**Marine Statistics
Theory and Practice**

Marine Statistics
Theory and Practice

E M Goodwin MSc., PhD., MRIN., MIS.
Principal Lecturer in Management Science,
City of London Polytechnic

J F Kemp PhD., Extra Master, FRIN.
Head of the School of Navigation,
City of London Polytechnic

STANFORD MARITIME LONDON

Stanford Maritime Limited
Member Company of the George Philip Group
12–14 Long Acre London WC2E 9LP
Editor D Nicolson

First Published 1979
© E M Goodwin and J F Kemp 1979
Printed in Great Britain by
William Clowes (Beccles) Limited
Beccles and London

British Library Cataloguing in Publication Data
Goodwin, E M
  Marine statistics.
  1. Mathematical statistics
  2. Shipping—statistical methods
  I. Title II. Kemp, John Frederick
  519.5'02'46238 QA276
  ISBN 0 540 07379 2

Preface

This book has been written as an introduction to statistical methods for anyone concerned with the operation of ships. It is intended for those who have to understand information presented in statistical form and for those who wish to use statistical methods to help in analysing problems they are faced with.

The knowledge of mathematics required to follow the book is not very great and the emphasis throughout is on a practical understanding of the various concepts involved. Exercises are provided at the end of each chapter and for the most part there are two exercises on each of the main points in a chapter. Full notes on the methods of solution are also given and it is hoped that this feature will be of use to students working on their own as well as to those who work in a group.

The authors would like to thank all the various people who have helped to make this book possible.

Elisabeth M Goodwin
John F Kemp

Contents

1	Introduction	9
2	Tabular and Graphical Representation of Numerical Data	13
3	Measures of Location	40
4	Measures of Dispersion	57
5	Probability	75
6	Discrete Probability Distributions	95
7	Continuous Probability Distributions	119
8	Sampling Distributions and Estimation	141
9	Hypothesis Testing	171
10	Tests Based on the χ^2 Distribution	199
11	Non-parametric Tests	225
12	Analysis of Variance	243
13	Regression and Correlation	278
	Summary of Notation and Formulae	315
	Tables	317
	Index	333

Chapter 1
Introduction

1.1 *Why Statistics for Mariners?*

The business of the mariner has been variously described as an art and a science. Probably it is both of these things since the efficient mariner has a feel for his work which may partly derive from natural aptitude and the intelligent use of experience and partly from an understanding of the processes he is required to control.

Certainly, mariners have never been afraid to use mathematical results as a basis for tackling problems, and then to use their experience to interpret those results in formulating sensible and seamanlike decisions. Thus a position fix may be calculated with trigonometrical precision, but the mariner will be aware that observational inaccuracies are such that it is only an estimate of the true position, and he will incorporate a reasonable margin of safety if he uses the fix to lay off a course for his next landfall. Similarly, a calculation of the safe working load for a wire rope gives an exact numerical answer, but the mariner may apply an additional safety factor if the wire shows signs of wear or if it is to be used for some particularly critical purpose.

There are many such examples where quite definite numerical results are available but where decisions have to be made in the face of further uncertainties. In the past, the mariner has acquired a feeling for these uncertainties during the course of his work and has allowed realistic margins of error, but technological advances have, in many cases, made this approach inadequate. The introduction of automatic and semi-automatic equipment has allowed reductions in manning standards, but it has also had the effect of often making the mariner more remote from the processes which he may be controlling. Thus, for example, in taking an astro observation a mariner is able to acquire a feel for the accuracy of his resulting position line through noting the appearance of the horizon, by taking account of any unsteadiness in holding his sextant due to weather conditions, etc. In taking a digital reading from a radio navigation aid, however, the navigator acquires no such feel for the accuracy of the corresponding position line and he has no way of directly sensing factors such as ionospheric disturbances which may affect that accuracy. Nevertheless, the mariner still requires information concerning the confidence which he can properly assign to his position line and this can be provided in numerical form from tables or from the navigation aid itself. The numerical

Marine Statistics

treatment of uncertainties is very much the business of statistics. As this aspect of the mariner's work progresses from an art in which subjective and often subconscious assessments of margins of safety are used, to a science in which numerate assessments are made necessary by the technology of the systems he has to operate, so does the mariner have an increasing and urgent need for an understanding of statistical methods.

We can draw similar conclusions from consideration of the second example, by looking at the safe working load of a wire rope. When derricks and winches are used for handling cargo, the running gear is continuously under observation and the experienced mariner can replace wires when he feels that they are becoming unsafe to operate. When ships are fitted with cranes, the wires are not so completely exposed to view during normal working, and more formal methods of ensuring safety are required. A statistical approach again becomes necessary.

1.2   Specific Examples

The examples discussed may be put in a more concrete form by considering a series of measurements made for test purposes and then the kinds of questions which might be asked and for which statistical analysis may provide answers.

Thus, if we wish to assess the accuracy of a navigation aid such as the Consol system, we might make a series of signal counts while the ship is in a known position for which the correct count is calculated to be 11 dots. The result of such an experiment, conducted for a number of different observers over a 24 hour period, is summarised in Table 1.1.

There are a number of questions which might be answered from a study of the information in Table 1.1. We might wish to know whether, on this

Table 1.1. Results of 40 counts of Consol signals for ship at anchor over 24 hour period

Count of dots	Number of observations
9	1
10	3
11	6
12	15
13	9
14	4
15	2
Total	40

Introduction

evidence, we should apply a fixed correction to other observations of the Consol signal made in the same general area. In addition, we might wish to know whether we can make useful statements about the accuracy of a subsequent signal count made under operational conditions.

Similarly, for the second example, we might recognise a need to 'formalise' the safe management of the wire runners used in a ship's crane and, for this purpose, to make use of the results of an experiment in which samples of wire rope were tested under simulated working conditions until they failed. A series of such results for 32 test pieces is illustrated in Table 1.2.

Table 1.2. Survival times for 32 samples of wire rope tested to destruction

Number of hours of test	Number of samples tested
Less than 1,000	0
1,000 but less than 2,000	0
2,000 but less than 3,000	2
3,000 but less than 4,000	6
4,000 but less than 5,000	12
5,000 but less than 6,000	8
6,000 but less than 7,000	3
7,000 but less than 8,000	1
8,000 and over	0
Total	32

Again, we may be interested in a number of questions. On safety grounds we may ask how many working hours the wire should be used before it is renewed, if we require that there should only be a 0·1% chance of it failing in service. Also, on economic grounds, we may wish to know how many wires with a remaining useful life of over 1,000 hours we will be condemning if we choose a fixed maximum working life on safety grounds. Is the economic penalty so great that we should investigate the possibility of using an alternative method of management such as a planned programme of inspections?

Statistical methods which can provide the answers to these and to many other types of problem are introduced in this book. They are discussed mainly in terms of examples which are drawn from the marine field, but the statistical techniques are perfectly general and have very wide applications.

1.3 *Purpose*

It is not the intention of this book to transform the readers into statisticians but it does have two important aims.

Firstly, as the technology of ship systems increases and as the function of ships' officers becomes more and more managerial, so they are presented with more and more information in statistical form. An important purpose of this book is to enable the mariner to accept such information critically and to use it with understanding and confidence for making his professional decisions.

Secondly, the intention is that the reader should become sufficiently acquainted with some of the simpler techniques that he can apply these to make his own analysis of some of the more straight-forward problems with which he is faced and for which information in a suitable form can be collected or may otherwise be available. At the same time, the limitations of the various techniques are emphasised so that the reader will appreciate where more advanced methods may be required and, perhaps, where the advice of a statistician may be necessary.

1.4 Summary

In general, the business of statistics, as presented in this book, is to take the evidence contained in a numerical record of certain entities and then to quantify particular features of the record which can be of use in a management context. Clearly, the starting point is to consider the form of the numerical information which is the raw material for statistical analysis, and this is the subject matter of the next chapter.

Chapter 2
Tabular and Graphical Representation of Numerical Data

2.1 Introduction

```
11.20  11.35  11.37  11.50  12.03  12.04  12.06  12.15
12.20  12.30  12.35  12.50  12.53  13.08  13.16  13.25
13.31  13.58  14.00  14.12  14.35  14.45  14.49  14.50
14.53  15.04  15.25  15.55  16.05  16.14  16.15  16.19
16.30  16.38  17.04  17.05  17.10  17.10  17.10  17.15
17.25  17.45  17.55
```

The above figures show the times at which ships passed the Felixstowe Ledge Buoy off Harwich during a survey of marine traffic on a day in July from 11.00 to 18.00. This set of data is typical of the raw material with which a statistician has to deal and on its own is of rather limited use. Immediately we look for ways of summarising it so that useful facts may be brought out. Thus, for example, we may want to know that in all 43 ships passed the buoy in 7 hours, or that 4 ships passed in the first hour, 9 ships in the second hour, etc. The particular method that is used to summarise a data set depends on the purpose for which it is to be used but it is evident that some form of summarisation is needed. The first task is to reduce the overwhelming mass of numbers so that (i) the amount of detail is reduced and (ii) the data is brought into a form whereby the significant features stand out prominently. In this chapter different aspects of the presentation of statistical data will be considered.

2.2 Definitions

Before proceeding further a few of the important basic statistical definitions will be considered as like most subjects statistics has plenty of its own jargon, which provides a useful shorthand as the topics develop.

2.2.1 **A *population* is the group from which data are to be collected,** e.g. ships over 100 grt registered in the UK. It is important always to define the population of interest very carefully and unambiguously.

2.2.2 **A *sample* is a subsection of a population.**

Marine Statistics

It is important in later chapters to know if a data set has been collected from a population or from a sample as a representative of a population but for the purpose of summarising data which is our present task the distinction is not so important.

2.2.3 A *variable* **is a feature characteristic of any member of a population yet differing in quality or quantity from one member to another.**

2.2.4 **A variable differing in quantity is called a** *quantitative variable* **or simply** *variable*, e.g. in the population of ships over 100 grt registered in the UK, the length of each ship measured in metres would be a quantitative variable.

2.2.5 **A variable differing in quality is called a** *qualitative variable* **or** *attribute*, e.g. in the same population as above the port of registry would be a qualitative variable.

2.2.6 **A** *discrete variable* **is one which can only take certain values in its range**, e.g. in the same population the number of men aboard each of the ships is a discrete variable since it must take the value of a whole number.

2.2.7 **A** *continuous variable* **is one which can take any value within its range**, e.g. the length of a ship may be measured to any required accuracy.

Often an arbitrary decision has to be made whether to treat a variable discretely or continuously. Thus length of ship measured to the nearest 100 metres would best be treated discretely, whereas length of ship measured to the nearest centimetre would best be treated continuously.

2.3 *Arrays*

Returning to the problem of how to summarise raw data, the first step might be to arrange the values in some sort of order. If the variable is quantitative then an arrangement starting with the smallest value and finishing with the largest would be suitable. If the variable is qualitative then an arbitrary ordering may be necessary, say alphabetical in the example of ports of registry. Such an arrangement is known as an *array* and would at least be a little easier to understand than would the raw data on length of ship or port of registry of ships over 100 grt registered in the UK, which would be in a very haphazard order as it was taken straight from Lloyd's Register of Shipping. However, an array in no way condenses the data. Our original data set of times of ships passing a buoy is in array form and we still have our original number of 43 items.

2.4 *Frequency Distributions*

The most popular form of tabular presentation of statistical data is the use of a *frequency distribution*. This is a table which shows, for each separate value or group of values of the variable, the number of times which it occurs. Consider the following set of figures which show, for Liverpool, the number of days each month for one year with rainfall of 0·3 mm or more:

January	18	July	15
February	13	August	15
March	13	September	15
April	13	October	17
May	13	November	17
June	13	December	19

This information could be alternatively shown in a frequency distribution showing the number of months in which a given number of rainy days occurred. The best way of preparing a frequency distribution is using a tally system, whereby the data items are taken in turn and an appropriate mark | made against the category to which it belongs. Groups of 5 are denoted as |||| . Thus the example on rainy days would be tabulated on a working sheet as follows:

Number of rainy days per month	Number of months
13					
14	
15				
16	
17			
18		
19		

The resulting frequency distribution would then be presented as shown in Table 2.1.

With this table we have lost precise details on what happens in each individual month but it is easier to get an appreciation of the likely number of rainy days in any one month. Here it was possible to show the individual recorded values of the variable separately, since they only ranged from 13 to 19. However in many cases, such as the original example on the time of ships passing a buoy, this would still give too much detail. In that data only one time (17.10) is repeated so that there would still be many different times with no consistent pattern between them, e.g. the first gap is 15 min, the next 2 min,

Marine Statistics

Table 2.1. *Frequency distribution of the number of days in a month with 0·3 mm or more of rain*

Number of days in a month with 0·3 mm or more of rain	Number of months
13	5
14	0
15	3
16	0
17	2
18	1
19	1
Total	12

etc. In these sorts of examples values of the variable are grouped together, wherever possible, in equal intervals. Thus a suitable frequency distribution for these data is given in Table 2.2.

Table 2.2. *Frequency distribution of the number of ships passing the Felixstowe Ledge buoy per hour in a survey from 11.00 to 18.00*

Time	Number of ships
11.00 but before 12.00	4
12.00 but before 13.00	9
13.00 but before 14.00	5
14.00 but before 15.00	7
15.00 but before 16.00	3
16.00 but before 17.00	6
17.00 but before 18.00	9
Total	43

A distribution of this type, where groups of values of the variable are taken, is often referred to as a *grouped frequency distribution*. Further examples of frequency distributions are given in the first chapter. Table 1.1 is an example of a distribution where the variable is not grouped and Table 1.2 is an example of a grouped frequency distribution.

2.5 Classes

If a variable has to be grouped then the groupings are referred to as *classes*. In the example above the first class was all times between 11.00 and 12.00, including 11.00 but not 12.00. The variable, time, is a continuous one and hence care had to be taken with the wording of the classes, so that the end points of each class appear in one class only. The size of the class, which is known as the *class interval*, is the length of the class measured on a continuous scale, so in this case it is 60 min. This definition is however a little less clear when the variable is a discrete one.

Consider Table 2.3 which shows a discrete frequency distribution, since the figure 29·5° for instance is not permitted by the wording of the classes, presumably because the measuring accuracy was not sufficient.

Table 2.3. Number of stars with given sidereal hour angles

Sidereal hour angles	Number of stars of magnitude less than 3
0°–29°	3
30°–59°	3
60°–89°	4
90°–119°	6
120°–149°	7
150°–179°	7
180°–209°	3
210°–239°	4
240°–269°	6
270°–299°	7
300°–329°	4
330°–359°	6
Total	60

The second class is written as 30°–59° and has a class interval of 30°, since to be measured on a continuous scale it would need to extend from $29\frac{1}{2}°$ to $59\frac{1}{2}°$. Alternatively the answer may be reached by counting the number of angles within the class, which gives 30 again. The class interval will be of use later on when we want to calculate values from a frequency distribution.

Obviously the choice of classes depends on the particular circumstances but there are a few general hints.

(i) Make the classes of equal size whenever possible.
(ii) Make the class intervals multiples of numbers in which people are used to thinking, such as 5, 10, 100, etc., and avoid numbers such as 13, 17, 7, etc.

Marine Statistics

(iii) The number of classes should usually be between 5 and 20.

(iv) In calculations involving frequency distributions, the midpoint of each class is chosen to represent the class, so if there is any clustering of frequencies about a particular value this should be a midpoint rather than an endpoint.

This rule however can often be particularly difficult to adhere to. For instance in the Lloyd's Register of Shipping tables of ships by gross registered tonnage, 499 grt is chosen as the endpoint of one class since this is a limiting value for equipment and construction standards, and on these grounds a ship of 499 grt is very different from one of 500 grt. However for this reason there are many ships of 499 grt built. In this example the main criterion was to keep ships which are alike because of another consideration altogether.

(v) If there are a few extreme values at either end of a distribution it is better to put them together in an open-ended class rather than having a series of classes with very few observations in them. Thus for the frequency distribution of annual incomes of employees of a shipping company, classes of £1,000 width from £3,000 to £10,000 might be shown separately but the first class would be 'All incomes up to £3,000', and the final class 'All incomes of £10,000 and over'. In future work we will assume that an open-ended class is the same size as the class immediately next to it.

It has already been suggested that on several occasions it will be necessary to break these rules since the statistician's main aim is to bring out as fairly as possible the principal features of a distribution. However in many cases they at least provide a useful guide for tackling the problem of tabulating a large set of raw data.

2.6 *Tabular Presentation in General*

It has been shown that a frequency distribution provides the most convenient means of summarising a set of statistical data when one variable is being measured. In a later chapter, bivariate frequency distributions are considered whereby for each member of the sample two variables are recorded. However there are a few basic rules when any form of tabular presentation is to be used. It is evident that a tabular presentation is more efficient and easier than narrative alone when summarising a set of data. The advantages may be listed as:

(i) The required figures may be located more readily.
(ii) Comparisons can be made more easily.
(iii) Patterns may be revealed.
(iv) Tables are easier for further computations.
(v) Tables take up less space.

It should however also be borne in mind that when presenting a report some narrative should accompany it, pointing out the most significant features.

Tables and Graphs

The general rules to bear in mind with any tabular presentation are as follows:

(i) Have the purpose of the table clearly in mind from the start. Typical aims might be:
    (a) to show the original figures in an orderly manner
    (b) to show a distinct pattern in the figures
    (c) to summarise the figures
    (d) to publish statistics other people may want to use, e.g. Government statistics or Lloyd's statistics
(ii) Aim at simplicity so that the table is easy to understand.
(iii) Always give the table a comprehensive explanatory heading.
(iv) State the units of measurement.
(v) State the source of the data.
(vi) Make sure that the row and column headings are unambiguous so that there is no doubt as to the category in which any item has been included.
(vii) It is often useful to include the total number of items in a distribution.
(viii) It can be helpful to show the items as actual numbers and also as percentages of the total. If this is done the base of the percentages should be clearly indicated.
(ix) Notes following a table can be used to explain any points about the table in general or individual figures which need further clarification.

An example illustrating some of these rules is shown in Table 2.4. The original purpose of the table was to show the growth in the number and tonnage of

Table 2.4. The number and gross tonnage of trading vessels of over 100 gross registered tons in 'flag of convenience' fleets, 1954 and 1976

Country of registration	1954 Number	1954 Gross tonnage in million tons	1976 Number	1976 Gross tonnage in million tons
Liberia	245	2·4	2,600	73·5
Panama	595	4·1	2,680	15·6
Honduras	130	0·4	57	0·1
Cyprus	—	—	765	3·1
Singapore	—	—	722	5·5
Lebanon	—	—	136	0·2
Somalia	—	—	255	1·8
Total	970	6·9	7,215	99·8
World total	32,358	97·4	65,887	372·0
Percentage of the world total	3	7	11	27

Source: Lloyd's Register of Shipping

Marine Statistics

vessels in the main countries considered as offering the facility of flag of convenience over the period 1954–1976.

2.7 *Graphical Presentation in General*

A quick general impression may be obtained from a set of data if some form of visual presentation is used in addition, or instead of, a tabular form. Obviously a table must be used if one needs to see a variety of figures whereas a graph will only be able to emphasise the most salient features. There is a tremendous variety of visual forms which may be employed varying in their degree of complexity, but as with tabular forms there are a few general rules.

These are:

(i) Unless one has the deliberate intention of misleading it is important to give the correct visual impression and to avoid optical illusions. Thus if a graph is drawn using a pair of axes, then it is important to state clearly the scale used on each axis.

(ii) It is sometimes convenient to have a vertical scale, say, which does not start at 0 but at some other point, e.g. if a set of figures ranged from 500–600 units.

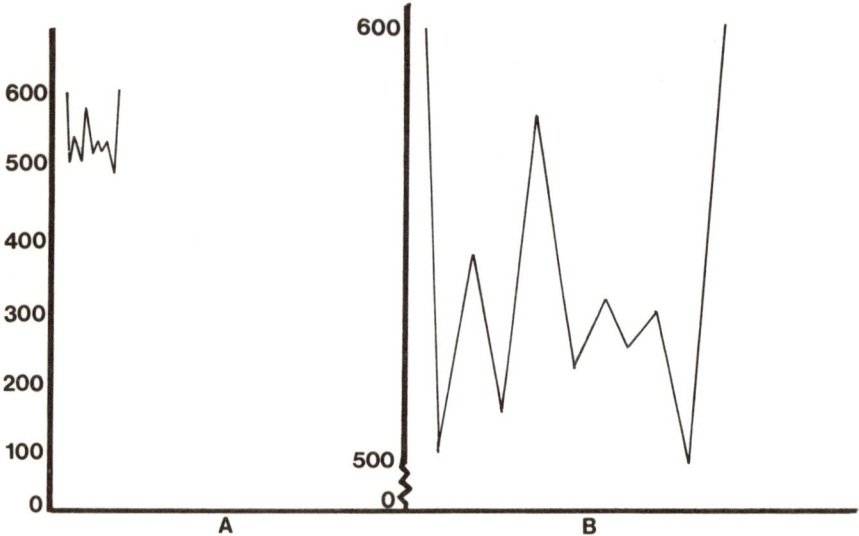

Fig. 2·1 Graphs showing a set of data on different scales.

Figure 2.1(A) gives the figures using a scale from 0 to 600 with the problem that the relevant part is very compressed, whereas in Fig. 2.1(B) by indicating the break from 0 to 500, the required portion may be shown more clearly.

(iii) The graph should have a clear comprehensive title.

Tables and Graphs

(iv) The variable measured (the independent variable) should be shown horizontally and the results of the measurement (the dependent variable) vertically. Thus in a frequency distribution the frequencies are displayed vertically.

(v) It is possible to display two sets of data on the same diagram in which case a double vertical scale may be employed but clearly indicated.

(vi) The axes used should be labelled.

(vii) The source of the graph should be given.

(viii) It is important to keep any form of visual representation simple, so that overcrowding is avoided.

It is difficult to classify all the forms of visual presentation of data which are used, since many of them are the result of much creative thinking. It is however possible to consider some of the more usual forms and in particular those which have been developed along strict theoretical lines for the analysis of frequency distributions.

2.8 *Histograms*

These are used to represent frequency distributions, and consist of a series of touching rectangles with bases on a horizontal axis. Each rectangle represents a value or class in the distribution and its height is drawn so that the area of the rectangle is proportional to the frequency it represents. Thus the total area of the *histogram* represents the total frequency in the distribution.

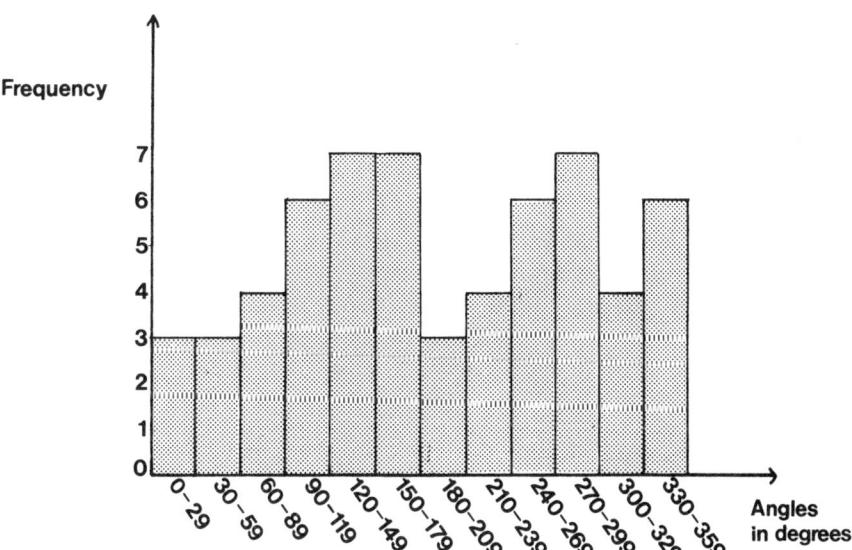

Fig. 2·2 **Histogram of the distribution of stars of magnitude less than 3 with given sidereal hour angles**

Marine Statistics

Figure 2.2 shows the histogram of the sidereal hour angle data given in Table 2.3. The labelling of the classes is important since this is a discrete frequency distribution and yet the histogram demands a series of touching rectangles drawn on a continuous scale. It is as though each rectangle extended $\frac{1}{2}°$ in both directions, thus the second one is drawn from $29\frac{1}{2}°$–$39\frac{1}{2}°$. In a continuous distribution there is no need for this extension and it is most usual to label the ends of the classes. Hence Fig. 2.3 shows the histogram for the ships' arrival pattern given in Table 2.2.

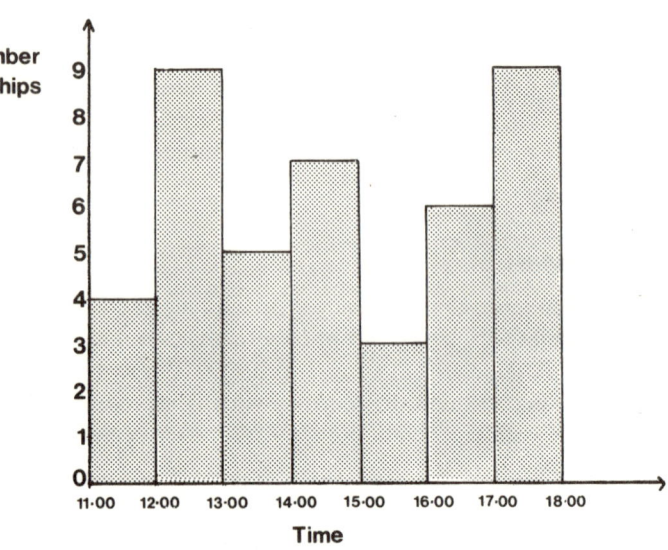

Fig 2·3 Histogram of the distribution of the number of ships passing the Felixstowe ledge buoy per hour in a survey from 11.00–18.00.

If, for some reason, the final two classes had been combined so that all that was known was that 15 ships passed between 16.00 and 18.00, the base of the final rectangle would be twice as large as the others, and hence to compensate the height would be $7\frac{1}{2}$, the result of dividing 15 by 2. By doing this the total area would still be proportional to 43 as shown in Fig. 2.4.

2.9 *Frequency Polygons*

Instead of a histogram, which can be rather tedious to draw, it is often easier to see the shape of a distribution by drawing a straight line graph called a *frequency polygon*. The first step is to mark off where the midpoint of the top

Tables and Graphs

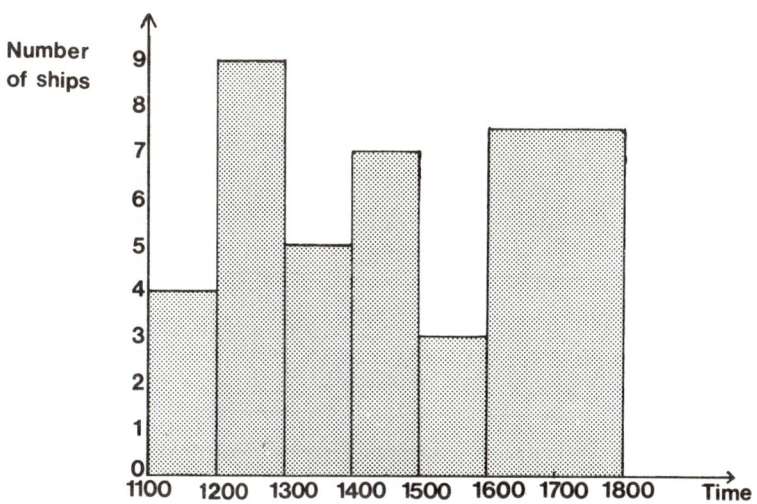

Fig. 2·4 The histogram for the distribution shown in Fig. 2·3 but with the two final classes combined.

of each rectangle in the histogram would come. Adjacent midpoints are then joined with straight lines. The effect of this is that the area of the triangle cut off from each rectangle of the histogram is equal to the area of the triangle included under the curve, if the classes are of equal size. To preserve the idea that the total area under the curve should represent total frequency, at either end a class interval is marked on the X axis equal to that of the adjacent class, and the midpoint of these joined to the ends of the curve to make a closed curve. If class intervals of unequal size are used, the property of total area representing total frequency is not strictly true. The only way this could be achieved would be to imagine the larger class intervals broken into single units and the upper midpoints of the single units joined, but this is seldom done. Figure 2.5 shows the frequency polygon corresponding to Fig. 2.3.

There are two points to note in connection with this figure. Firstly, the curve simply denotes the shape of the distribution, and intermediate points have no meaning at all. Thus no reading must be taken between dots. Secondly, it is not a good idea to label the horizontal axis beyond the extent of the observed values, because it gives the impression that 0 ships were observed between 10.00 and 11.00 whereas in fact no observations were taken. Frequency polygons obviously have to be handled with care but are useful for comparing two distributions. They are also useful in later theoretical work particularly if there are a large number of classes. In this situation it becomes easier to draw a continuous smooth frequency curve rather than a disjointed frequency polygon, since the limit of any polygon, as the number of sides increases infinitely, is a curve.

Marine Statistics

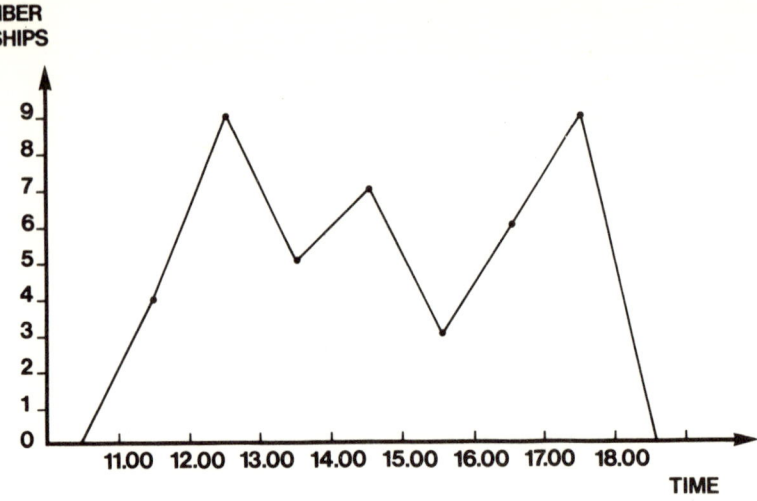

Fig 2·5 Frequency polygon of the distribution of the number of ships passing the Felixstowe ledge buoy per hour in a survey from 11.00–18.00

2.10 Relative Frequency Histograms and Polygons

Instead of using actual frequencies when preparing a distribution relative frequencies may be used and the resulting distribution is known as a *relative frequency distribution*. The relative frequency with which a particular value or group of values occur is equal to the ratio of the actual frequency to the total frequency observed. Thus the relative frequency distribution corresponding to Table 2.2 is given in Table 2.5.

Table 2.5

Time	Relative frequency
11.00 but before 12.00	0.09 (=4/43)
12.00 but before 13.00	0·21
13.00 but before 14.00	0·12
14.00 but before 15.00	0·16
15.00 but before 16.00	0·07
16.00 but before 17.00	0·14
17.00 but before 18.00	0·21
Total	1·00
Number of ships	43

Tables and Graphs

The relative frequency histogram and relative frequency polygon are obtained by plotting the relative frequency distribution obeying similar rules as for the ordinary frequency distribution. The area under each of the relative frequency curves is proportional to 1.

2.11 Ogives

An *ogive* or cumulative frequency polygon is the graph obtained by plotting a cumulative frequency distribution against its variable. As the name suggests the cumulative frequency distribution is formed by accumulating the frequencies, either from the smallest value of the variable, when it is termed a *'less-than'* distribution, or from the largest value of the variable, when it is termed a *'more-than'* distribution. The 'less-than' distribution tends to be the more common of the two. Table 2.6 shows the cumulative frequency distribution of the 'less-than' variety corresponding to the sidereal hour angle data of Table 2.3.

Table 2.6

Sidereal hour angles	Cumulative number of stars of magnitude less than 3
Less than 29°	3
Less than 59°	6
Less than 89°	10
Less than 119°	16
Less than 149°	23
Less than 179°	30
Less than 209°	33
Less than 239°	37
Less than 269°	43
Less than 299°	50
Less than 329°	54
Less than 359°	60

The first entry 3 is obviously the number in the first class, whereas the second entry 6 is the number in the first class together with the number in the second class (3 + 3) and so on. The final entry should be equal to the total frequency.

When plotting the ogive the same problem arises as with the histogram in that this is a discrete frequency distribution which has to be plotted on a continuous scale. Thus for strict accuracy the class limit should be taken as the midpoint between the end of one class and the start of the next. Hence the

Marine Statistics

first value 3 is plotted against $29\frac{1}{2}°$ and so on until the final value 60 is plotted against $359\frac{1}{2}°$. Successive points are linked by a continuous smooth curve since it makes sense to read off intermediate points. Since there are 0 observations at the start of the distribution the curve is drawn to the start of the first class on the horizontal axis, in this particular example to $-\frac{1}{2}°$. Figure 2.6 shows the completed ogive.

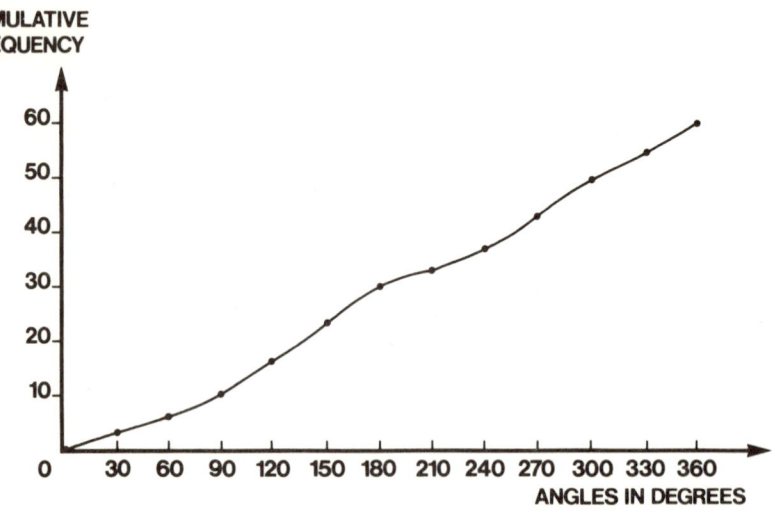

Fig 2·6 Ogive or cumulative frequency curve for the sidereal hour angle date in table 2·3

The ogive is a useful means of displaying one cumulative frequency curve but there is little advantage in displaying two such curves on the same graph. The ogive of the one with the larger total frequency will simply tower over the other. The only feasible way for comparison is to plot cumulative relative frequency curves on the same axes.

The three types of graph described so far, the histogram, frequency polygon and ogive, are of use in analysing frequency distributions rather than in simply portraying data and hence must be constructed in strict accordance with the rules. The other types to be discussed are more for visual impact and hence are not so strictly defined.

2.12 *Historigram or Time Series Graph*

Numerical data which have been recorded at intervals of time form a time series. A *historigram* is simply a time graph showing the level of the variable at the different time points, successive points being joined by straight lines. It is usually nonsensical to read off intermediate points. Figure 2.7 is an example.

26

Tables and Graphs

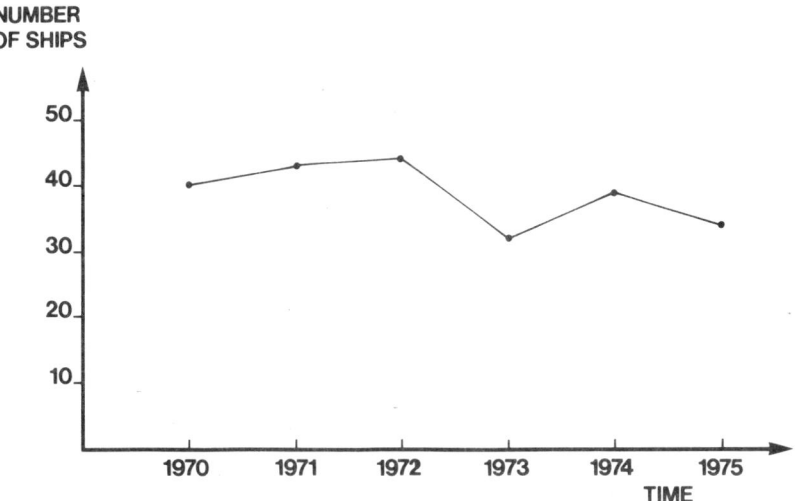

Fig 2·7 Annual incidence of total losses of merchant ships of 100 g.r.t and over due to collision 1970–1975

Sometimes a graph of this type is used for analysis work to determine if there is any underlying movement, known as a trend, in the data. For this purpose it is often a good idea to draw a smooth continuous curve. Another analysis which may be required is a forecasting analysis to predict future events, in which case the time variable must be interpreted continuously and again a smooth curve drawn. If one is plotting totals they should be plotted at the end of each period, but if the figures are averages they should be plotted in the middle of the period.

2.13 *Semilogarithmic Graphs*

There are occasions when it is sensible to choose a straight-forward scale on one axis, usually the horizontal axis, but to plot the logarithm of values on the vertical axis. It is possible to use logarithmic scales on both axes but the *semilog graphs* are the more usual. There are two main reasons for using a semilog graph.

(i) A great range of values may be shown on the vertical axis, e.g. $\log_{10} 100 = 2$ and $\log 100{,}000 = 5$, hence the range 100–100,000 may be depicted easily in 3 units.

(ii) It is easier to compare percentage growth rates or any rates of change, e.g. suppose firm *A* grows in turnover from £1,120,000 in 1970 to £1,680,000 in 1975 and firm *B* grows in turnover from £492,000 in 1970 to £738,000 in 1975. Then firm *A* increases turnover at a rate of £112,000 per annum and

Marine Statistics

firm *B* at a rate of £49,000 per annum. Thus on an ordinary graph the slope of the line for *A* would be much steeper than that for *B*. However the turnover for *A* for 1975, expressed as a percentage of turnover in 1970, is 150% and similarly for *B* and on a semilogarithmic graph the curves would appear as parallel lines as shown in Fig. 2.8.

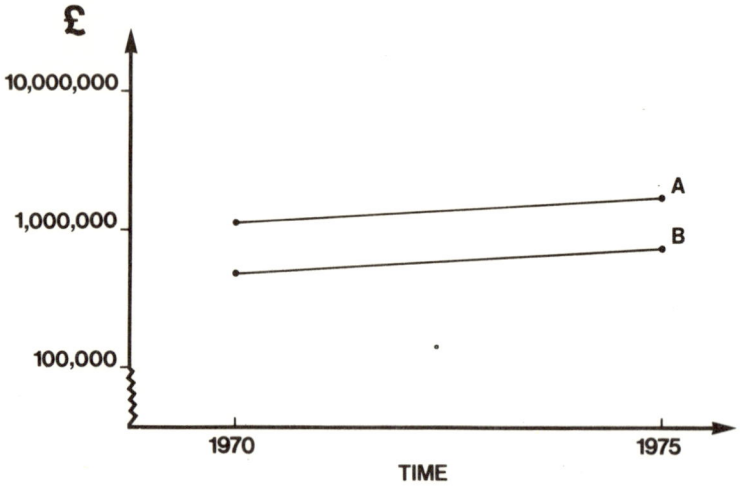

Fig 2·8 Figure to compare the growth of firms A and B over the period 1970-1975

The conclusion from this graph is that both firms are performing equally well in relation to size. Special graph paper is available with a logarithmic scale already on the vertical axis but the graph may be plotted easily on ordinary paper by plotting the logarithm of the variable.

2.14 Lorenz Curves

Lorenz curves are used to show how even the distribution of an item may be over a range of objects. Consider the data on flag of convenience fleets for 1976 given in Table 2.4 and arranged in ascending order of size of fleet in terms of numbers. The cumulative percentages, in terms of numbers and tonnage, are also given in Table 2.7.

By plotting the pairs of cumulative percentage points and joining them with a straight line from the origin to the point (100%, 100%) shows the line of equal distribution, i.e. all the points would lie on this line if the tonnage owned by each country was proportional to the number of ships owned. The disparity between the actual curve and this line gives an indication of the difference in size of ships with respect to numbers owned in the different countries.

Tables and Graphs

Table 2.7

Country of registration	Number	Cumulative percentage	Tonnage in million tons	Cumulative percentage
Honduras	57	0·1	0·1	0·1
Lebanon	136	2·7	0·2	0·3
Somalia	255	6·2	1·8	2·1
Singapore	722	16·2	5·5	7·6
Cyprus	765	26·8	3·1	10·7
Liberia	2,600	62·9	73·5	84·3
Panama	2,680	100·0	15·6	100·0
Totals	7,215		99·8	

Either variable on its own provides a different impression of the relative importance of the countries with respect to flag of convenience fleets, and this curve helps to show how different the results of using each would be.

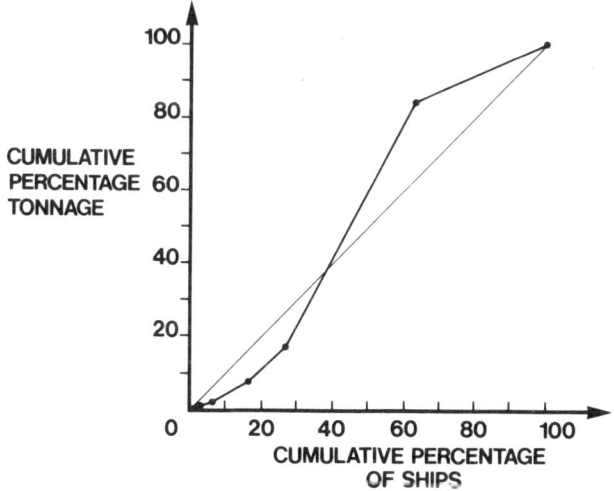

Fig 2·9 Lorenz curve to show the numbers and gross tonnage of ships in flag of convenience fleets

2.15 Pictograms

Here pictures are used to represent the data. The best plan is to adopt a symbol which is used to represent a given number of items, and then the same picture is repeated as many times as necessary together with a fraction of the

Marine Statistics

picture to complete the total. As an example consider the data on the number of trading vessels of over 100 grt in flag of convenience fleets in 1954 given in Table 2.4. This can be represented by a *pictogram* as shown in Fig. 2.10.

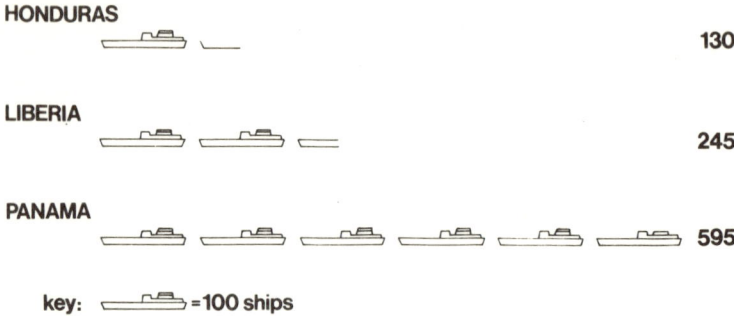

Fig 2·10 Number of trading vessels over 100 g.r.t. in flag of convenience fleets in 1954

A pictogram provides a very striking visual presentation but the main problem is that accuracy is very difficult. It is extremely hard to judge how much of a ship is drawn when a fractional symbol is used. A major trap which some people fall into is to attempt to use larger symbols to show larger quantities. Thus if it is required to show that the export of beer from one country *B* is twice that of country *A* they represent it thus

Tables and Graphs

To keep the barrel for country *B* in proportion, in doubling the height, all dimensions are doubled. Hence the size of the finished symbol is not just twice that of the original but eight times it. Unfortunately if one tried to keep the volume of *B* only twice that of *A*, there would still be doubts if it were height or volume that was to be considered. Despite this problem such a device is often used as it is eyecatching, but must be treated with care.

2.16 Bar Charts

In a bar chart data are represented by a series of bars, either vertical or horizontal, with equal bases but the length of each bar indicating the size of figure represented. Thus, instead of the pictogram for the size of the flag of convenience fleets in 1954 shown in Fig. 2.10, Fig. 2.11 gives a bar chart.

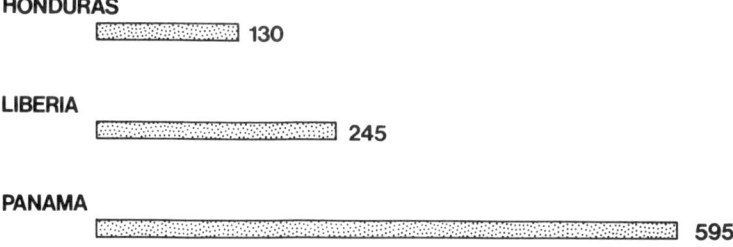

Fig 2·11 The number of trading vessels over 100 g.r.t in flag of convenience fleets in 1954

The two most obvious differences between a bar chart and a histogram are that in a bar chart it is the height of the bar which is proportional to frequency and the bars do not touch each other.

Various degrees of complexity may be introduced into bar charts. For instance component bar charts are where the total bar is shown subdivided into component parts. Multiple bar charts on the other hand show the components of a total as a series of bars next to each other. Thus if one wished to compare, for the Dover Strait and the Baltic Sea, the number of sea collisions and the respective proportion that occurred in restricted visibility over a period of years, then Fig. 2.12(a) shows the comparison using a compound bar chart and Fig. 2.12(b) the comparison using a multiple bar chart.

Component bar charts are obviously more suitable when one wishes to compare totals and one component only; multiple bar charts are better when one wishes to compare all the components and not totals. However, if there are more than three or four components bar charts are not the most suitable form of representation and it is best to use a pie chart.

Marine Statistics

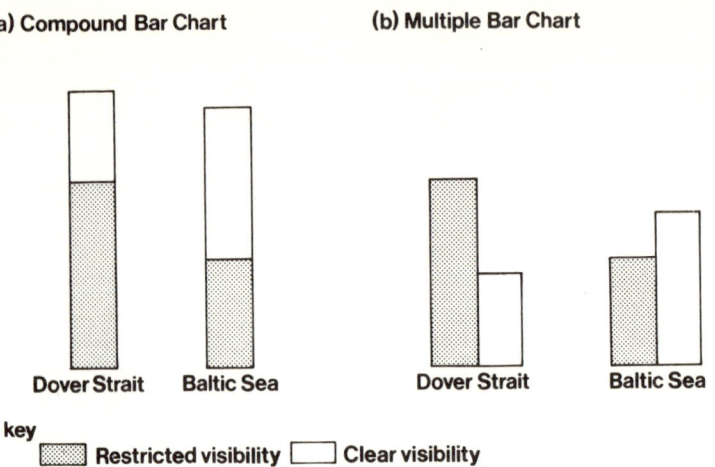

Fig 2·12 Comparison of the number of sea collisions in the Dover Strait and Baltic Sea in resticted and clear visibility

2.17 *Pie Charts*

A pie chart is a circle divided by radial lines into sections, like slices of a cake or pie. The area of each section is made proportional to the size of the figure represented and it therefore provides a convenient method of showing the relationship between the sizes of component parts and the overall total. Two distributions may be compared using different circles. Some people recommend making the radius of each circle proportional to the total frequency in each distribution, but this must be treated very carefully for the same reasons as with pictograms. It is often simpler to take circles of the same radius.

The method of construction of the pie chart is first to calculate the central angle of each section as the same proportion of 360° (a complete circle) as the frequency it represents is of the total frequency. It can be shown that the angle at the centre of a section is proportional to the area of the section and hence

	Number of ships	Central angle
Honduras	130	$\frac{130}{970} \times 360° = 48°$
Liberia	245	$\frac{245}{970} \times 360° = 91°$
Panama	595	$\frac{595}{970} \times 360° = 221°$
	970	360°

32

Tables and Graphs

the areas will be in the right ratio. A circle can then be drawn and subdivided according to these ratios. The 1954 flag of convenience ship data can be used to illustrate this method, although the main point we would be trying to make this time by using it would be the relative importance of the three countries to each other rather than the absolute numbers of ships involved as in the pictogram and bar chart.

Fig 2·13 Pie chart to compare the number of trading vessels over 100 g.r.t in 'flag of convenience' fleets in 1954

2.18 Summary

In this chapter we have considered how a statistician sets about the task of reducing a set of raw data into a form whereby it can be easily understood by anybody else. Different methods of presentation of data, both tabular and graphical, have been considered and it only remains to emphasise again that the particular choice of method depends on the purpose for which the data are intended. In the next chapter we will consider how to summarise the data set even further, into single figures which can be used to describe the whole distribution.

Exercises

1. At a particular port a ship took aboard 48 loaded 20 ft containers. The weights of these containers, in tonnes (rounded down to the nearest tenth), in order of loading, were as follows:

Marine Statistics

11·5	4·6	5·4	13·3	3·6	3·6	20·4	11·8
2·4	10·7	6·8	4·0	8·8	3·2	12·2	14·3
9·7	14·2	20·7	10·9	12·5	10·8	7·4	12·2
12·6	9·7	15·1	14·2	6·3	16·4	17·3	18·1
3·3	19·5	1·8	11·4	21·6	10·3	13·5	23·2
20·5	11·3	7·4	18·6	11·7	8·2	9·8	5·7

Arrange the above data as an array.

2. Arrange the data from Ex. 1 as a frequency distribution with six classes. What are the class boundaries of the second class?

3. Construct a histogram from the data in Ex. 1.

4. Construct a frequency polygon for the data in Ex. 1.

5. Provide the histogram and frequency polygon constructed in Ex. 3 and Ex. 4 with a scale of relative frequency on the right-hand side.

6. Construct a 'less-than' cumulative frequency distribution for the data from Ex. 1 and hence draw the ogive.

7. A ship loads 48 × 20 ft containers with a total weight of 600 tonnes, 12 × 30 ft containers with a total weight of 200 tonnes and 20 × 40 ft containers with a total weight of 360 tonnes. Construct diagrams to show:
   (a) the comparison of numbers of containers of each class
   (b) the comparison of weight of cargo loaded in containers of each class

8. Over the six-year period 1974–1979, two shipping companies competing on a ferry service carried the following numbers of vehicle units:

	Vehicle numbers (thousands)		
	Company A	Company B	Total
1974	10	42	52
1975	12	45	57
1976	15	50	65
1977	16	52	68
1978	20	60	80
1979	25	68	93

(a) Construct a historigram for the above data showing each company's record on the same coordinates.
(b) Also, plot the same data as semilogarithmic graphs, and explain the apparent difference in the appearance.

Tables and Graphs

9. The managers of the port used by the ferry services in Ex. 7 wish to publicise the increasing use of the port. Suggest how this could be illustrated in pictogram form.

10. The total figures given in Ex. 8 can be broken down into commercial and private vehicles as follows:

	Private vehicles	Commercial vehicles	Total (thousands)
1974	36	16	52
1975	38	19	57
1976	39	26	65
1977	33	35	68
1978	40	40	80
1979	45	48	93

Show this information in the form of a component bar chart.

Answers

1. Data arranged as an array in order of weight:

1·8	4·0	7·4	9·8	11·4	12·5	14·3	19·5
2·4	4·6	7·4	10·3	11·5	12·6	15·1	20·4
3·3	5·4	8·2	10·7	11·7	13·3	16·4	20·5
3·3	5·7	8·8	10·8	11·8	13·5	17·3	20·7
3·6	6·3	9·7	10·9	12·2	14·2	18·1	21·6
3·6	6·8	9·7	11·3	12·2	14·2	18·6	23·2

2.

Weight in tonnes	Number of containers
0·0 but under 4·0	6
4·0 but under 8·0	8
8·0 but under 12·0	14
12·0 but under 16·0	10
16·0 but under 20·0	5
20·0 but under 24·0	5

Class boundaries are: 4·0–8·0.

Marine Statistics

3. (and 5.)

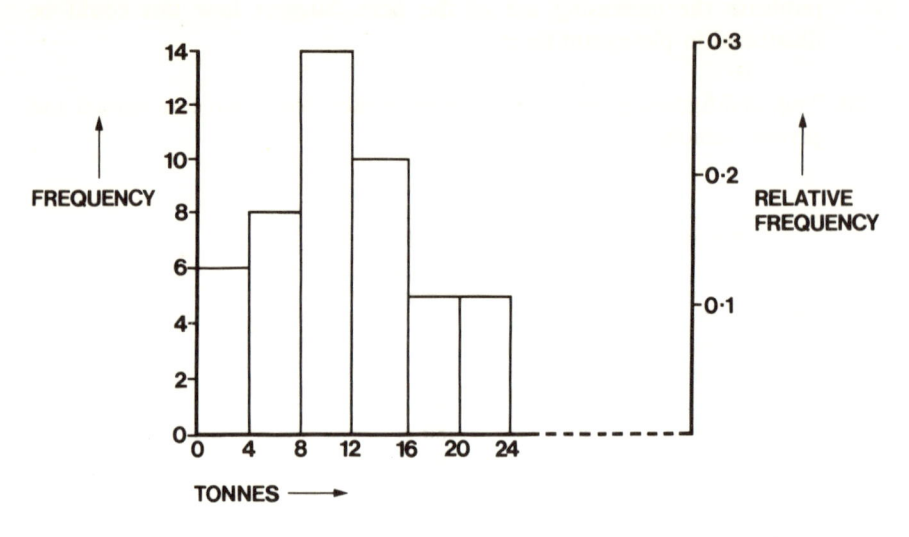

[Histogram: Frequency vs Tonnes, with bars at 6, 8, 14, 10, 5, 5 for intervals 0-4, 4-8, 8-12, 12-16, 16-20, 20-24. Relative frequency scale 0.1, 0.2, 0.3 on right axis.]

4. (and 5.)

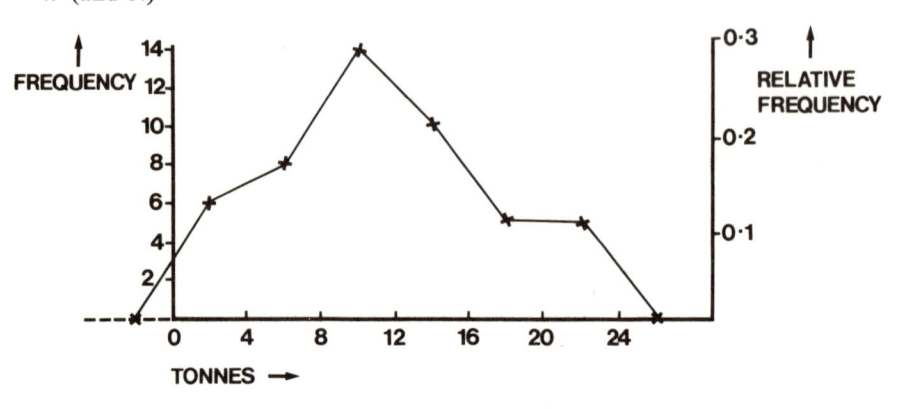

[Frequency polygon plotting points at tonnes 2, 6, 10, 14, 18, 22 with frequencies 6, 8, 14, 10, 5, 5, tailing to zero at both ends.]

5. See answers 3 and 4 above.

6.

Weight in tonnes	Cumulative number of containers
Less than 4	6
Less than 8	14
Less than 12	28
Less than 16	38
Less than 20	43
Less than 24	48

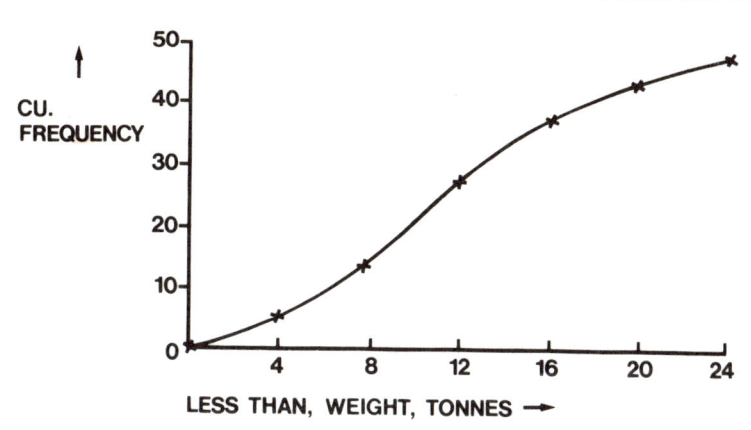

Tables and Graphs

↑
CU. FREQUENCY

50
40
30
20
10
0
   4   8   12   16   20   24
LESS THAN, WEIGHT, TONNES →

7.   A. NOS. OF UNITS   B. TOTAL WEIGHTS
                                        OF UNITS

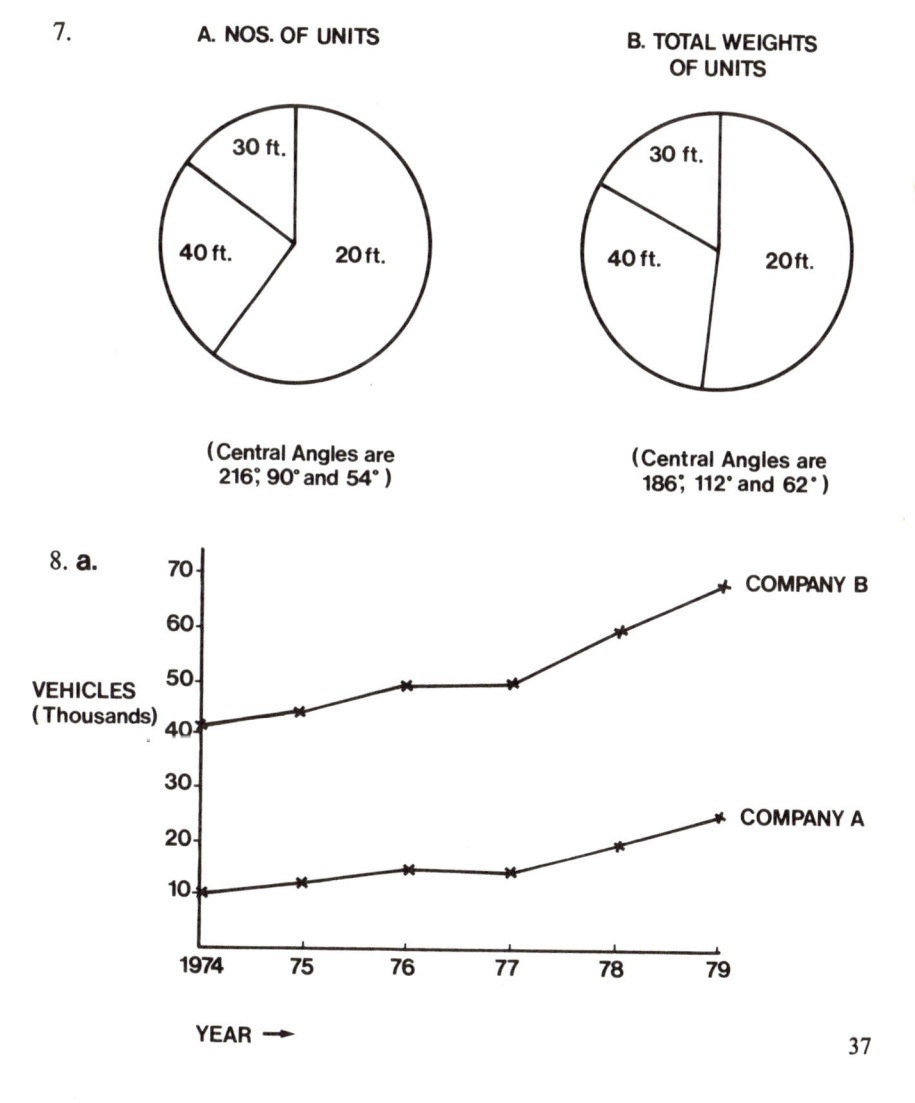

   30 ft.   30 ft.
   40 ft.   20 ft.   40 ft.   20 ft.

(Central Angles are   (Central Angles are
 216°, 90° and 54°)   186°, 112° and 62°)

8. a.

70
60
VEHICLES
(Thousands) 50
40
30
20
10
   1974 75 76 77 78 79

COMPANY B
COMPANY A

YEAR →

37

Marine Statistics

b.

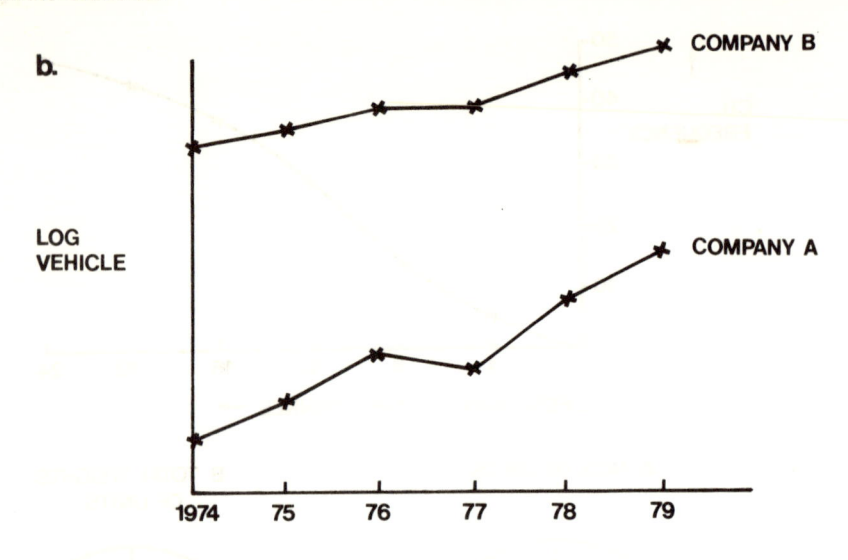

[Graph showing LOG VEHICLE on y-axis and years 1974-79 on x-axis, with two lines: COMPANY B (upper) and COMPANY A (lower, rising more steeply)]

In graph (a) the time series curves are diverging, indicating that company B is generally increasing its number of vehicles carried per annum by more vehicles than company A. In graph (b) the curves are converging, indicating that, in relation to its size, company A is generally increasing its throughput of vehicles at a greater rate than company B.

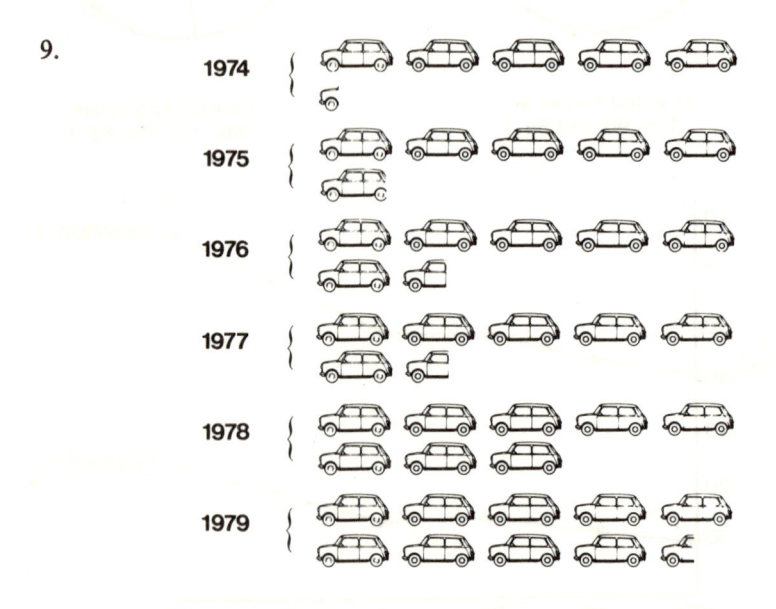

9.

1974 { [6 cars]

1975 { [7 cars]

1976 { [7 cars]

1977 { [7 cars]

1978 { [9 cars]

1979 { [10 cars]

EACH SYMBOL REPRESENTS 1000 VEHICLES

Tables and Graphs

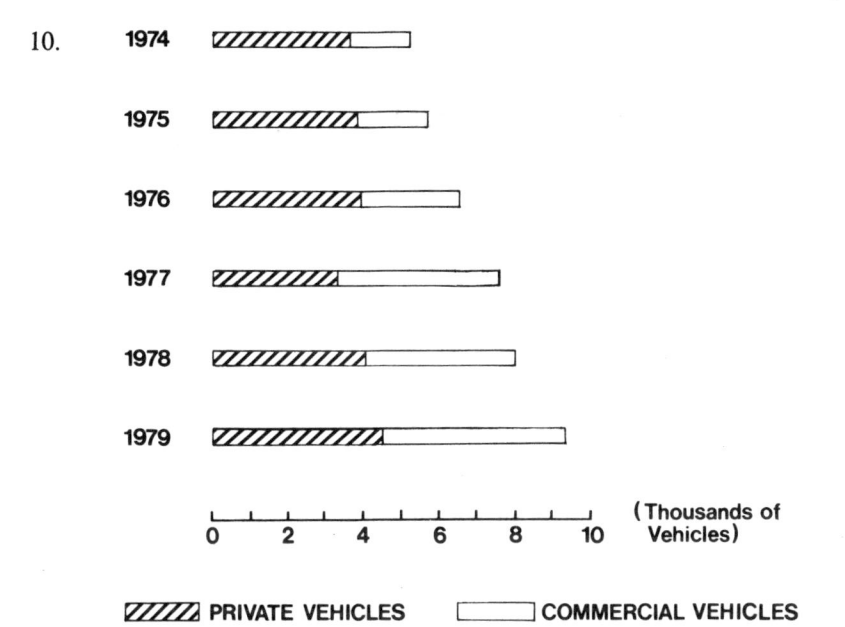

10.

Year	
1974	
1975	
1976	
1977	
1978	
1979	

(Thousands of Vehicles)

▨ PRIVATE VEHICLES   ☐ COMMERCIAL VEHICLES

Chapter 3
Measures of Location

3.1 Introduction

'All the crew on this ship have more than the average number of legs for the population' remarked the chief engineer who was doing a course in statistics. Needless to say this statement was greeted with amazement by everyone else who did not think that the crew of the ship were abnormal in any way. Yet the engineer was perfectly correct in his statement, because in the population although most had two legs, there may have been one with only one leg or possibly none at all. However, if he were to read a little further into the chapter on measures of location in his book he might have phrased his statement a little differently as we will soon find out.

Let us continue with our aim of reducing a mass of raw data into a form whereby the significant features stand out. In the last chapter we considered tabular and graphical presentations of raw data but we were still left with having to take in several pieces of information at the same time. Whenever one reads a book or an article one remembers the author's central theme even if the detail is gone and similarly, with numerical data, it is useful to have some form of central measure. Again if we want to compare two distributions then it is easiest done through a single number for each distribution. Hence our aim now is to find single numbers to sum up a distribution, and for this we will find that the measures of central tendency which locate central points in the distribution are the most useful.

3.2 Arithmetic Mean

Probably the best known of the measures of central tendency is the *arithmetic mean* or the *average*. However, it is a good idea to try to avoid the word 'average' because it is often used in everyday language as synonymous with measure of central tendency and hence can refer to one of the other measures. There is also a measure known as the *geometric mean* but whenever there is no danger of confusion, the arithmetic mean is referred to simply as the *mean*. In this book in future if the term mean is used it may be taken to be the arithmetic mean.

3.2.1 The *arithmetic mean* or *mean of a set of numbers* **is the sum of the numbers divided by the number of contributions to the total.**

Measures of Location

As an example suppose there is a part cargo of 2,000 bags of similar shape and weight to be loaded. Ten bags are chosen at random from the cargo and their weights are 48·7, 49·8, 51·2, 47·3, 52·4, 48·6, 49·2, 50·1, 50·6 and 48·1 kg respectively. Then the mean weight of the ten bags is

$$\frac{(48·7 + 49·8 + 51·2 + 47·3 + 52·4 + 48·6 + 49·2 + 50·1 + 50·6 + 48·1)}{10}$$

$$= \frac{496}{10} = 49·6 \text{ kg}$$

Hence, if we can assume that any one bag has a mean weight of 49·6 kg, the total weight of the 2,000 bags will be estimated to be 2,000 × 49·6 kg = 99·2 tonnes. Having established the idea of a mean it may be instructive if we have a short break from it to discuss some very useful mathematical notation.

The sign $\sum$, pronounced sigma, is the capital letter in the Greek alphabet for the sound s, but mathematically it means *sum* or *add up*. It can be used in a number of slightly different ways. Suppose we have n readings on a variable x, which we can call $x_1, x_2, \ldots, x_n$. Then $\sum x$ means add up all the values of x. $\sum_i x_i$ is slightly more formal and is saying give x_i each of the values of i, in this case 1 to n inclusive and then add them up. $\sum_{i=1}^{i=n} x_i$ is the most formal of all and is saying take the values $x_1, x_2, \ldots, x_n$ and add them all up. This is mathematically the most correct form but can often look rather formidable so if there is no possible reason for ambiguity one of the two previous forms is often used.

However, suppose we only wanted to add up the values of x from the third to the penultimate one then we would have to use the last form, viz.:

$$\sum_{i=3}^{i=n-1} x_i \quad \text{or} \quad \sum_{i=3}^{i=n-1} x_i$$

The $\sum$ notation enables us to produce a formula for the calculation of the mean which is therefore much quicker than having to write down a long verbal description each time. This is in fact generally true—that a mathematical formula is simply a shorthand for what can be a lengthy process to describe. There are two alternative notations used for the mean of a set of data of measurements on a variable x. The symbol $\bar{x}$, pronounced x-bar, is used if the data we are referring to is sample data taken from a larger population. The symbol μ, pronounced mu, is used if the data we are referring to is population data. At this stage it does not matter which we use, but it will become important when we reach the later chapters and discussion of how to use the results from a sample, to generalise about a population.

It is now time to rewrite the definition of the mean in mathematical terms.

3.2.2 The *mean* of a set of numbers is given by the formula

$$\bar{x} \text{ (or } \mu) = \frac{\sum x}{n}$$

where n is the number of terms.

Marine Statistics

Thus in our example on the bags x would be the weight of one bag and $n = 10$. Although this definition and formula are correct in any situation it is easier if we rewrite them for the case of a frequency distribution.

Consider first of all a situation whereby we know the maximum air temperature recorded in a certain place on every day in the month of June during one year. The frequency distribution would be as given in Table 3.1.

Table 3.1. The maximum daily air temperature recorded throughout June in degrees Centigrade

Temperature: x	Number of days: f
11	1
12	1
13	2
14	3
15	3
16	4
17	3
18	4
19	2
20	2
21	2
22	2
23	1
Total	$\Sigma f = 30$

The variable which we are measuring each day is the maximum air temperature and hence we can call this x. The other column gives the number of times each value of x occurs and we can call this f for frequency. To find the mean maximum daily air temperature we must take into account that whereas a value of 11°C occurred only once, a value of 16°C occurred four times and hence we want a contribution of $4 \times 16°$ to the total temperature sum, thus:

$$\mu = \frac{[1 \times 11 + 1 \times 12 + 2 \times 13 + 3 \times 14 + 3 \times 15 + 4 \times 16 + 3 \times 17 + 4 \times 18 + 2 \times 19 + 2 \times 20 + 2 \times 21 + 2 \times 22 + 1 \times 23]}{30}$$

$$= \frac{510}{30} = 17°C$$

We can write this result in formula form as

$$\bar{x} \text{ (or } \mu) = \frac{\sum_i (f_i x_i)}{\sum_i f_i} \quad \text{or just} \quad \frac{\Sigma (fx)}{\Sigma f}$$

Measures of Location

In other words what we must do to get the mean, is to multiply each value of x by the number of times, f, it occurs, to get each fx and then add all these products together dividing the final total by the number of values in the distribution, obtained by adding up all the frequencies. In earlier books on statistics a lot of attention was paid to methods of calculating the mean and other statistical measures which cut down the amount of mental arithmetic involved. The poor student therefore, not only had to learn the basic definition and formulae for the mean, but also all the other formulae which acted as aids to computations. In this present age of computers and electronic calculators, these alternative methods just make for another burden so they will be omitted here. Many calculators today have a memory which makes it possible to compute the sum of a number of two factor products, such as $\sum fx$, without writing down the individual products. However, if this is not available it is a good idea to include a third column, for working, alongside the frequency distribution, which shows the individual values of each product, fx. The next example will illustrate this point.

Table 3.2 shows the frequency distribution for the length of time that a particular component, of a certain make of VHF transmitter, lasted although it had a guaranteed life of 2 yr. The time was recorded for 40 similar components to the nearest tenth of a year.

Table 3.2. Frequency distribution of components' lives

Length of life	Number
1·0–1·4	2
1·5–1·9	1
2·0–2·4	4
2·5–2·9	15
3·0–3·4	10
3·5–3·9	5
4·0–4·4	3
Total	40

Looking at the data we can see that no component failed in the first year, presumably because they were given an initial test in the factory and the immediate duds were sorted out then. We can also see that none of them lasted more than 4·4 yr, although we have no ideas as to the exact length of time the longest one lasted. We are interested however in finding the mean life of one of these components. As we have no information on the exact length of life of the component within the range of values given we have to make an assumption, and what is usually done is to assume that all values falling within a group or class take the midpoint of that group, the *class mark*. It was suggested in the last chapter, when considering how to form frequency

Marine Statistics

distributions, that if there were any points of clustering then these should be made midpoints of classes for this reason. In any case, assuming that the values falling within a class are reasonably evenly spread, then the midpoint is a good representative value to take. Of course in most cases when we are presented with a ready formed frequency distribution we have no way of knowing how the values are spaced.

We can now prepare a working table for the calculation of the mean component life, which is given in Table 3.3.

Table 3.3. Calculation of mean component life

Length of life	Midpoint: x	Frequency: f	fx
1·0–1·4	1·2	2	2·4
1·5–1·9	1.7	1	1·7
2·0–2·4	2·2	4	8·8
2·5–2·9	2·7	15	40·5
3·0–3·4	3·2	10	32·0
3·5–3·9	3·7	5	18·5
4·0–4·4	4·2	3	12·6
Totals		$\Sigma f = 40$	$\Sigma fx = 116·5$

The second column gives the midpoint of each class which, as we have discrete data, is the mean of the two end points of the class. The same rule applies if we have continuous data. The midpoints give the values of x and the third column shows the frequency, f, with which each x-value occurs. The fourth column, which may be omitted, shows the individual products fx. The mean is then calculated as before:

$$\mu = \frac{\Sigma fx}{\Sigma f} = \text{sum of the fourth column divided by the sum of the third column}$$

$$= \frac{116.5}{40} \simeq 2·91 \text{ yr}$$

It should be noticed that 2·91 yr is given as the result because, as a very general rule, the result of a statistical calculation is quoted to one more significant figure than the original data shown.

3.2.3 The *mean* of a *frequency distribution* is given by the formula

$$\bar{x} \text{ (or } \mu\text{)} = \frac{\sum_i (f_i x_i)}{\sum_i f_i}$$

Measures of Location

where x_i is the ith value of x or, in a grouped distribution, the midpoint of the ith class, and f_i the frequency associated with x_i.

The arithmetic mean is not however the only way to sum up a set of numbers as we mentioned at the beginning, so we will describe the other two major ones and then compare them all.

3.3 Median

3.3.1 The *median* of a set of numbers is that number which lies in the middle when the set is arranged in ascending order.

Suppose nine trainee navigators were asked to estimate the distance of another ship from their own. Their estimates were: 3·0, 2·5, 1·75, 2·0, 0·5, 4·0, 1·75, 6·0 and 1·25 nautical miles. If we calculated the mean of these estimates we would get about 2·5 nautical miles. However this is likely to be rather high for the true reading as one of the values, 6·0 nautical miles, is very much larger than the others. There is one extreme small value of 0·5 nautical miles but there is a limit anyway of 0 as to how small this could be, but the large values could be of any size depending on how wild the guess was.

If we arrange the estimates in ascending order, thus: 0·5, 1·25, 1·75, 1·75, 2·0, 2·5, 3·0, 4·0, 6·0 then we can see that the middle value is 2·0 nautical miles, which is the median of this distribution. This value seems more representative in the situation than the mean which was distorted by extreme values. As a general rule it is more satisfactory to use the median whenever there are extreme values in a distribution.

The notation for the median of a set of readings on a variable x is $\tilde{x}$ (pronounced x-tilde). If there had been an even number of values, then we would have calculated the median as the midpoint of the two middle values. Thus in the example above, if we did not have the reading of 0·5, we would be left with eight readings, the middle two being 2·0 and 2·5. The median would therefore be taken to be 2·25.

To calculate the median of a frequency distribution we make use of the associated cumulative frequency distribution. Consider first the data on maximum daily temperatures in June whose cumulative frequency distribution is shown in Table 3.4.

Strictly speaking the median of 30 items is the value of the $15\frac{1}{2}$th item but for large amounts of data of n items, where n is 30 or more, we usually say the median is the value of the $n/2$th item. Hence in this case we want the value of the 15th item. This falls against the entry '17 and less' and so as we are ascending in 1° steps, we conclude that $\tilde{x}$, the median, is 17°C. This was the same value as we obtained for the mean, which will happen whenever the distribution is reasonably symmetrical.

Consider now the data on components' lives for which the cumulative frequency distribution is shown in Table 3.5.

Marine Statistics

Table 3.4. Cumulative frequency distribution of maximum daily temperature in June

Temperature °C	Cumulative frequency
11	1
12 and less	2
13 and less	4
14 and less	7
15 and less	10
16 and less	14
17 and less	17
18 and less	21
19 and less	23
20 and less	25
21 and less	27
22 and less	29
23 and less	30

Table 3.5. Cumulative frequency distribution of length of life of components

Length of life in years	Cumulative frequency
0·9 and less	0
1·4 and less	2
1·9 and less	3
2·4 and less	7
2·9 and less	22
3·4 and less	32
3·9 and less	37
4·4 and less	40

The median observation is observation number 20 which lies in the class 2·4–2·9, which we can call the median class. To find its exact value we can either use the ogive and read off the value of observation number 20 or we can use linear interpolation in the class 2·4–2·9, which we shall do here.

As we have a discrete variable, on a continuous scale the class stretches from 2·45 to 2·95 yr, an interval of 0·5 yr. At the start of this class we have only got to observation number 7, but at the end of it we have reached observation number 22, and we want to know where observation number 20 is. Assuming that the 15 observations are evenly spread in this interval, then we want to know where the 13th of them will lie. The answer is given by calculating $2·45 + 13/15 \times 0·5 = 2·45 + 0·4333 \simeq 2·88$ yr.

We add onto the starting point of the class a fraction of the class interval, where the fraction is the relative position of the median observation to all the other observations in the median class.

3.4 Mode

3.4.1 The *mode* of a set of numbers is the value which occurs most frequently.

Suppose four people read the temperature of the room from a thermometer and three of them said that it was 21°C and one of them said 19°C, then the conclusion that we would probably draw is that 21°C is the most suitable value to take for the temperature of the room. We would in this situation be swayed by the majority figure and suppose that the 19°C represented a value not as typical as the others, presumably because the thermometer had been wrongly read. Of the four figures, 21°C represents the mode or modal value and is the most typical, or loosely speaking the best 'average' we can get. The mean of these figures, at $20\frac{1}{2}$°C, is probably a distortion and, in many cases such as this, it is unnecessary to get people to understand the concept of the median if the mode provides a suitable answer.

If we have a set of qualitative data, say the frequency with which a variety of code numbers for cargo markings arise, then it is impossible to do any form of calculation and the mode (the code number which occurs most frequently) is the only representative value we can quote. We can now return to our chief engineer's statement of the number of legs in the crew. As far as a trouser manufacturer is concerned, the most meaningful central measure is the mode, and he will make all his trousers with two legs. However, the numerical value of the mean for this distribution will be probably 1·99··· legs per person. The crew's confusion would have been avoided if the engineer had not used that overworked word 'average'.

Having discussed some situations where the mode is a very useful concept, we must set the balance straight by considering some situations where it is not applicable. If we take the temperature data given in Table 3.1 then immediately we can see that the mode is not always uniquely defined. On four occasions each, temperatures of 16° and 18°C were recorded, which is an example of a bimodal distribution. It is also worth noting that the frequencies associated with all the recorded values were very similar, so for this reason as well, the mode is not a very suitable measure in this case.

If we are dealing with a grouped frequency distribution it is impossible to define the mode at all. The distribution of the components' lives given in Table 3.2 helps to illustrate this. The largest frequency of 15 is associated with the class 2·5–2·9 yr, so we might term this the 'modal class' but we have no way of knowing that the mode itself lies in this class. For instance in the original ungrouped data we might have had a frequency of 3 associated with

Marine Statistics

each of the values 2·5, 2·6, 2·7, 2·8 and 2·9 making a total of 15 in the 2·5–2·9 class, but a frequency of 10 associated with the value 3·0 making this the actual mode. Some statisticians suggest a graphical method of determining where the mode might lie, on the assumption that it does lie in the modal class. The method consists of drawing part of the histogram for the modal class and the two adjacent classes on either side. This is done in Fig. 3.1 for

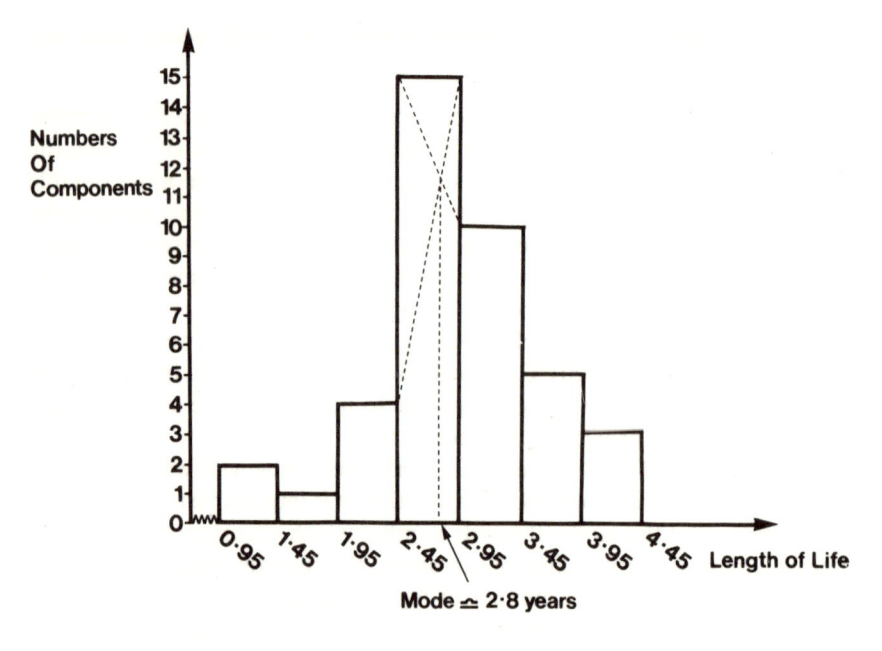

Fig. 3·1 Graphical method for finding the mode for the components data

the components' data. The tops of the rectangles are then joined in the manner shown in the diagram, i.e. the top left corner of the modal class of components is joined to the top left corner of the class above, and the top right corner of the modal class is joined to the top right corner of the class below. The intersection of these lines is then read off on the continuous scale below, remembering that if the distribution is discrete, as in this case, care must be taken to link adjacent classes. Thus the lower end of the modal class should be read as 2·45 yr, giving the mode as approximately 2·8 yr. This result may be checked as the effect of the construction is to divide the class interval in a ratio determined by the difference in frequencies on either side of the modal class. The difference between the frequency in the modal class and the class above it is 15 − 10 = 5, where on the other side the difference is 15 − 4 = 11. Thus the mode is located 11/16 of the distance along the modal class interval. However, it is debatable whether this sort of calculation is really necessary or even justifiable since the mode may not even lie in the modal class. Another

Measures of Location

suggestion often adopted is to take the midpoint of the modal class as an approximation to the mode. For the components data this gives a value of 2·7 yr and certainly makes little difference in this case. In most situations, however, it is sufficient to use the modal class as an indication of where the greatest frequency lies, rather than attempting to get a single value.

3.5 *Other Measures of Central Tendency*

There are two other measures of central tendency which will be mentioned briefly, but their use is very limited in comparison with the three previously considered. The first of these is the geometric mean and if there are n readings on a variable x, this is defined as the nth root of the product of the n readings.

Thus if there is a sample consisting of values $x_1, x_2, \ldots, x_n$, the geometric mean of these values

$$= \sqrt[n]{x_1 . x_2 . x_3 . \ldots . x_n}$$

For example, if there is a set of four numbers, 2, 3, 12, 18, then the geometric mean of these numbers is $\sqrt[4]{2.3.12.18} = \sqrt[4]{1{,}296} = 6$.

The second is the harmonic mean and if there are n readings on a variable x this is defined as the reciprocal of the arithmetic mean of the reciprocals of the values. Thus if there is a sample consisting of values $x_1, x_2, \ldots, x_n$, the harmonic mean of these values is

$$\frac{1}{\frac{1}{n}\left[\frac{1}{x_1} + \frac{1}{x_2} + \frac{1}{x_3} + \cdots + \frac{1}{x_n}\right]}$$

For the same set of numbers as above then the harmonic mean is

$$\frac{1}{\frac{1}{4}\left[\frac{1}{2} + \frac{1}{3} + \frac{1}{12} + \frac{1}{18}\right]} = \frac{4}{\frac{35}{36}} = 4 \cdot 11$$

The geometric mean is used mainly when the individual values are ratios or rates of change. The harmonic mean is of very limited use apart from some special situations. For instance if a ship steams at 15 knots for an outward 10 nautical miles and at 11 knots for the return 10 nautical miles then the speed which will have been averaged for the whole journey is the harmonic mean of 15 and 11 not the arithmetic mean. This may be shown as follows: since the time for the first leg is 10/15 hr and for the second leg is 10/11 hr, the total time is 10/15 + 10/11 hr to go a distance of 2 × 10 nautical miles.

$$\text{Speed over whole journey} = \frac{2 \times 10}{\frac{10}{15} + \frac{10}{11}} = \frac{1}{\frac{1}{2}\left[\frac{1}{15} + \frac{1}{11}\right]} = 12 \cdot 7 \text{ knots}$$

which is the harmonic mean of 15 and 11.

3.6 *Comparison of the Mean, Median and Mode*

To conclude the work on measures of central tendency, a comparison of the various properties of the most important measures may make it easier to decide which measure is the most suitable in a particular situation. Most of the points made will have been discussed at the appropriate point earlier in the chapter so will not be elaborated on here. It is not worth including the geometric mean and harmonic mean as their use is so limited.

The Mean

(1) Every value in the distribution is used in calculating the mean, therefore it is very representative.

(2) An extreme value in the data can distort the mean badly.

(3) It is a simple concept and one that is widely understood.

(4) It is calculated purely arithmetically and the concept may be expressed in a simple algebraic formula. This makes it very suitable for advanced mathematical treatment.

(5) The value obtained for the mean will often be a value which is not meaningful in the context of the data, e.g. mean number of people per household in a survey was found to be 3·2, whereas in no household was this number actually found!

The Median

(1) Extreme values in the data have no distorting effect on the median.

(2) It is an unfamiliar concept to many people although not a difficult one.

(3) It is not calculated by an arithmetic process and hence cannot easily be described algebraically.

(4) It will probably be an actual value in the distribution.

The Mode

(1) Extreme values in the data have no distorting effect on the mode.

(2) It is an extremely easy concept.

(3) It is not calculated by an arithmetic process and hence cannot easily be described algebraically.

(4) It will be an actual value of the distribution.

(5) It is sometimes not defined uniquely.

(6) It is difficult in a grouped frequency distribution to assign a value to the mode.

(7) If the data are qualitative it is the only representative value which can be given.

These points should be borne in mind if one is faced with having to decide which measure to use for a certain set of data or if one is having to evaluate

Measures of Location

someone else's choice of measure. However, there is no doubt that the choice of measure will depend on one's purpose for a set of data as the following example may illustrate. The earnings structure in a certain small shipping company is a reasonably simple one as the table below shows in that there are only a few salary levels.

Annual salary (£)	Number of employees
2,500	4
4,000	20
5,000	10
6,000	9
8,000	4
10,000	2
20,000	1

The union negotiators choose the mode of £4,000 as their average salary, while the management choose the mean of £5,320 as their average salary. An outside arbitrator wishing to compare with other companies chooses the median of £5,000 as their average salary, since half the employees earn below that amount and half the employees earn above it and it thus provides a useful comparison.

3.7 Other Measures of Location

The chapter is headed 'measures of location' and so far we have only considered measures of central tendency, which are in fact the most frequently needed of the measures of location. However, it can also be useful to locate other points in the distribution, such as the values which divide the distribution into quarters, or tenths, or hundredths.

3.7.1 **The *lower* or *first quartile of a distribution* is the value such that one quarter of the distribution lies below it and three-quarters lies above it.** It is usually written as Q_1.

3.7.2 **The *upper* or *third quartile of a distribution* is the value such that three-quarters of the distribution lies below it and one quarter lies above it.** It is usually written as Q_3.

In line with these two definitions the median of a distribution may be thought of as the second quartile and can be written as Q_2.

3.7.3 **The *deciles of a distribution* divide the distribution into tenths.** Thus the lowest decile is the value such that 10% of the distribution lies below it and 90% lies above it. Similarly the highest or ninth decile is the value with 90% of the distribution below it and 10% above. These two are the most frequently used of the deciles, but obviously the median may be interpreted as the fifth decile.

Marine Statistics

3.7.4 The *percentiles of a distribution* divide the distribution into hundredths. Thus the 34th percentile for example is the value such that 34% of the distribution lies below it and 66% lies above it.

3.7.5 In general the terms *fractiles* or *quantiles* may be used to describe the values which divide the distribution into a given fraction.

Any of these measures may be calculated in a similar way to that used for the median. The first step is to locate the number of the observation corresponding to the required fraction. Thus if we wanted the first decile of a distribution with 50 observations we would want to know the value of observation number 5. In general if we have n observations and we want to know the number of the observation such that $1/k$th of the distribution lies below it then the required number is n/k.

The second step is to read off the value of the observation from the ogive or to use linear interpolation within the class in which the observation lies. As an example suppose we want the upper quartile of the components' data. There are 40 items in the distribution so we want the value of observation number $30 = \frac{3}{4} \times 40$. This lies in the class 3·0–3·4 in which 10 observations in all lie.

$$Q_3 = 2\cdot 95 + \frac{8}{10} \times 0\cdot 5 \simeq 3\cdot 35 \text{ yr}$$

Alternatively we could have got this from the ogive shown in Fig. 3.2.

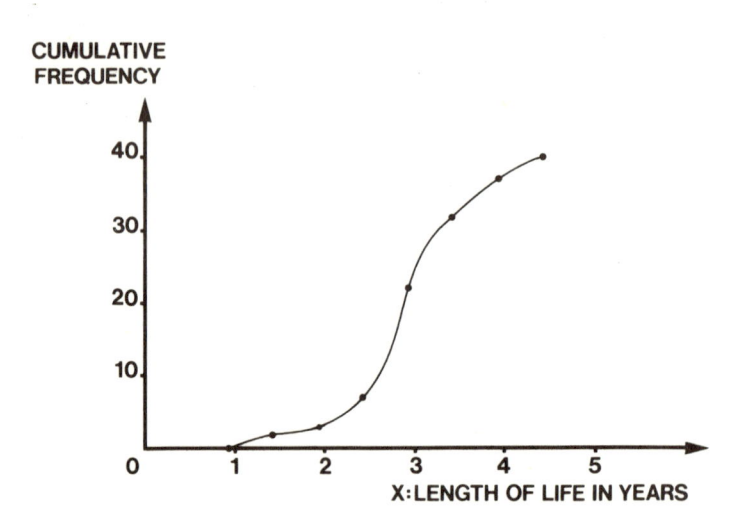

Fig 3·2 Ogive of the components data

As a second example we can see from the ogive that the lower quartile, the value of observation number 10, is 2·55 yr.

3.8 Summary

We have seen in this chapter how a set of data may be reduced to one single representative value. Various alternative definitions have been considered for a suitable single value but the final choice depends on the particular set of circumstances. Apart from the *mean*, the *median* and the *mode* which are the three most important ways of summing up a distribution, some less well-known methods have been discussed together with measures for locating other key points in the distribution. In the next chapter we will consider ways of calculating additional values which help to give a more comprehensive picture of a distribution.

Exercises

1. Using the weights of the containers given in Ex. 1 of Chapter 2:
   (a) Calculate the mean weight of a container.
   (b) Find the median weight.
   (c) Explain why the mode is not a suitable measure for these data.

2. Fifty samples of a wire rope were tested to destruction and the breaking loads were measured to the nearest tenth of a tonne. The results are summarised in the frequency distribution below. Find (a) the mean and (b) the median breaking load.

Breaking load (tonnes)	Number of samples
4·0–4·9	4
5·0–5·9	8
6·0–6·9	22
7·0–7·9	13
8·0–8·9	3

3. At a particular port, the delays to 100 ships in berthing were noted precisely and summarised in the following frequency distribution.

Delay limits (hr)	Number of ships
0 but less than 12	11
12 but less than 24	20
24 but less than 36	24
36 but less than 48	17
48 but less than 60	13
60 but less than 72	9
72 but less than 84	4
84 and over	2

Marine Statistics

(a) Calculate the mean delay time.
(b) Find the median delay time.
(c) What is the modal class?
(d) How would you explain the fact that the mean and the median are appreciably different in this example whereas in Ex. 1 they were almost the same?

4. A consignment of general cargo is tallied into a hold as follows:

Cargo mark	Tally
LOND/4	⊮ ⊮ ‖
◊-LA	⊮ ‖‖
$\dfrac{CP}{B}$	⊮ ⊮ ⊮ ⊮ ‖‖
19-14-32	⊮ ⊮ ⊮
EXPO-10	⊮ ‖
ADAMS	‖‖
9 ⊛ 5	⊮

What measure of location is suitable for describing this distribution?

5. At the port in Ex. 3 it is policy that additional berths should be constructed if more than one tenth of the ships using the facilities have to wait for more than 3 days. What is the upper decile delay time for the sample given and how does this relate to the port management policy?

Answers

1. (a) Mean, $\bar{x} = \sum x/n = 542.5/48 = 11.30$ tonnes.
   (b) Median, $\tilde{x} = n/2$th value = 24th value = 11.3 tonnes. Alternatively, interpolate between 24th and 25th value to give $\tilde{x} = 11.35$ tonnes.
   (c) The mode is not a suitable measure for these data because several weights occur twice, i.e. 3.3 tonnes, 3.6 tonnes, 7.4 tonnes, 9.7 tonnes, 12.2 tonnes and 14.2 tonnes, and none occurs more frequently. Quoting these multi-modal values is not a very useful way of describing the data.

Measures of Location

2. (a)

Breaking load (tonnes)	Class mark x	Frequency (f)	fx
4·0–4·9	4·45	4	17·8
5·0–5·9	5·45	8	43·6
6·0–6·9	6·45	22	141·9
7·0–7·9	7·45	14	104·3
8·0–8·9	8·45	2	16·9
		$\Sigma f = 50$	$\Sigma fx = 324·5$

Mean, $\bar{x} = \Sigma fx / \Sigma f = 6·49$.

(b)

Breaking load (tonnes)	Cumulative frequency
4·9 or less	4
5·9 or less	12
6·9 or less	34
7·9 or less	48
8·9 or less	50

Median, value of 25th observation:

$\tilde{x} = 5·95 + 1·0 \times \frac{9}{22}$
$= 6·36$ tonnes

3. (a)

Delay limits (hr)	Class mark x	Frequency (f)	(fx)
0 but less than 12	6	11	66
12 but less than 24	18	20	360
24 but less than 36	30	24	720
36 but less than 48	42	17	714
48 but less than 60	54	13	702
60 but less than 72	66	9	594
72 but less than 84	78	4	312
84 and over	90	2	180
		$\Sigma f = 100$	$\Sigma fx = 3,648$

Mean, $\bar{x} = \Sigma fx / \Sigma f = 36·48$ hr.

Marine Statistics

(b)

Delay time (hr)	Cumulative frequency
Less than 12	11
Less than 24	31
Less than 36	55
Less than 48	72
Less than 60	85
Less than 72	94
Less than 84	98
Unlimited	100

The median class (containing the 50th ship) is '24 but less than 36 hours' and, by linear interpolation:

$$\tilde{x} = 24 + 12 \times \tfrac{19}{24}$$
$$= 33.5 \text{ hr}$$

(c) The modal class is '24 but less than 36 hours'.

(d) The appreciable difference between the mean and the median is typical of a non-symmetrical (skewed) distribution.

4. The mode is the only appropriate measure and is clearly equal to CP/B since this is the most numerous class.

5. The cumulative frequency table given in answer 2(b) indicates that the upper decile lies between the limits '60 but less than 72 hours'. Linear interpolation gives the decile as:

$$D_9 = 60 + 12 \times \tfrac{5}{9}$$
$$= 66.67 \text{ hr}$$

On this evidence, new berths are not yet required under the port management policy.

Chapter 4
Measures of Dispersion

4.1 Introduction

At a certain port there are two cargoes, waiting to be loaded onto a ship, each of which consists of 1,000 bags of similar size. The bags in the two cargoes are the same shape and size and on first glance there appears to be little difference between them. A sample of five bags is taken from the first cargo and when weighed the readings are: 48·3, 49·4, 49·6, 50·2 and 50·5 kg, giving a mean value of 49·6 kg. A sample of five bags is then taken from the second cargo and their weights are found to be 38·4, 42·5, 49·6, 56·8 and 60·7 kg, again giving a mean value of 49·6 kg. However, comparison of the two sets of weights suggests that although the two samples have the same mean weight, there is a considerable difference in the variability in weights within the samples. The first cargo would appear to be much more homogeneous than the second. By summarising data into one single statistic, such as the mean, there may be too great a simplification of the picture, and especially when comparing sets of figures there is a need for an additional statistic to denote the amount of variability in the data. In this chapter we are concerned with various statistics which do this and which are termed *measures of dispersion*, or *spread* or *scatter*. As with the measures of location there are several measures of dispersion which can be used, the most suitable one for any situation depending on the particular nature of the data.

4.2 Range

The most simple of all these measures is the *range* which is the difference between the highest and smallest values in the data. Thus, considering the cargoes we have been talking about, the range in the first sample is 50·5 − 48·3 kg = 2·2 kg, and the range in the second sample is 60·7 − 38·4 kg = 22·3 kg. Immediately the main difference between the two samples is highlighted and for this particular example the range provides a very suitable measure of scatter. If we are dealing with a grouped frequency distribution then there is some disagreement as to the most suitable definition of the range. Some statisticians argue that the range should be taken as the difference between the top limit of the final class and the bottom limit of the first class,

Marine Statistics

whereas others argue that it should be taken as the difference between the midpoints of the two end classes. As the convention for the calculation of the mean is to consider that all values in a class take the midpoint of the class, the second definition will be used here. Although it may well give a slight under-estimate for the range, the error is likely to be less than with the over-estimate that would occur with the first definition.

4.2.1 **The *range of a set of data* is the difference between the highest and lowest values in the data.** For a grouped frequency distribution this is interpreted to be the difference between the midpoints of the two end classes.

As an example of the range of a grouped frequency distribution, we can consider the components' data given in Section 3.2. Using the definition given above the range would be calculated as 4·2 − 1·2 yr, giving a value of 3·0 yr.

Although the range is straight-forward to calculate and easy to understand, it does present other problems as a suitable measure of dispersion in some circumstances. We have already considered one problem, that of definition for a grouped frequency distribution. The most serious problem is that the range is considerably affected by an extreme value in the data. Suppose we have a series of five numbers, 3, 4, 5, 6, 100, then the range would be 100 − 3 = 97, the result being dominated entirely by the figure 100. Extreme values in data can often arise quite naturally but in some circumstances one may suspect them of being errors of some sort. For instance in this example it might be just a recording error and the true value was in fact 10, but we usually have no justifiable reason for excluding an extreme reading. It would also seem good policy if a measure of dispersion could take into account the values of all the sample members. For instance the sets of numbers 3, 51, 53, 53, 100 and 3, 28, 53, 76, 100 both have the same ranges of 97 and the same means of 52 but there is still a considerable difference in the way the numbers are distributed between the extremes of 3 and 100. Another useful point would be if the measure of dispersion was related to a central point of the distribution, to show dispersion around the central point. In calculating the range no attention is paid to any central point such as the mean of the distribution. A further disadvantage of the range is that there is no mathematical formula which can be used to summarise the method of calculation and as we noted with the measures of location, this limits its use in further theoretical work. However, despite all these limitations which mean that more sophisticated measures must usually be used in detailed work, there is no doubt that the range is very useful for giving an immediate impression of a set of data. When faced with a new set of data it is often the first thing that a statistician will calculate to get a feel for the information. As a final point it must be stressed that there are situations where the range, because it concentrates on extremes, is exactly the statistic we require, as the following example shows:

A bulk carrier has to be designed to carry cargoes of differing stowage factors, so that its stability is at all times satisfactory. These are:

Measures of Dispersion

Coal	stowage factor	1·3 m³ per tonne
Light grain	stowage factor	1·7 m³ per tonne
Iron ores	stowage factor	0·4 m³ per tonne
Manganese ore	stowage factor	0·5 m³ per tonne
Wheat	stowage factor	1·4 m³ per tonne
Slurry	stowage factor	0·8 m³ per tonne
Nitrates	stowage factor	1·0 m³ per tonne
Concentrates	stowage factor	0·5 m³ per tonne

The stability must be satisfactory when the ship is loaded with cargoes with the extreme values of stowage factor, i.e. for light grain with a stowage factor of 1·7 and iron ore with a stowage factor of 0·4 m³ per tonne and hence the range of 1·3 m³ is the measure of dispersion required here.

4.3 Standard Deviation

Having considered the most straight-forward of the measures of dispersion we will turn to what is probably the most complicated one but the one that is used most widely, viz. the *standard deviation*. In defining standard deviation, an attempt is made to overcome most of the various adverse comments that were made about the range. Verbally the standard deviation of a set of numbers may be defined as the square root of the arithmetic mean of the squared deviations of the observations about the arithmetic mean of the distribution. Probably the easiest way to understand this long string of words is to follow through a very simple example.

Suppose we have a set of five numbers, x, {5, 6, 7, 10, 12}. The first step is to find the arithmetic mean of these numbers. Thus

$$\bar{x} = \frac{\sum x}{5} = \frac{5 + 6 + 7 + 10 + 12}{5} = \frac{40}{5} = 8$$

The second step is to find for each number x_i, its deviation d_i, from the arithmetic mean. Thus

$$d_i = x_i - \bar{x}$$

and for

we get

x_i	5	6	7	10	12
d_i	−3	−2	−1	2	4

If we were simply to sum the deviations d_i we would get the value 0, so to get a measure of the magnitude of the deviations about the mean, as a third step we square each of the d_i giving d_i^2. Thus for

and

we get

x_i	5	6	7	10	12
d_i	−3	−2	−1	2	4
d_i^2	9	4	1	4	16

Marine Statistics

The fourth step is to sum these squared deviations giving

$$\sum d_i^2 = 9 + 4 + 1 + 4 + 16 = 34$$

The fifth step is to divide the sum of the squared deviations by the number of observations to give the arithmetic mean of the squared deviations. Thus

$$\frac{\sum d_i^2}{5} = \frac{34}{5} = 6 \cdot 8$$

The statistic which we have calculated at this stage is termed the variance of the set of numbers and is often denoted by the symbol σ^2, pronounced sigma squared. This is the lower case symbol in the Greek alphabet corresponding to the symbol $\sum$ we have met previously. However, despite the same pronunciation the symbols are used in such different contexts that there is no confusion. Thus for this set of data $\sigma^2 = 6 \cdot 8$. However, the variance is measured in the square of the original units of the data, so to relate the measure of dispersion to the same units as the data, as a final step we take the square root of the variance to give the standard deviation of the data. This is often denoted by the symbol σ (pronounced sigma and the Greek letter corresponding to s). Thus for this set of data, the standard deviation is $\sigma = \sqrt{6 \cdot 8} = 2 \cdot 6$.

4.3.1 The *standard deviation* of a set of data is the square root of the mean squared deviation of the observations from their arithmetic mean.

This may be written algebraically as

$$\sigma = \sqrt{\frac{1}{n}\sum_{i=1}^{i=n}(x_i - \mu)^2}$$

where σ is the standard deviation, μ is the arithmetic mean, and n is the number of observations, $x_1, x_2, \ldots, x_n$.

4.3.2 The *variance* of a set of data is the square of its standard deviation.

This may be written algebraically as

$$\sigma^2 = \frac{1}{n}\sum_{i=1}^{i=n}(x_i - \mu)^2$$

with the symbols as defined above.

From the discussion of standard deviation so far, it can be seen that every value in the distribution is taken into account in the calculation and hence the dispersion of all the values is considered. The dispersion is measured about a central point, the arithmetic mean of the data and most important of all the standard deviation has an algebraic definition which means that either the standard deviation, or the variance especially, are easily incorporated in advanced theoretical work. Since every value in the distribution is considered, then extreme values in the data will produce a distorting effect which can

Measures of Dispersion

sometimes be a nuisance but the effect is less marked than with the range. Apart from this disadvantage the most serious problem is that people find it a rather hard concept to understand and the calculation is nearly always very long and tedious especially if it is done by hand. Pocket calculators help considerably in this respect and many now incorporate a feature which automatically calculates the standard deviation of a set of input data. For very large data sets an electronic computer will probably be needed but again this is a very standard output using a prewritten package.

Ability to interpret the significance of the magnitude of a standard deviation comes largely with practice, although in some cases there is a special use of the standard deviation as we shall see later in the book. Measures of dispersion are often required for comparison purposes so if we compare the set of data discussed above, having values of 5, 6, 7, 10, 12 a mean of 8 and a standard deviation of 2·6, with another set having values of 6, 7, 8, 9, 10, a mean of 8 and a standard deviation of 1·4 we have in the standard deviation a means of numerically summarising the fact that the first set has more spread in it than the second. The standard deviation is always measured in the same units as the original data, which may cause problems of comparison, if one set is measured in pounds say, and the other in kilograms. To overcome this it is possible to define a coefficient of variation which is the standard deviation expressed as a percentage of the mean.

4.3.3 The *coefficient of variation* for a distribution is the ratio of the standard deviation to the mean multiplied by 100.

Thus

$$C = \frac{\sigma}{\mu} \times 100$$

With frequency distributions the formulae for the calculation of the standard deviation or the variance have to be adapted slightly, since each value of x does not occur once usually but with a frequency f. This means that after the third stage when each of the deviations of the individual values from the arithmetic mean have been squared to produce d_i^2, an extra stage must be added when each of these terms must be multiplied by its associated frequency f_i to produce $f_i d_i^2$. It is these values of $f_i d_i^2$ which are then summed and the arithmetic mean of the squared deviations then found, by dividing by the total number of observations, i.e. the sum of the frequencies f_i.

4.3.4 The formula for the standard deviation of a frequency distribution is

$$\sigma = \sqrt{\frac{\sum_i f_i (x_i - \mu)^2}{\sum_i f_i}}$$

Marine Statistics

where μ is the arithmetic mean of the distribution and f_i is the frequency associated with the value x_i.

If the distribution consists of single values of x, such as in the maximum air temperature example of the previous chapter, then the values of x taken are simply those as written. However, if it is a grouped frequency distribution then the same convention is adopted as for the calculation of the mean whereby the midpoint of each group is chosen as the representative value of the class. Again no attention will be paid to most of the methods devised for reducing the amount of arithmetic which used to fill statistics books, but an alternative version of the formula above will be considered as it is the most suitable for use on an electronic calculator. The algebraic proof of the equivalence of the two formulae will not be given here but it is reasonably straightforward for those who enjoy manipulations of this type.

4.3.5 The standard deviation of a frequency distribution is most easily calculated from the formula

$$\sigma = \sqrt{\frac{1}{n}\left[\sum_i f_i x_i^2 - \frac{(\sum_i f_i x_i)^2}{n}\right]}$$

where

$$n = \sum_i f_i$$

To see this formula in action let us consider the components' data from the previous chapter which provides an example of a grouped frequency distribution and the working is set out in Table 4.1.

Table 4.1. Calculation of the standard deviation of component life

Length of life in years	Midpoint: x	Frequency: f	fx	fx^2
1·0–1·4	1·2	2	2·4	2·88
1·5–1·9	1·7	1	1·7	2·89
2·0–2·4	2·2	4	8·8	19·36
2·5–2·9	2·7	15	40·5	109·35
3·0–3·4	3·2	10	32·0	102·40
3·5–3·9	3·7	5	18·5	68·45
4·0–4·4	4·2	3	12·6	52·92
		$\sum f = 40$	$\sum fx = 116·5$	$\sum fx^2 = 358·25$

Measures of Dispersion

Then

$$\sigma^2 = \frac{1}{40}\left[358{\cdot}25 - \frac{116{\cdot}5 \times 116{\cdot}5}{40}\right]$$

$$= \frac{1}{40}[358{\cdot}25 - 339{\cdot}31]$$

$$= 0{\cdot}4736$$

Thus

$$\sigma = 0{\cdot}69 \text{ yr}$$

The column fx is calculated by multiplying the x value by the appropriate f value and it is helpful if a note is made of each of these fx values together with their sum which is $\sum fx$. The column fx^2 can then be calculated by multiplying each x value by its corresponding fx value and this time there is no need to make a note of the individual results as a memory facility on a calculator can be used to accumulate them, to produce $\sum fx^2$. Alternatively the fx^2 column may be calculated by taking each value of x, squaring it and then multiplying by the corresponding frequency. The sum of the fx^2 column, in this case 358·25, is put in as the first term in the formula and the second term is the square of the sum of the fx column, i.e. 116·5 × 116·5, divided by the number of observations, 40. The difference between these two terms is then calculated and divided by the number of observations. This gives a value for the variance of 0·47 and the square root of this gives the standard deviation of 0·69 yr.

To conclude the discussion on standard deviation and variance, a brief mention should be made of notation.

If we are talking about data comprising a population then the mean is usually given the symbol μ and the standard deviation the symbol σ. However, if our data is to be treated as a sample from a population then the mean is usually given the symbol $\bar{x}$, as discussed in the previous chapter and the standard deviation the symbol s. Unfortunately, for reasons which we will discuss later, if we are going to use the standard deviation of the sample to draw numerical conclusions about the population then a very slightly different formula has to be used to calculate s:

$$s = \sqrt{\frac{1}{n-1}\left[\sum_i f_i(x_i - \bar{x})^2\right]}$$

or

$$s = \sqrt{\frac{1}{n-1}\left[\sum_i f_i x_i^2 - \frac{\left(\sum_i f_i x_i\right)^2}{n}\right]}$$

Marine Statistics

The essential difference is that we divide by one less than the number of observations in this situation. If we are dealing with a large sample the numerical difference is minimal. Again, if we are treating a set of data purely descriptively, then we can use the basic definition of standard deviation and call it σ since we can consider any set of data to be a population. It is not unusual to see the symbols $\bar{x}$ and σ used together on the same set of data but it is suggested that this should be avoided whenever possible, to help make the situation clearer when we are in a situation of using sample data to draw conclusions about the population. Having said that the standard deviation is the most widely used measure of dispersion, we should now consider an alternative measure which is sometimes preferable in descriptive work.

4.4 *Semi-interquartile Range*

We saw in the previous chapter that there are some situations for which the median is a better measure of central tendency than the mean. These tend to be situations where the distribution has a long tail of extremes in one direction, such as the earnings data we considered, and hence these extremes distort the mean to a certain extent. The same remark is true for the standard deviation and so it is often more suitable to calculate a measure of dispersion based on the quartiles of the distribution.

    4.4.1 **The *semi-interquartile range* of a distribution is one half of the difference between the upper and lower quartiles.**

i.e.   Semi-interquartile range = $\frac{1}{2}(Q_3 - Q_1)$

where Q_3 is the upper quartile and Q_1 is the lower quartile.

    We calculated the quartiles of the components' data in the last chapter to be $Q_3 = 3\cdot35$ yr and $Q_1 = 2\cdot55$ yr. Thus the semi-interquartile range is $\frac{1}{2}(0\cdot80) = 0\cdot40$ yr. This measure is not affected by extreme values at all and is indirectly linked to a central measure, the median, as it is the mean of the distances of the two quartiles from the median. It has the same disadvantages as the median in that it has no simple algebraic formulation and is also introducing a new concept which most people are unfamiliar with. However, whenever the median of a set of data is quoted, it is quite usual to quote the interquartile range with it.

4.5 *Mean Deviation*

In some older statistics textbooks it is possible to come across another measure of dispersion known as the mean deviation about the mean. The principle behind it is to measure the deviation of each observation from the arithmetic mean of the distribution, but instead of squaring these results simply to take the size of them, ignoring the sign, to give the absolute deviations. The mean of these absolute values is then taken. Thus for the set of values 5, 6, 7, 10, 12 the calculation would be as follows:

x_i	5	6	7	10	12	Mean 8
$d_i = x_i - \mu$	-3	-2	-1	$+2$	$+4$	
$\|d_i\|$	3	2	1	2	4	

$$\sum_i |d_i| = 12$$

$$\frac{\sum_i |d_i|}{5} = \frac{12}{5} = 2 \cdot 4$$

$|d_i|$ is the symbol for the size of d_i

Hence the mean deviation about the mean for this data set is 2·4.

Algebraically the formula for the mean deviation about the mean for a frequency distribution is:

$$\frac{\sum_i f_i |x_i - \mu|}{\sum_i f_i}$$

Another variation is to measure the deviations about the median and then proceed in the same way. By ignoring the signs of the deviations, one obtains their sizes, which is an alternative procedure to squaring the deviations. However, although the calculation may be expressed in algebraic formula, absolute values, or moduli as they are also called, are not very easy to manipulate especially in further theoretical work. The improvement in calculating aids has meant that the increased arithmetic involved in squaring values to calculate the standard deviation has become of little importance and hence the mean deviation about the mean is very rarely used now as it had no other advantage over the standard deviation. It is included here purely for reference.

4.6 Comparison of the Range, Standard Deviation, and Semi-interquartile Range

As in the previous chapter a short summary of the various properties of the three most common measures of dispersion may make it easier to decide which measure is the most suitable in a particular situation.

The Range

(1) It is affected badly by extreme values.
(2) The distribution of items between the extremes is not considered.
(3) It is not related to a central measure.

Marine Statistics

(4) It has no easy algebraic formulation.
(5) It is very easy to understand as a concept.
(6) It is easily and quickly calculated.

The Standard Deviation

(1) Every value in the distribution comes into its calculation.
(2) It measures dispersion about the mean of the distribution.
(3) Extreme values can distort it to some extent.
(4) It has an algebraic formulation.
(5) It is rather difficult to understand as a concept.
(6) Calculation without any aid can be long and tedious.

The Semi-interquartile Range

(1) It is not affected by extreme values.
(2) It measures indirectly dispersion about the median of the distribution.
(3) It is only based on the middle half of the observations.
(4) It has no exact algebraic formulation.
(5) It is not a well-known concept.

As a general rule the mean and the standard deviation are the most common pair of statistics to calculate given a set of data. However, the mean and range may often be used together, or the mode and range or, if the median is calculated, then this is usually paired with the semi-interquartile range.

4.7 Skewness

As a final section in this chapter, a brief mention will be given of a third way in which a distribution may be described. The measure of central tendency and the measure of dispersion are the most important but occasionally we need to introduce a third description which is that of *skewness* and concerns the overall shape of a distribution. It is possible to calculate a measure of skewness but this is rarely done, and it is often sufficient to describe a distribution as skew or its opposite symmetrical.

4.7.1 **A distribution is said to be *symmetrical* if there is a central line which divides a distribution into two mirror images.**

4.7.2 **A distribution is said to be *skew* if it has a long tail of extremes in one direction only.**

Comparison of the various frequency curves in Fig. 4.1 illustrates these definitions.

Measures of Dispersion

(a) A unimodal symmetrical distribution
(b) A bimodal symmetical distribution
(c) A negatively skewed distribution
(d) A positively skewed distribution
(e) A symmetrical 'U' shaped distribution

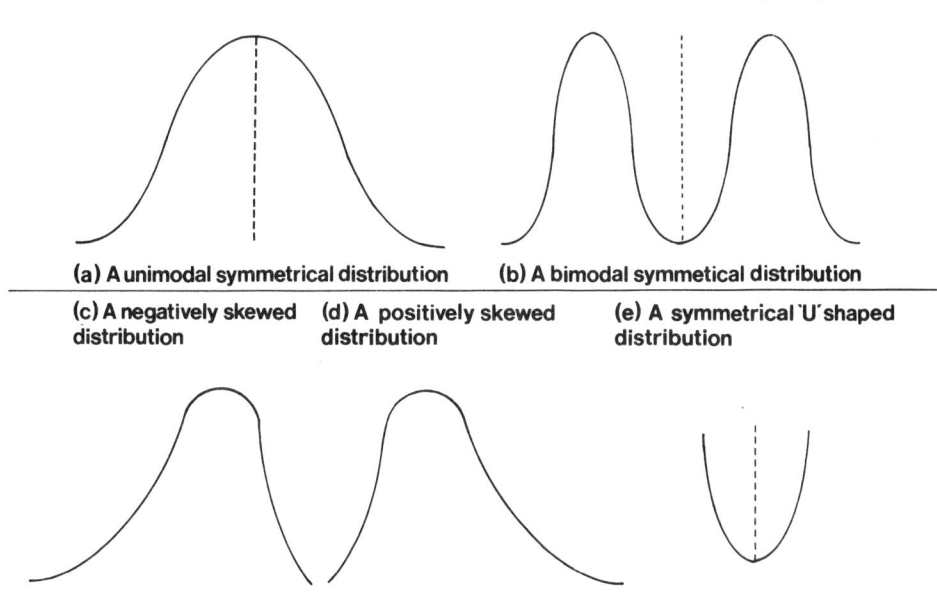

Fig. 4.1 Comparison of different shapes of distributions

The distribution in Fig. 4.1(a) is a symmetrical distribution and in this example the mean, the median and the mode all coincide. Similarly the distribution in Fig. 4.1(b) is also symmetrical but this is a bimodal distribution and only the median and the mean coincide. Figure 4.1(c) shows a distribution with a long tail to the left and is an example of a negatively skewed distribution. The median of this distribution will be greater than its mean. An example of a positively skewed distribution is given in Fig. 4.1(d) where, this time, the tail is to the right. In this case the median would be less than the mean. Another example of a symmetrical distribution is given in Fig. 4.1(e) and this characteristic shape is usually known as a U-shaped distribution. If one does actually want to put a numerical value on skewness then the usual measure is the *Pearsonian coefficient of skewness*.

4.7.3 The Pearsonian coefficient of skewness $= \dfrac{3(\text{mean} - \text{median})}{\text{standard deviation}}$

For symmetrical distributions the mean and the median are equal, hence the value of this coefficient is 0. The more skew the distribution, the larger the distance between the two measures, and the effect of dividing by the standard deviation is to remove the effect of the unit of measurement which was the same technique adopted with the coefficient of variation.

4.8 *Summary*

This chapter is the last one dealing purely with the descriptive aspects of numerical data. In it we have considered methods of summarising the data

Marine Statistics

in terms of the spread of the observations and finally we have talked briefly about the shape of a distribution. After three chapters of describing the raw material with which a statistician works we are now ready to consider some of the tools of analysis which are available.

Exercises

1. With the ship in a surveyed position, a reading from a radio aid was observed on 12 occasions. The displacements, in cables, of the resulting position lines from the true position were noted as follows, a northerly displacement being named positive and a southerly displacement negative:

0·5	2·7	−1·3	3·8	−6·3	0·3
−1·8	8·2	−4·3	−2·1	5·4	1·3

    (a) Find the range of the displacements.
    (b) Find the standard deviation.

2. In a survey, the speeds of 60 ships were recorded as they passed through a traffic separation scheme. The results are given in the frequency distribution below:

Speed (knots)	Number of ships
0–4	6
5–9	14
10–14	24
15–19	10
20–24	4
25–29	2

    (a) Find the range of speeds.
    (b) Calculate the standard deviation.

3. A cargo of oil is loaded into the 9 compartments of a tanker and the temperature is measured in each compartment as follows in degrees Celsius:

    30° 34° 32° 28° 31° 35° 26° 31° 33°

    (a) Find the mean temperature.
    (b) Find the standard deviation of the temperature.

4. The times taken for 8 ships to transit a channel between two reporting points were recorded as follows:

    Hours: 2·5 3·6 2·1 2·7 4·2 2·5 9·2 3·4

Measures of Dispersion

(a) Calculate the standard deviation.
(b) Find the range.
(c) Find the semi-interquartile range.
(d) Comment on the comparative virtues of these three measures for describing this particular set of data.

5. Using the data given in Ex. 4, calculate: (a) the coefficient of variation and (b) the Pearsonian coefficient of skewness.

6. At an estuarial port, the density of the water was measured every hour through a tidal cycle of 13 hr with the following results.

Hours after low water	Relative density
1	1·010
2	1·008
3	1·012
4	1·013
5	1·010
6	1·015
7	1·018
8	1·021
9	1·020
10	1·016
11	1·012
12	1·010
13	1·010

(a) Find the standard deviation.
(b) Find the range.

7. Using the data for ship delay times given in Ex. 3 of Chapter 3, calculate:
(a) The variance.
(b) The semi-interquartile range.

8. Sketch a histogram of the data for ship delay times given in Ex. 3 of Chapter 3.
(a) Is this distribution positively or negatively skewed?
(b) Calculate the Pearsonian coefficient of skewness.

Answers

1. (a) Range $= 8\cdot2 - (-6\cdot3)$
    $= 14\cdot5$ cables

   (b) $\sigma = \sqrt{\frac{1}{n}\left[\sum x^2 - \frac{(\sum x)^2}{n}\right]} = \sqrt{\frac{1}{12}\left[187\cdot68 - \frac{6\cdot4^2}{12}\right]}$
   $= 3\cdot92$

Marine Statistics

x	x^2
0·5	0·25
2·7	7·29
−1·3	1·69
3·8	14·44
−6·3	39·69
0·3	0·09
−1·8	3·24
8·2	67·24
−4·3	18·49
−2·1	4·41
5·4	29·16
1·3	1·69
$\Sigma x = 6·4$	$\Sigma x^2 = 187·68$

2. Range = highest class mark − lowest class mark

    = 27 − 2

    = 25 knots

Speed (knots)	Class mark x	f	fx	fx^2
0–4	2	6	12	24
5–9	7	14	98	686
10–14	12	24	288	3,456
15–19	17	10	170	2,890
20–24	22	4	88	1,936
25–29	27	2	54	1,458
			$\Sigma fx = 710$	$\Sigma fx^2 = 10,450$

$$\sigma = \sqrt{\frac{1}{n}\left[\Sigma fx^2 - \frac{(\Sigma fx)^2}{n}\right]}$$

$$= \sqrt{\frac{1}{60}\left[10,450 - \frac{710^2}{60}\right]}$$

$$= 5·84 \text{ knots}$$

3.

Temp	Temp²
30°	900
34°	1,156
32°	1,024
28°	784
31°	961
35°	1,225
26°	676
31°	961
33°	1,089
280	8,776

$$\mu = \frac{\sum x}{n} = \frac{280}{9} = \underline{31 \cdot 1°}$$

$$\sigma = \sqrt{\frac{1}{n}\left[\sum x^2 - \frac{(\sum x)^2}{n}\right]}$$

$$= \sqrt{\frac{1}{9}\left[8,776 - \frac{280^2}{9}\right]}$$

$$= \underline{2 \cdot 7°}$$

4. (a)

Time	Time²
2·5	6·25
3·6	12·96
2·1	4·41
2·7	7·29
4·2	17·64
2·5	6·25
9·2	84·64
3·4	11·56
30·2	151·00

$$\sigma = \sqrt{\frac{1}{n}\left[\sum x^2 - \frac{(\sum x)^2}{n}\right]}$$

$$= \sqrt{\frac{1}{8}\left[151 - \frac{30 \cdot 2^2}{8}\right]}$$

$$= \underline{2 \cdot 15 \text{ hr}}$$

(b) Range = $9 \cdot 2 - 2 \cdot 1 = \underline{7 \cdot 1 \text{ hr}}$

(c) $\left.\begin{array}{l}2 \cdot 1 \\ 2 \cdot 5\end{array}\right\}$ $Q_1 = 2 \cdot 5$

$\left.\begin{array}{l}2 \cdot 5 \\ 2 \cdot 7 \\ 3 \cdot 4 \\ 3 \cdot 6\end{array}\right\}$ $Q_3 = \dfrac{3 \cdot 6 + 4 \cdot 2}{2} = 3 \cdot 9$

Interquartile range = $3 \cdot 9 - 2 \cdot 5$
$= 1 \cdot 4$

$\left.\begin{array}{l}4 \cdot 2 \\ 9 \cdot 2\end{array}\right\}$ Semi-interquartile range = $\dfrac{1 \cdot 4}{2}$
$= \underline{0 \cdot 7 \text{ hr}}$

(d) Generally, the variation in transit times of ships is quite small, but both the range and, to a lesser extent, the standard deviation are heavily influenced by the single, very large value of 9·2 hr, which might have been an untypical occurrence due perhaps to an engine breakdown en route. The semi-interquartile range is not affected by this extreme value.

Marine Statistics

5. Referring to the data array in answer 4(b), we find the median:

$$\tilde{x} = \frac{3 \cdot 4 + 2 \cdot 7}{2} = \underline{3 \cdot 05 \text{ hr}}$$

Referring to the tabulation in answer 4(a), we find the mean:

$$\bar{x} = \frac{\Sigma x}{n} = \frac{30 \cdot 2}{8} = \underline{3 \cdot 77 \text{ hr}}$$

(a) Coefficient of variation (C) is given by:

$$C = \frac{\sigma}{\mu} \times 100$$

$$= \frac{2 \cdot 15}{3 \cdot 77} \times 100$$

$$= \underline{57}$$

(b) Pearsonian coefficient of skewness is given by

$$\frac{3(\text{mean} - \text{median})}{\text{standard deviation}}$$

$$= \frac{3(3 \cdot 77 - 3 \cdot 05)}{2 \cdot 15}$$

$$= \frac{3 \times 0 \cdot 72}{2 \cdot 15}$$

$$= \underline{1 \cdot 00}$$

6. x = rel. density (in rank order)

x	x^2
1·008	1·0161
1·010	1·0201
1·010	1·0201
1·010	1·0201
1·010	1·0201
1·012	1·0241
1·012	1·0241
1·013	1·0262
1·015	1·0302
1·016	1·0322
1·018	1·0363
1·020	1·0404
1·021	1·0424
13·175	13·3524

Measures of Dispersion

Mean, $\bar{x} = \dfrac{\Sigma x}{n} = \dfrac{13\cdot175}{13} = \underline{1\cdot0135}$

S.D., $\sigma = \sqrt{\dfrac{1}{n}\left[\Sigma x^2 - \dfrac{(\Sigma x)^2}{n}\right]}$

$= \sqrt{\dfrac{1}{13}\left[13\cdot3524 - \dfrac{13\cdot175^2}{13}\right]}$

$= \underline{0\cdot002}$

Range $= 1\cdot021 - 1\cdot008 = \underline{0\cdot013}$

7.

Delay limits (hr)	x	f	fx	fx^2
0 but less than 12	6	11	66	396
12 but less than 24	18	20	360	6,480
24 but less than 36	30	24	720	21,600
36 but less than 48	42	17	714	29,988
48 but less than 60	54	13	702	37,908
60 but less than 72	66	9	594	39,204
72 but less than 84	78	4	312	24,336
84 and over	90	2	180	16,200
		100	3,648	176,112

(a) Variance $= \sigma^2 = \dfrac{1}{n}\left[\Sigma fx^2 - \dfrac{(\Sigma fx)^2}{n}\right]$

$= \dfrac{1}{100}\left[176,112 - \dfrac{3,648^2}{100}\right]$

$= \underline{430\cdot33}$

(b) In this example, the quartiles are taken as the 25th and 75th delay times. The 25th delay time is within the interval 12–24 hr and may be estimated by linear interpolation as:

$Q_1 = 12 + \dfrac{12 \times 14}{20} = \underline{20\cdot4 \text{ hr}}$

Similarly: $Q_3 = 48 + \dfrac{12 \times 3}{13} = \underline{50\cdot8 \text{ hr}}$

Semi-interquartile range $= \dfrac{50\cdot8 - 20\cdot4}{2}$

$= \underline{15\cdot2 \text{ hr}}$

Marine Statistics

8. (a) Positively skewed.
   (b) Taking the median as the 50th delay time, this falls in the interval 24–36 hr and may be found by interpolation as:

   $$\tilde{x} = 24 + \frac{12 \times 19}{24} = \underline{33.5 \text{ hr}}$$

   The mean has previously been found as:

   $$\bar{x} = \underline{36.48 \text{ hr}}$$

   The variance has previously been found as:

   $$\sigma^2 = \underline{430.33}$$

   Hence the standard deviation is:

   $$\sigma = \underline{20.74 \text{ hr}}$$

   The Pearsonian coefficient of skewness $= \dfrac{3(\bar{x} - \tilde{x})}{\sigma}$

   $$= \frac{3(36.48 - 33.50)}{20.74}$$

   $$= \underline{\underline{0.43}}$$

Chapter 5
Probability

5.1 Introduction

'What is the probability that the ship will have engine trouble on this voyage?'

'What are the odds that the second mate will make a mistake in the next position fix?'

'What is the chance, as I look at my bridge hand, that my partner has the ace of spades?'

'What is the likelihood that we will have beef stew again for supper this evening?'

All these hypothetical questions have one thing in common. In each one of them the speaker is trying to make an assessment of uncertainty. If we knew that a particular event was going to happen then we could make contingency plans to cope with it. For instance if one had a strong dislike of beef stew then one could find out the answer from the cook and not turn up for supper if one really wanted to. However, one would probably have to feel very strongly to take this action and in any case in most situations of uncertainty we have no way of finding out. Yet again in many uncertain situations it is useful to have some sort of assessment, preferably numerical, as to whether or not an event will occur. When the owners of a ship are insuring it against a variety of different unwelcome events, the amount of premium they are charged will depend on the monetary consequences of the event, such as foundering, together with how likely the foundering is to occur.

In the initial questions four different terms were used to express the idea of how likely an event was to occur, but in statistics we tend to use probability most frequently. For a statistician who wants to draw conclusions from his raw data the theory of probability is the basic tool which he has at his disposal. In this chapter we will consider first how we can assess the probability of an event happening; we will then go on to consider the laws governing the way in which we can produce probabilities of compound events.

5.2 Basic Approaches to Probability

1. Theoretical Approach

In some situations, or experiments as they are often called in theoretical terms, it is possible to enumerate all the equally likely events which can occur.

Marine Statistics

Having done this the probability of a particular occurrence may be defined as the ratio of the number of equally likely events giving the particular occurrence to the number of equally likely events which can occur. If we let p stand for probability, then informally we may write

$$p = \frac{\text{number of possibilities in which event occurs}}{\text{total number of possibilities}}$$

To be slightly more formal as there may be several events of interest arising from any one situation then we may write $p(A)$ as the probability that an event A occurs.

To illustrate this definition consider the probability of tossing a head when a perfectly fair coin is tossed. Since the coin has two sides, a head and a tail, and we are told the coin has no bias in it to either side, then in this experiment there are two equally likely events, one of which gives us what we require. Thus

$$p(\text{head}) = \frac{1}{2}$$

As a second example consider the probability of drawing a spade from a normal well-shuffled pack of playing cards. There are 52 cards, 13 of which give the required event and all of which are equally likely to be drawn. Thus

$$p(\text{spade}) = \frac{13}{52} \quad \text{or} \quad \frac{1}{4}$$

This approach can also be used in any situation for which we can enumerate all the equally likely probabilities. For instance on a ship with five navigation lights, i.e. red and green sidelights, two masthead lights and a stern light, the red sidelight is considered the most important. All lights are fitted with similar lamps. What is the probability that it is the red sidelight which has failed if the navigation light alarm buzzer sounds to indicate a fault? If all lights are equally likely to fail then we have five equally likely events, one of which is of concern to us, hence

$$p(\text{red light fails}) = \frac{1}{5}$$

The main thing to guard against when using this approach is to ensure that the events enumerated are equally likely. Thus if we consider the probability of a ship having a collision on a particular voyage, then two things can happen, either the ship collides or it does not. However, very few of us would believe that a ship has an even chance of colliding or not especially if the next voyage was across the Atlantic, and the reason is that the event of no collision is much more likely than the event of collision. Just how much more likely is impossible to answer this way, which is why we need other approaches to probability. As a second example of this trap consider an experiment of tossing two fair coins. The outcomes may be listed as two heads, two tails and one of each, and for many years several famous mathematicians argued that as there were three events the probability of occurrence for each of them was 1 in 3. However if we consider how each of these events occurs we see that these odds are wrong. The simplest way is to draw up a table showing what each of the two coins could show.

Coin 1	Coin 2	Event
Head	Head	Two heads
Tail	Tail	Two tails
Head	Tail	One of each
Tail	Head	One of each

Thus we see that the event, a head and a tail can occur in two ways whereas the other events can only occur in one way. Hence the true odds should be: $p(2 \text{ heads}) = \frac{1}{4}$, $p(2 \text{ tails}) = \frac{1}{4}$, $p(1 \text{ head and 1 tail}) = \frac{1}{2}$. In this case we have still been able to solve the problem but in many cases, this symmetrical approach to probability will not work.

2. Relative Frequency Approach

The relative frequency or experimental approach to probability enables us to come up with a solution on other occasions. In this approach we use historical evidence and define the probability of an event as the ratio of the number of times the event has occurred to the number of times the experiment was repeated. As we are using empirical evidence we obviously will not obtain an exact result but if we base our estimate on as large a sample as we can, we will obtain as close a result as possible. This point may be illustrated by considering the tossing of a fair coin to estimate the probability of getting a head. If we toss the coin only once, then our answer will be 0 or 1, both of which are very far from the truth. If we toss the coin 10 times we might well get 6 heads giving a result of 6/10, 100 tosses might well yield 55 heads giving a result of 55/100 or 11/20 and 1,000 tosses might well yield 490 heads giving a result of 490/1,000 or 49/100. In the long run if the coin is fair we will eventually converge on to the value 1/2. In most situations we just have to accept what evidence we have but we must be aware that the result is based purely on practical evidence. As an example we can consider the following situation. Over a period of time it was found that 20 cases of fruit were damaged during loading operations out of a total of 10,000 cases loaded. In a subsequent loading operation, under similar conditions, what is the probability that a particular case will be damaged?

There were 10,000 cases loaded of which 20 were damaged so the proportion of damaged cases is 20/10,000 = 0·002. Hence assuming all conditions similar

Marine Statistics

the probability that a particular case will be damaged on a subsequent loading operation is 0·002. Obviously if any of the loading methods were changed or the type of packing case used changed then this result would be less valid, although if no other information were available it would still be the best answer we could give.

In some situations it is still difficult to get a figure for the denominator in the probability definition, i.e. the number of times the experiment is repeated. Suppose we were interested in the probability that an oil tanker will be involved in a collision. The first problem is one of definitions and the answer will depend on the purpose for which the probability is required. For instance do we mean a collision with another ship only, or are we including fixed structures such as dock walls? Let us suppose we require the probability that an oil tanker of over 1,000 grt will be involved in an open sea collision anywhere in the world. We can get from reports such as Lloyds' casualty lists the number of collisions in this category over a certain time period, but then we have to know the number of oil tankers of over 1,000 grt exposed to risk of collision over this period.

Exposed to risk is a standard phrase in this sort of work and is used by demographers in calculating human mortality rates. In our context, this means that from a knowledge of the number of tankers registered in this period, we should exclude those which are laid up for the whole period. Those which are laid up for a fraction of the period should only be counted for the fraction of time they are not laid up. Some utilisation factor should be applied for the length of time spent in ports loading and unloading. Similarly adjustments should be made for ships that were scrapped during the period or had their maiden voyages during the period. Obviously in most cases it is impossible to get an accurate figure but the more adjustments that can be made the more realistic the final result will be. In any case the result will only be applicable to present-day ships if there have not been too many changes since the time the data related to.

The other point which comes out of this discussion is that it is possible for several different answers to be given for the probability of an event, all of which have been calculated perfectly correctly using a relative frequency approach. They have differed only in the extra conditions which have been taken into account. This is the basis of the idea of conditional probability which we will meet more formally later in this chapter.

3. Subjective Approach

Suppose we are asked the question 'What is the probability that it will rain today?' One answer could be given using the relative frequency approach and referring to a meteorological table showing the number of rainy days over a period of time. However another answer would be given if we looked out of the window before answering. Suppose there were no clouds in the sky and the weather records for the time of year suggested settled weather then we would probably answer 0. This answer is reached from our own personal experience of the situation. Similarly if we were shown a coin and asked the probability that if it were tossed we would get a head we would probably answer $\frac{1}{2}$ since experience has taught us that coins tend to come down an equal number of times heads or tails. It is thus important to realise that subjective assessment can often play an important role in assigning probabilities. Experience of a situation can be a great help and often explains why a businessman places greater confidence on one deal succeeding rather than another. The main reason for understanding how our probabilities for individual events are obtained is so that we know what reliance we can put on them, but however they are obtained there are some basic rules for manipulation which we will now consider.

5.3 *Mathematical Properties of Probability*

The probability of an event A, which is written as $p(A)$, may be thought of as a mathematical quantity about which various things are true.

(1) The smallest value that $p(A)$ can take is 0 and if $p(A) = 0$ then we say that A is an impossibility. This comes from the basic theoretical definition of probability and the only way in which $p(A) = 0$ is if there are no outcomes of the experiment which yield A. Thus if a die is tossed with 6 faces numbered 1 to 6 inclusive, then the probability that a 7 is thrown is 0. Alternatively if using the relative frequency approach we get $p(A) = 0$ then we have to ask ourselves if this is realistic. For instance at the time of writing, the probability of a nuclear-powered ship being in a collision would be calculated, the relative frequency way, as 0. However the probability of a large bulk carrier foundering was 0 by this method, until one did (the *Berg Istrol*).

(2) The largest value that $p(A)$ can take is 1 and if $p(A) = 1$ then we say that A is a certainty. This again comes from the basic theoretical definition when all equally likely events of an experiment give A. Thus if the same die is tossed the probability that a number between 1 and 6 inclusive is thrown is 1. Again care has to be taken if the relative frequency approach yields $p(A) = 1$.

(3) If the probability that an event A happens is p, then the probability that it does not happen is $1 - p$. This is known as complementary probability.

We write A' or $\overline{A}$ to represent the complementary event to A. Thus $p(A') = 1 - p(A)$.

This can again be shown from the basic definition because the number of outcomes which give A together with the number of outcomes that do not give A must make the total number of outcomes of the experiment.

This can be a useful property when looking at a complex situation, as the probability of the complement occurring may sometimes be easier to calculate.

Marine Statistics

We will see an example of this later. As a simple example we can see that if the probability of a magnetic compass failing during a forthcoming voyage is estimated to be 1/1,000 then 999/1,000 is the estimated probability that it will not fail during the forthcoming voyage.

As a final point in this section it should be emphasised that if the probability of an event is assessed as -0.4 or 1.2 say, then the result must be closely questioned because it is impossible for a probability to be outside the range 0 to 1 inclusive.

5.4 *Addition Law of Probability*

One way in which we may want to compound probabilities is to find the probability of one thing or another happening, and for this we need the addition law of probability.

5.4.1 **The addition law of probability** states that: **the probability of at least one of event A or event B happening is the probability that event A happens plus the probability that event B happens less the probability that events A and B happen together.**

We use the symbol $A \cup B$ to denote the event of at least one of A or B. This symbol comes from the algebra of sets and is used to denote what is termed the union of A and B. **We use the symbol $A \cap B$ to denote the event of both A and B.** This symbol again comes from the algebra of sets and is used to denote what is termed the intersection of A and B.

Another useful device from set theory is the Venn diagram which can be used to denote sets. We draw a circle to represent set or event A and similarly for set or event B, in Fig. 5.1. The intersection of A and B, $A \cup B$, is repre-

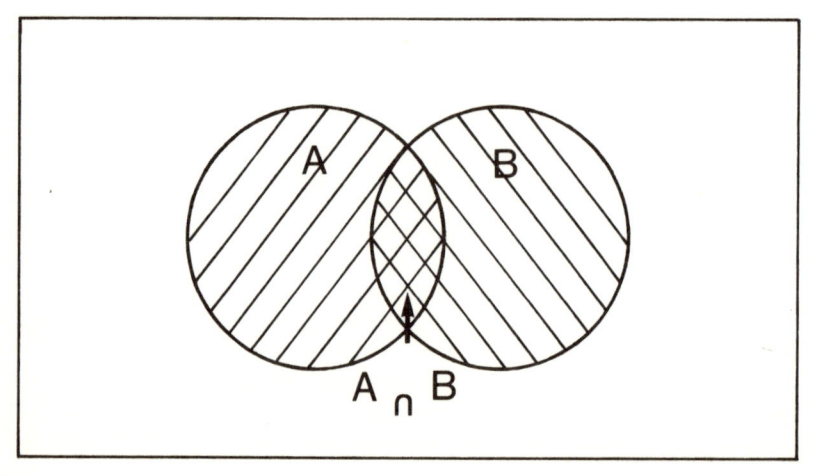

Fig 5.1 The Intersection and Union of Two Sets

Probability

sented by the common area between the sets, and the union of A and B, $A \cup B$, is represented by the whole shaded area. The rest of the rectangle in which the shaded diagram is enclosed represents the results of the experiment which do not give A or B. We can now write the addition law of probability symbolically as

$$p(A \cup B) = p(A) + p(B) - p(A \cap B)$$

A simple example may help to make the law clearer.

Let A denote the event drawing an ace from a normal pack of cards and let S denote the event drawing a spade.

$$p(A) = \frac{4}{52} = \frac{1}{13} \quad \text{and} \quad p(S) = \frac{13}{52} = \frac{1}{4}$$

But the probability of drawing an ace or spade,

$$p(A \cup S) = p(A) + p(S) - p(A \cap S)$$

$p(A \cap S)$, is the probability of drawing an ace and a spade together, i.e. the ace of spades = 1/52. Thus

$$p(A \cup S) = \frac{1}{13} + \frac{1}{4} - \frac{1}{52} = \frac{16}{52} = \frac{4}{13}$$

From first principles we could have derived this answer since there are only 16 separate cards which satisfy this criterion of ace and spade. Let us consider a second more practical example. From past records, the probability of fog occurring at the Varne lightvessel during the course of a day is found to be 1/80 and the probability of fog occurring at the South Goodwin lightvessel is found to be 1/50. As they are reasonably close the probability of fog occurring at both lightvessels together must be considered and is found to be 1/100. What is the probability that fog will occur at at least one of these two lightvessels on a particular day? If we let $p(V)$ denote the probability of fog at the Varne and $p(S)$ denote the probability of fog at the South Goodwin then we require $p(V \cup S)$.

$$p(V \cup S) = p(V) + p(S) - p(V \cap S)$$

$$= \frac{1}{50} + \frac{1}{80} - \frac{1}{100}$$

$$= \frac{9}{400}$$

5.4.2 Two events are said to be *mutually exclusive* if they cannot happen together.

5.4.3 If two events are mutually exclusive then the probability of them happening together is 0.

Marine Statistics

If we consider an ordinary pack of 52 playing cards then the events drawing a heart and drawing a spade are mutually exclusive since there is no card that is both a heart and a spade. However the events drawing a spade and drawing an ace are not mutually exclusive since the ace of spades represents their intersection.

On a Venn diagram two mutually exclusive events would be represented by non-overlapping circles as shown in Fig. 5.2.

Fig 5.2 Two Mutually Exclusive Sets

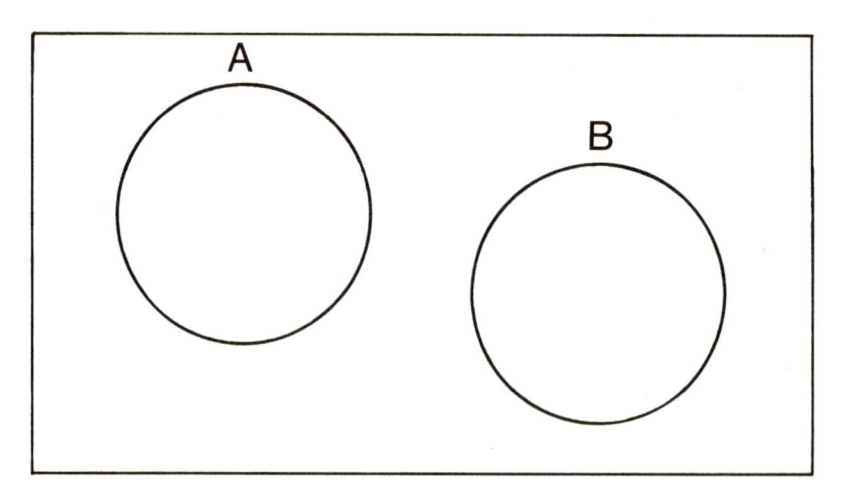

Since the events have no overlap, the probability of the intersection $A \cap B$ occurring is 0; hence in this case the addition law becomes modified.

5.4.4 If A and B are mutually exclusive events then the probability of A or B happening is the sum of the probabilities of A and B.

Thus

$$p(A \cup B) = p(A) + p(B)$$

Consider the following example. A piece of signalling equipment is comprised of components A, B, C and D. Past experience indicates that, when a failure occurs, the probability that it is due to a fault in one of these components is respectively:

$$p(A) = \tfrac{1}{4} \quad p(B) = \tfrac{1}{8} \quad p(C) = \tfrac{1}{2} \quad p(D) = \tfrac{1}{8}$$

and it is never due to a fault in two or more simultaneously. If a fault occurs, what is the probability that it is (i) due to a fault in components A or B? (ii) due to a fault in components other than D?

(i) We require $p(A \cup B) = p(A) + p(B) = \tfrac{3}{8}$.
(ii) We require $p(D') = 1 - p(D) = \tfrac{7}{8}$.

Probability

This modified addition law is easily extended to cover several mutually exclusive events. Thus if $A, B, C, D, \ldots$ are a series of mutually exclusive events then the probability of A or B or C or $D \cdots$ happening is the sum of the individual probabilities. Hence

$$p(A \cup B \cup C \cup D \cdots) = p(A) + p(B) + p(C) + p(D) + \cdots$$

Thus in the second part of the last example we could have obtained the answer by adding $p(A)$, $p(B)$ and $p(C)$ giving $\frac{7}{8}$ as before. However this is an example of how the complementary probability may be the easier to find in some situations.

The general addition law when the events are not mutually exclusive can also be extended but the form gets progressively more complicated. For instance in mathematical symbols for three events,

$$p(A \cup B \cup C) = p(A) + p(B) + p(C) - p(A \cap B) - p(B \cap C) \\ - p(A \cap C) + p(A \cap B \cap C)$$

5.5 Multiplication Law of Probability

The other way in which we may want to compound probabilities is to find the probability of two events happening together. We write this mathematically as $p(A \cap B)$, which of course we have been using as part of the addition law. However before defining the multiplication law of probability, which we use to find $p(A \cap B)$, we must first discuss the concept of conditional probability.

5.5.1 The *conditional probability* **of event *A* on event *B* is the probability that event *A* happens given that event *B* has happened.** It is written as $p(A|B)$ and is read as the probability of *A* given *B*.

Similarly the probability that *B* happens given that *A* has happened is the conditional probability of *B* on *A* and is written as $p(B|A)$.

An example may help to clarify these ideas. Suppose a shipping company has 90 deck officers with either a Master's or a First Mate's certificate. If we categorise these officers by certificate and by marital status, married or unmarried, the following table is obtained.

	Master	First Mate
Married	40	10
Unmarried	16	24

Let *C* be the event having a Master's certificate, *F* be the event having a First Mate's certificate, *M* be the event being married, and *U* be the event being unmarried.

Marine Statistics

Then $p(M)$, the probability of being married $= \dfrac{50}{90} = \dfrac{5}{9}$

$p(M|C)$, the probability of being married given that

a man has a Master's certificate $= \dfrac{40}{56} = \dfrac{5}{7}$

$p(C|M)$, the probability of having a Master's certificate

given that a man is married $= \dfrac{40}{50} = \dfrac{4}{5}$

If we compare $p(M)$ and $p(M|C)$ we can see an illustration of an earlier point, that the probability of an event can change given different circumstances. To calculate $p(M)$ we took the number of all married officers in the company as a fraction of all the officers whereas to calculate $p(M|C)$ we took the number of married people with Master's certificates as a fraction of all the officers with Master's certificates. $p(C|M)$ was calculated by taking the number of married people with Master's certificates as a fraction of all the married officers. Thus $p(M|C)$, the probability of being married given that the man has a Master's certificate, is a very different thing from $p(C|M)$, the probability that a man has a Master's certificate given that he is married. Having defined conditional probability we are now in a position to define the multiplication law of probability.

5.5.2 The *multiplication law* of probability states that the probability that two events A and B happen together is the probability that event A happens, multiplied by the probability that event B then happens, given that event A has happened, or vice versa.

Thus in symbols

$$p(A \cap B) = p(A).p(B|A)$$

or

$$p(A \cap B) = p(B).p(A|B)$$

In our example on the shipping company the probability that a man is married and has a Master's certificate is

$$p(M \cap C) = p(M).p(C|M)$$

Thus

$$p(M \cap C) = \dfrac{50}{90}.\dfrac{40}{50} = \dfrac{40}{90} = \dfrac{4}{9}$$

Similarly

$$p(M \cap C) = p(C).p(M|C)$$
$$= \dfrac{56}{90}.\dfrac{40}{56} = \dfrac{4}{9}$$

Probability

The multiplication law can also be used in reverse to find conditional probabilities. Thus

$$p(A|B) = \frac{p(A \cap B)}{p(B)}$$

and

$$p(B|A) = \frac{p(A \cap B)}{p(A)}$$

As a second example on the multiplication law consider the probability of drawing two spades from a pack of cards when two cards are drawn in succession without replacing the first.

Let A be the event the first card is a spade, and B be the event the second card is a spade. Then we require $p(A \cap B) = p(A).p(B|A)$.

$$p(A) = \frac{13}{52}$$

$$p(B|A) = \frac{12}{51}$$

Thus

$$p(A \cap B) = \frac{13}{52} \cdot \frac{12}{51} = \frac{1}{17}$$

The probability of the event that the second card was a spade depended on whether or not the first card was a spade. Thus the two events are called non-independent events.

5.5.3 The events A and B are independent if $p(B|A) = p(B)$ and $p(A|B) = p(A)$, i.e. if the probability of one happening is not dependent on the other happening or not.

In the example above we had two non-independent events, but the two events, getting a head on one coin followed by getting a head on a second coin, would be independent events. In some cases we have to make arbitrary decisions as to whether or not two events shall be assumed to be independent. Thus we would probably assume that the event that one member of the crew was injured in an accident during a voyage was independent of the event that another member of the crew was injured, even though they could both be injured in the same accident. This is a usual assumption in any sort of insurance work, whereby for instance, it is always assumed that a man and his wife are independent with respect to dying.

If we have two independent events A and B then the multiplication law of probability becomes modified.

5.5.4 If A and B are two independent events then the probability of A and B happening together is found by multiplying the individual probabilities that A and B happen. Thus, $p(A \cap B) = p(A).p(B)$.

Marine Statistics

As a first example we can consider the probability of tossing two heads on two throws of a coin.

Let A be the probability of a head on the first throw, and let B be the probability of a head on the second throw. Then

$$p(A) = \frac{1}{2} \text{ and } p(B) = \frac{1}{2}$$

and

$$p(A \cap B) = p(A).p(B) = \frac{1}{2} \cdot \frac{1}{2} = \frac{1}{4}$$

As a second example we can take the following situation. A ship is equipped with two independent radar systems A and B. From past experience it is estimated that the probability of system A failing during a forthcoming two-week voyage is 1/10 and the probability of system B failing is 1/7.

(i) What is the probability that both systems will fail during the voyage?
(ii) What is the probability that one or the other of the two systems will fail during the voyage?

Let A be the event that system A fails during the voyage and B be the event that system B fails. Then (i) we require $p(A \cap B) = p(A).p(B)$ since the systems are independent. Thus

$$p(A \cap B) = \frac{1}{10} \cdot \frac{1}{7} = \frac{1}{70}$$

(ii) we require $p(A \cup B) = p(A) + p(B) - p(A \cap B)$. Thus

$$p(A \cup B) = \frac{1}{10} + \frac{1}{7} - \frac{1}{70} = \frac{16}{70} = \frac{8}{35}$$

We can note that by having the two systems on board we have greatly reduced the risk of a complete radar blackout.

The modified multiplication law is also easily extended to cover several independent events. Thus if $A, B, C, D, \ldots$ are a series of independent events then the probability of them all happening together is the product of the individual probabilities. Hence

$$p(A \cap B \cap C \cap D \cdots) = p(A).p(B).p(C).p(D) \cdots$$

The general multiplication law can also be extended to a series of non-independent events but again the format gets progressively more difficult. Thus we have, symbolically, for three events:

$$p(A \cap B \cap C) = p(A).p(B|A).p(C|A \cap B)$$

Probability

5.6 Bayes' Rule

It is sometimes useful to work out the probability of an initial cause given the final result. For instance it may be desired to know why a man has decided to leave the sea and take a shore job. Unfortunately the particular man had gone before he could be asked so the only information available was from a large study. Among all dissatisfied seamen, it was found that 20% are dissatisfied mainly because they dislike the work, 50% because they felt they are underpaid, and 30% because they found the long periods away from home unsettling. It was also found that the probability that a man will leave, given that he is in one of the above categories, is 0·6, 0·4 and 0·9 respectively. We know that the man left and we want to know the probability that he left because he disliked the work.

To answer a problem like this we need *Bayes' rule*. In its simplest form this rule takes account of the fact that:

$$p(A \cap B) = p(A).p(B|A) = p(B)p(A|B)$$

Thus

$$p(B|A) = \frac{P(B).p(A|B)}{P(A)}$$

for instance. However we have a more complicated situation where event A, say, has several causes $B_1, B_2, \ldots, B_K$.

5.6.1 Bayes' rule. If event A has several mutually exclusive causes B_1, $B_2, \ldots, B_K$ then,

$$p(B_i|A) = \frac{p(B_i)p(A|B_i)}{p(B_1)p(A|B_1) + p(B_2)p(A|B_2) + \cdots + p(B_K)p(A|B_K)}$$

for $i = 1, 2, \ldots, K$

Bayes' rule has aroused much criticism because one is arguing from effect to cause. It is also mainly used in more advanced statistical calculations so for the general reader is perhaps not of so much interest, as nothing else in this book depends on its use. However, for readers who are still staying the course and have not jumped to the next chapter, we will set about solving the case of the dissatisfied seaman.

Let A be the event the man leaves the sea, B_1 be the event he dislikes the work, B_2 be the event he thinks he is underpaid, and B_3 be the event he dislikes the long periods away. The information we are given is that

$$p(A|B_1) = 0.6 \quad p(B_1) = 0.2$$
$$p(A|B_2) = 0.4 \quad p(B_2) = 0.5$$
$$p(A|B_3) = 0.9 \quad p(B_3) = 0.3$$

Marine Statistics

We require $p(B_1|A)$. By Bayes' rule,

$$p(B_1|A) = \frac{p(B_1)p(A|B_1)}{p(B_1)p(A|B_1) + p(B_2)p(A|B_2) + p(B_3)p(A|B_3)}$$

$$= \frac{0\cdot2 \times 0\cdot6}{0\cdot2 \times 0\cdot6 + 0\cdot5 \times 0\cdot4 + 0\cdot3 \times 0\cdot9}$$

$$= \frac{12}{59}$$

5.7 Permutations and Combinations

As a final section in this chapter a very brief word will be given on permutations and combinations as they are useful mathematical means for counting how many ways events can occur in some situations. This section is again by no means essential for the rest of the book but will be quite useful.

5.7.1 **A *permutation* is an arrangement of all or part of a set of objects.**

5.7.2 **The number of permutations of *n* distinct objects is *n*!** (pronounced *n* factorial or *n* shriek).

$$n! = n \times n-1 \times \cdots \times 1$$
$$0! = 1$$
$$1! = 1$$
e.g. $3! = 3 \times 2 \times 1 = 6$

Example: If we have three letters *A*, *B*, *C*, then we can arrange them in 6 ways, viz.: *ABC, BAC, BCA, ACB, CAB, CBA*.

5.7.3 **The number of permutations of *n* distinct objects taken *r* at a time is $_nP_r = n!/(n-r)!$**

Example: If we have three letters *A*, *B*, *C* and we take them two at a time, then we can arrange them in $_3P_2 = 3!/1! = 6$ ways, viz.: *AB, BA, AC, CA, BC, CB*.

5.7.4 **The number of distinct permutations of *n* objects of which n_1 are alike, n_2 are alike, ..., n_K are alike is $n!/(n_1!\,n_2!\ldots n_K!)$**

Example: If we have five letters *A, A, B, B, B*, then the number of distinct permutations is $5!/2!\,3! = 10$, viz.: *AABBB, ABBBA, ABBAB, ABABB, BBBAA, BBAAB, BABAB, BBABA, BABBA, BAABB*.

5.7.5 **The number of ways in which *r* objects can be selected from *n* objects without regard to order is called the number of combinations of *r* objects from *n* objects. It is written $\binom{n}{r}$ and $= \frac{n!}{r!(n-r)!}$. $\binom{n}{r}$ is read as *n c r*.**

Example: If we have five letters A, B, C, D, E, the number of combinations of 3 letters from the 5 is

$$\binom{5}{3} = \frac{5!}{3!\,2!} = 10$$

viz.: $ABC, ABD, ABE, ACD, ACE, ADE, BCD, BDE, BCE, CDE$.

As a final example consider the following situation. How many four-man work teams can be chosen from a pool of 12 deckhands?

We want the number of combinations of 4 objects out of 12

$$= \binom{12}{4} = \frac{12!}{8!\,4!} = \frac{12.11.10.9}{4.3.2.1}$$
$$= 495$$

Thus the probability that a particular group of four men are put in a work-team together is 1/495.

5.8 Summary

In this chapter we have considered the general concept of probability and how the probability of a particular event may be assessed in three different ways. We then considered the mathematical properties of probability and the laws for calculating the probabilities of compound events. In the next chapter we will extend these ideas to consider how probability is distributed over all the different outcomes of a particular situation.

Exercises

1. A growler paints on a radar screen with the same brilliance as wave returns, but can be distinguished by an observer if it paints in the same position on two successive sweeps. If the probability that the growler will paint on a particular sweep is 0·5 and the observer watches for three sweeps, what is the probability that the growler can be detected? Tabulate the possible outcomes and derive the probability from first principles.

2. From a climatic chart, a navigator finds that the percentages of winds from different directions at a certain locality are as follows:

Direction	%
N	15
NE	10
E	10
SE	5
S	15
SW	20
W	15
NW	10

Marine Statistics

The navigator's ship is anchored with good shelter except from southerly winds. What is the probability that, on a particular occasion, the wind will blow:
(a) from a direction from SE through S to SW?
(b) from a direction other than S?

3. At a certain port, the probability of a stoppage of at least a day during the loading of cargo is estimated as 1/10 due to rain and 1/20 due to labour disputes. What is the probability that a stoppage will occur due to these causes?

4. On a particular trade, log book records were analysed for 60 voyages to investigate the relationship between weather conditions and cargo claims. The following results were deduced:
   (i) The probability of experiencing winds of gale force and above during a voyage was 0·4.
   (ii) The probability of a cargo claim, given that a wind of gale force or above was experienced, was 0·75.
   (iii) The probability of a voyage being associated with a cargo claim but no wind of gale force or above was 0·2.
      (a) Find the probability that a voyage was associated with a cargo claim and a wind of gale force or above.
      (b) Find the probability of a cargo claim occurring given that no wind of gale force or above was recorded.

5. The probability of a magnetic compass failing during a forthcoming voyage is estimated as 0·001 and the probability of a gyro compass failing is estimated as 0·01. What is the probability that both compasses will fail during the course of the voyage?

6. In an analysis of case cargo damaged over a number of voyages, the following results were obtained:
   (i) 60% were damaged due to inadequate ventilation and, for these, there was a probability of 0·2 that they would be rejected.
   (ii) 10% were damaged due to weather effects and, for these, there was a probability of 0·4 that they would be rejected.
   (iii) 30% were damaged during loading or discharging operations and, for these, there was a probability of 0·6 that they would be rejected.

   If a particular damaged case was rejected without a cause being stated, what is the probability that this was due to inadequate ventilation?

7. In how many ways can three of the 26 international code alphabetical flags be selected:
   (a) If order is important, as in a signal?
   (b) If no regard is paid to order?

8. In order to enter a particular port, a ship has to cross a bar while the height of tide is greater than 2 m and pass through a channel marked

by unlit buoys which is not navigable by night. Consulting tide tables indicates that the height of tide is greater than 2 m for 5 hr in each tidal cycle, and consulting the Nautical Almanac indicates that, for the time of year in question, daylight extends for 14 hr in each 24 hr period. If the ship arrives at a random time, what is the probability it will be able to enter port?

9. A port navigation service finds that, over a period of time, for vessels negotiating a particular section of channel, the probability of an encounter with another ship on a reciprocal course is 0·6, the probability of an overtaking encounter is 0·1 and the probability of a crossing encounter with a ferry is 0·2. What is the probability that a vessel entering the channel will have an encounter of some kind?

10. A ship's steering system may be divided into two parts. The first part comprises completely duplicated control circuits and steering actuators of which each of the duplicated sub-systems has a probability of failure of 0·01 during a month of operation. The second part consists of the rudder and rudder-head assembly which is estimated to have a probability of failure of 0·0002 during a month of operation. What is the probability that a ship will be without a steering capability during a forthcoming month?

11. A commuter in a big city has to catch a ferry to work in the morning, and he always catches one of two, A or B. If he catches A his probability of being late for work is 0·03 while if he catches B his probability of being late for work is 0·42. On 4 out of 5 mornings he catches ferry A. What is the overall probability of his being late for work?

12. There is a fifty-fifty chance that company A will bid for a new freight route. Company B submits a bid and the probability that it will get the contract is $\frac{2}{3}$ provided company A does not bid; if company A submits a bid, however, the probability that company B will get the contract is only $\frac{1}{5}$. If company B gets the contract, what is the probability that company A did not bid?

Answers

1. Table of outcomes
   p p p
   n p p
   p n p
   n n p
   p p n
   n p n
   p n n
   n n n

   (p = paint, n = no paint) Consideration of the table indicates that there are 8 equally likely possible outcomes of which 3 contain at least two successive paints. The probability is thus $\frac{3}{8}$ that the growler could be detected.

Marine Statistics

2. Since we are dealing with mutually exclusive events, the probability of a wind from SE to SW is given by:

   (a) $p(\text{SE, S, SW}) = p(\text{SE}) + p(\text{S}) + p(\text{SW})$
   $= 0.05 + 0.15 + 0.20$
   $= \underline{0.4}$

   (b) $p(\text{other than S}) = p(\text{S}')$
   $= 1 - p(\text{S})$
   $= 1 - 0.15$
   $= \underline{0.85}$

3. The overall probability of a stoppage $p(S)$ in terms of the separate probabilities of a stoppage due to rain $p(R)$ and due to labour disputes $p(D)$ is given by:

   $p(S) = p(R \cup D)$
   $= p(R) + p(D) - p(R \cap D)$
   $= \dfrac{1}{10} + \dfrac{1}{20} - \dfrac{1}{10} \cdot \dfrac{1}{20}$ (assuming stoppages due to rain and labour disputes are independent)
   $= \underline{0.145}$

4. $p(G)$ = probability of gale or stronger wind
   $p(C)$ = probability of cargo claim
   $p(B)$ = probability of below gale force wind

   (a) $p(C \cap G) = p(C|G) \cdot p(G)$
   $= 0.75 \times 0.4$
   $= \underline{0.3}$

   (b) $p(C|B) = \dfrac{p(C \cap B)}{p(B)}$
   $= \dfrac{0.2}{1 - 0.4}$
   $= \underline{0.33}$

5. Since these events may be assumed independent:
   $p(F) = p(M) \cdot p(G)$
   $= 0.001 \times 0.01$
   $= \underline{0.00001}$

Probability

6. Using Bayes' rule:

$$p(A|B_1) = 0{\cdot}2 \qquad p(B_1) = 0{\cdot}6$$
$$p(A|B_2) = 0{\cdot}4 \qquad p(B_2) = 0{\cdot}1$$
$$p(A|B_3) = 0{\cdot}6 \qquad p(B_3) = 0{\cdot}3$$

$$p(B_1|A) = \frac{p(B_1).p(A|B_1)}{p(B_1).p(A|B_1) + p(B_2).p(A|B_2) + p(B_3).p(A|B_3)}$$

$$= \frac{0{\cdot}6 \times 0{\cdot}2}{0{\cdot}6 \times 0{\cdot}2 + 0{\cdot}1 \times 0{\cdot}4 + 0{\cdot}3 \times 0{\cdot}6}$$

$$= \underline{0{\cdot}35}$$

7. (a) Permutation $\quad _nP_r = {_{26}P_3} = \dfrac{26!}{23!} = \underline{15{,}600}$

   (b) Combination $\quad \dbinom{n}{r} = \dbinom{26}{3} = \dfrac{26!}{3!\,23!} = \underline{2{,}600}$

8. Probability of a suitable tide, $P(T) = \dfrac{5}{13}$

   Probability of daylight, $\qquad P(F) = \dfrac{14}{24}$

   Since these are independent events, the probability that they both occur together, $P(B)$ is given by:

   $$P(B) = P(T).P(F)$$
   $$= \frac{5}{13}.\frac{14}{24}$$
   $$= \underline{0{\cdot}224}$$

9. We let probability of reciprocal course encounters $= P(R)$
   probability of overtaking course encounters $= P(O)$
   probability of crossing course encounters $\ = P(C)$
   Overall probability of an encounter $\qquad = P(T)$
   Then, since these probabilities are not exclusive, we have:

   $$P(T) = P(R \cup O \cup C) = P(R) + P(O) + P(T) - P(R \cap O)$$
   $$- P(R \cap C) - P(O \cap C)$$
   $$+ P(R \cap O \cap C)$$
   $$= 0{\cdot}6 + 0{\cdot}1 + 0{\cdot}2 - 0{\cdot}6 \times 0{\cdot}1$$
   $$- 0{\cdot}6 \times 0{\cdot}2 - 0{\cdot}1 \times 0{\cdot}2$$
   $$+ 0{\cdot}6 \times 0{\cdot}2 \times 0{\cdot}1$$
   $$= \underline{0{\cdot}712} \quad \text{(assuming encounters are independent)}$$

Marine Statistics

10. Since the duplicated sub-systems should be entirely separate, their failure may be considered independent events and so the probability $P(D)$ that they will both fail is given by:

    $$P(D) = 0.01 \times 0.01 = 0.0001$$

    Failure of the duplicated section $P(D)$ and failure of the rudder assembly $P(R)$ are not mutually exclusive events and so the probability that at least one will occur, $P(F)$, is given by:

    $$P(F) = P(D) + P(R) - P(D \cap R)$$
    $$= 0.0001 + 0.0002 - 0.0001 \times 0.0002$$
    $$\simeq 0.0003$$

11. Let L be the event late. Then

    $$p(L|A) = 0.03 \qquad p(L|B) = 0.42$$
    $$p(A) = \frac{4}{5} \qquad p(B) = \frac{1}{5}$$
    $$p(L) = p(A \cap L) + p(B \cap L)$$
    $$= p(A)p(L|A) + p(B)p(L|B)$$
    $$= \frac{4}{5} \times 0.03 + \frac{1}{5} \times 0.42$$
    $$= 0.024 + 0.084$$
    $$= 0.108$$
    $$\simeq 0.11$$

12. A — A bids. A' — A does not bid. B — B wins.

    $$p(A) = p(A') = \frac{1}{2} \qquad p(B|A) = \frac{1}{5} \qquad p(B|A') = \frac{2}{3}$$

    By Bayes' rule

    $$p(A'|B) = \frac{p(B|A')p(A')}{p(B|A)p(A) + p(B|A')p(A')}$$
    $$= \frac{\frac{2}{3} \times \frac{1}{2}}{\frac{1}{5} \times \frac{1}{2} + \frac{2}{3} \times \frac{1}{2}} = \frac{0.333}{0.433} = \underline{0.77}$$

Chapter 6
Discrete Probability Distributions

6.1 *Introduction*

Over a period of time before 1977, in the approaches to a port, it was found that on average 1 out of every 1,000 ships entering or leaving the port touched the ground at some stage. In 1977 a new system of buoyage was introduced and subsequently, with similar traffic patterns, it was found that 5 ships out of 2,000 touched the ground. The port authority wants to know if the new buoyage scheme is effective or not, so their tame statistician suggests that he should calculate the probability of getting this number of groundings or more, occurring if the buoyage changes had no effect on navigation in the area. If this calculated probability is very small, then conversely, the complementary probability is very large that under the old system there would have been a better record than 5 groundings out of 2,000 movements, i.e. only 0 or 1 or 2 or 3 or 4 groundings out of 2,000 movements.

The statistician needs to calculate the probability of each of these separate number of groundings, and since it is certain that there must be some number of groundings, albeit 0, the sum of the probabilities for all possible numbers of groundings from 0 to 2,000 must be 1. Thus the probability of 5 or more groundings could be calculated by subtracting the sum of the probabilities for 0 to 4 groundings inclusive from 1. In fact what this statistician is really interested in is the distribution of probability over all possible happenings. This idea is known as a probability distribution. In this chapter we are concerned with probability distributions over a discrete variable, first of all in general and secondly some standard distributions which provide useful models if practical situations can be fitted to them.

6.2 *Discrete Probability Distributions in General*

In Chapter 2 we considered how a relative frequency distribution could be formed from a frequency distribution and in Chapter 5 we saw that in practical situations the best way to calculate the probability that something will occur is to use the relative frequency with which it has occurred.

Let us consider an example in which a navigator is interested in the number of times a buoy paints on a radar screen. He observes the number of times the

Marine Statistics

buoy paints in three successive sweeps of the screen and repeats this observation 16 times. The distribution he obtains is as follows in Table 6.1.

Table 6.1

Number of paints in 3 sweeps	Frequency	Relative frequency
0	6	0·38
1	7	0·43
2	2	0·13
3	1	0·06
Total	16	1·00

Hence it may be said that the empirical probability distribution is given by the relative frequency distribution. This is an example of a discrete probability distribution since the independent variable being measured, the number of paints of the buoy in 3 sweeps, can only take the values 0, 1, 2 or 3, and no intermediate values such as $2\frac{1}{2}$ or $3\frac{3}{4}$. It shows us how the probability of the outcome of a situation is scattered over the possible outcomes. We could get a graphical representation of the probability distribution by drawing a histogram in the same way as for a relative frequency distribution. Since one of the outcomes must occur the sum of the probabilities of the different mutually exclusive outcomes is 1. The other main point of interest is that because this is a distribution it has a mean and a standard deviation.

It will help to introduce some notation here so we call the independent variable x and $p(x)$ the function or equation which shows how the probability is distributed over the values of x. In this particular case $p(x)$ has no neat mathematical formula but in later examples we will consider situations where it has.

Then the mean of the frequency distribution would be given by the formula:

$$\mu = \frac{\sum_i f_i x_i}{n}$$

where f_i is again the frequency with which the value x_i occurs as in Chapter 3 and $n = \sum_i f_i$ (total frequency). This may be rewritten as

$$\mu = \sum_i \left(\frac{f_i}{n}\right) x_i$$

But $f_i/n = p(x = x_i)$, the probability that x takes the value x_i; hence

$$\mu = \sum_i x_i p(x_i)$$

In this example $\mu = 0 \times 0·38 + 1 \times 0·43 + 2 \times 0·13 + 3 \times 0·06 = 0·87$.

Discrete Distributions

It is quite common notation to write $\sum_x xp(x)$ as an alternative to $\sum_i x_i p(x_i)$, indicating that the summation is over the values of x. By a similar argument it may be shown that the variance of this distribution is given by the formula written in the alternative notation as

$$\sigma^2 = \sum_x x^2 p(x) - \left[\sum_x xp(x)\right]^2$$

Thus

$$\sigma^2 = 0 \times 0.38 + 1 \times 0.43 + 4 \times 0.13 + 9 \times 0.06 - (0.87)^2$$
$$= 1.49 - 0.76$$
$$= 0.73$$

As a final point it can be seen that to return from the probabilities of occurrence to the actual numbers observed in the distribution all that is required is to multiply each probability by 16. The results from this example can be summed up to give the first basic definitions of this chapter.

6.3 Definitions

6.3.1 A *random variable* is a variable that takes any one of its values by chance.

6.3.2 A *discrete probability distribution* shows how the probability of an outcome of a situation varies between the different values of the discrete variable x which describes the outcomes of a situation. The mathematical function or equation for the distribution of probability is often written as $p(x)$. It should be noted that (i) $0 \leqslant p(x_i) \leqslant 1$ for each x_i, (ii) $\sum_x p(x_i) = 1$.

6.3.3 The mean μ of a discrete probability distribution is given by the formula $\mu = \sum_x x_i p(x_i)$. It is alternatively written as $E(x)$ and is termed the *expectation* of x, or expected value.

6.3.4 The *variance*, σ^2, of a discrete probability distribution is alternatively written as $V(x)$ and is given by the formula

$$V(x) = \sigma^2 = \sum_x x_i^2 p(x_i) - \left(\sum_x x_i p(x_i)\right)^2$$

This may be alternatively written as

$$V(x) = E(x^2) - [E(x)]^2 \quad \text{where} \quad E(x^2) = \sum_x x_i^2 p(x_i)$$

6.3.5 In general the expectation of x^r for a discrete probability distribution is written as $E(x^r)$ and is defined mathematically by $E(x^r) = \sum x^r p(x_i)$. It is also called the *r*th moment of the probability distribution about the origin of measurement.

In all the above definitions the summation is taken over all values of x.

Marine Statistics

6.3.6 If a situation is observed N times then we would expect each value of x to occur $Np(x)$ times.

From these basic ideas of discrete probability distributions it is now possible to consider some of the more usual standard models which arise.

6.4 *Discrete Uniform Distribution*

Consider the following example. Ten seamen, designated $A, B, C, D, E, F, G, H, I, J$, join a ship at the commencement of a voyage. Three men are to be allocated to each of the three watches and the remaining one is allocated as a day worker. The allocation is performed simply by drawing the letters out of a hat. What is the probability for each man that he will become the day worker?

This situation is reasonably straight-forward, and we can assume without affecting the results that the day worker is drawn first. Since there are ten pieces of paper in a hat, and again it may be assumed that they are all the same size, the probability that the piece with an A on it is drawn first is $1/10$ and is the same for all the letters. Formalising this result we have a random variable x, which is the seaman drawn first who is to be the day worker, and x can take any of the values $A, B, \ldots, J$. Then $p(x) = 1/10$ for each of these values. We can note also that $0 < p(x) < 1$ for all x, and

$$\sum_x p(x) = \frac{1}{10} + \frac{1}{10} + \cdots + \frac{1}{10} = 10 \cdot \frac{1}{10} = 1$$

This is an example of a discrete uniform distribution.

6.4.1 In general if x is a discrete random variable which can take any one of n values $x_1, x_2, \ldots, x_n$ and if $p(x) = 1/n$ for each of these values, i.e. the probability of occurrence is the same for all the values, then x has a discrete uniform distribution.

The probability histogram for the example above is given in Fig. 6.1. This is typical of any discrete uniform distribution. A second example of a discrete

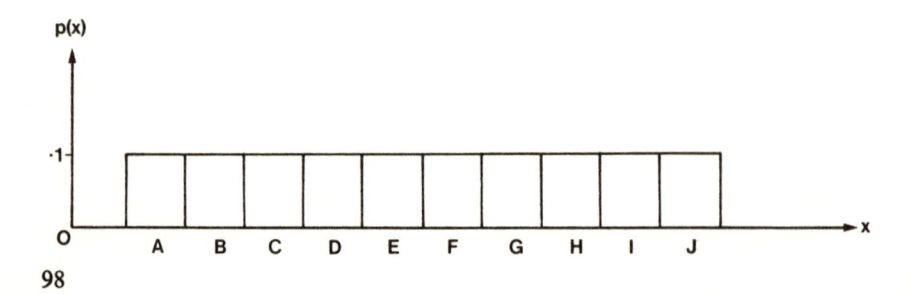

Fig 6.1 Probability Histogram for the Allocation of a Seaman to the Day Shift

Discrete Distributions

uniform distribution is given by considering the outcomes of tossing a fair six-sided die. The outcome, x, can be any one of the six numbers 1 to 6 inclusive and their probabilities of occurrence are equally likely. This may be summarised as follows:

x	1	2	3	4	5	6
$p(x)$	$\frac{1}{6}$	$\frac{1}{6}$	$\frac{1}{6}$	$\frac{1}{6}$	$\frac{1}{6}$	$\frac{1}{6}$

Our previous example involved a qualitative variable so no mean or variance could be calculated but in this situation they can.

Using the earlier results

$$\mu = \sum_x x_i p(x_i)$$

$$= 1 \times \frac{1}{6} + 2 \times \frac{1}{6} + 3 \times \frac{1}{6} + 4 \times \frac{1}{6} + 5 \times \frac{1}{6} + 6 \times \frac{1}{6}$$

$$= \frac{21}{6} = 3\tfrac{1}{2}$$

$$\sigma^2 = \sum_x x_i^2 p(x_i) - \left[\sum_x x_i p(x_i)\right]^2$$

$$= 1 \times \frac{1}{6} + 4 \times \frac{1}{6} + 9 \times \frac{1}{6} + 16 \times \frac{1}{6} + 25 \times \frac{1}{6} + 36 \times \frac{1}{6} - \frac{49}{4}$$

$$= \frac{91}{6} - \frac{49}{4} = \frac{35}{12}$$

Thus the mean score when a die is tossed is 3·5 with standard deviation $\sqrt{35/12} = 1\cdot7$.

6.5 Binomial Distribution

Let us again start by considering an example. A navigator makes an arithmetical error once in every 20 calculations for obtaining a position line from an astronomical observation. During a certain evening he obtains observations of six stars. What is the probability that: (i) one of the six reduction calculations will contain an arithmetical error, and (ii) more than half of the six reduction calculations will contain such an error?

We must assume first that he makes each calculation completely independent of the others and all are of the same complexity so that the chance error is 1/20 for all of them and hence the chance of success is 19/20 for all of them. Let C denote a correct calculation and E an erroneous one, and let the probability of success be $p = 19/20$ and the probability of error be $q = 1/20$.

If we consider the six calculations in order, then to answer part (i) we want

Marine Statistics

to know the probability of the mistake being on any one of the six calculations and the other five calculations being correct. Hence our sequence of events could be as follows:

```
Calculation number  1  2  3  4  5  6
                    E  C  C  C  C  C
                or  C  E  C  C  C  C
                or  C  C  E  C  C  C
                or  C  C  C  E  C  C
                or  C  C  C  C  E  C
                or  C  C  C  C  C  E
```

Any of these six sequences of results gives us the final situation of one mistake. The probability of any one of them is given by $p^5 q = 19^5/20^6$ using the multiplication law of probability. However using the addition law of probability, since we have mutually exclusive events, the probability of getting one mistake in 6 calculations is $6 \cdot 19^5/20^6 = 0.23$.

To answer part (ii) we must know the probability of getting 4 or 5 or 6 mistakes out of 6 calculations; 4 mistakes can occur in any of the ways mentioned below, of which there are 15 in all.

```
Calculation number  1  2  3  4  5  6     1  2  3  4  5  6
                    E  E  E  E  C  C     E  C  E  E  E  C
                    E  E  E  C  E  C     E  C  E  E  C  E
                    E  E  E  C  C  E     E  C  E  C  E  E
                    E  E  C  E  E  C     E  C  C  E  E  E
                    E  E  C  E  C  E     C  E  E  E  E  C
                    E  E  C  C  E  E     C  E  E  E  C  E
                                         C  E  E  C  E  E
                                         C  E  C  E  E  E
                                         C  C  E  E  E  E
```

In fact 15 is the number of ways in which 6 objects may be rearranged, 4 of which are alike and the other 2 of which are alike. This is known as the combination of 4 items out of 6 and has the symbol $\binom{6}{4}$. It is defined mathematically to be equal to $6!/4!\,2!$ where, e.g. 6! (called 6 factorial or 6 shriek) = $6 \times 5 \times 4 \times 3 \times 2 \times 1$. Thus

$$\binom{6}{4} = \frac{6!}{4!\,2!} = \frac{6 \times 5 \times 4 \times 3 \times 2 \times 1}{4 \times 3 \times 2 \times 1 \times 2 \times 1} = 15$$

By symmetry

$$\binom{6}{2} = \frac{6!}{2!\,4!} = 15$$

Discrete Distributions

Those readers who looked at the final section in Chapter 5 will already have met combinations there.

By a similar argument as before

$$p(x = 4) = 15\left(\frac{1}{20}\right)^4\left(\frac{19}{20}\right)^2$$

where x is the number of mistakes. Similarly

$$p(x = 5) = \binom{6}{5}q^5 p = \frac{6!}{5!\,1!}\left(\frac{19}{20}\right)\left(\frac{1}{20}\right)^5 = \frac{6.19}{20^6}$$

$$p(x = 6) = \binom{6}{6}q^6 p^0 = \frac{6!}{6!\,0!}\left(\frac{19}{20}\right)^0\left(\frac{1}{20}\right)^6$$

$$= \left(\frac{1}{20}\right)^6 \quad \text{since 0! is defined to be 1}$$

Using the addition law, the probability of more than half the calculations being wrong

$$= \frac{15.19^2}{20^6} + \frac{6.19^1}{20^6} + \frac{19^0}{20^6}$$

$$= \frac{(5{,}415 + 114 + 1)}{20^6} = \frac{5{,}530}{20^6} \simeq 0{\cdot}00$$

If we were to calculate the complete probability distribution for this example we would get the following results:

x: number of mistakes	0	1	2	3
$p(x)$	$\left(\frac{19}{20}\right)^6$	$6\left(\frac{19}{20}\right)^5\frac{1}{20}$	$15\left(\frac{19}{20}\right)^4\left(\frac{1}{20}\right)^2$	$20\left(\frac{19}{20}\right)^3\left(\frac{1}{20}\right)^3$

x: number of mistakes	4	5	6
$p(x)$	$15\left(\frac{19}{20}\right)^2\left(\frac{1}{20}\right)^4$	$6\left(\frac{19}{20}\right)\left(\frac{1}{20}\right)^5$	$\left(\frac{1}{20}\right)^6$

The terms in this probability distribution are the terms in what is called, mathematically, the binomial expansion of $(19/20 + 1/20)^6$. Hence this probability distribution is an example of what is termed a binomial distribution.

6.5.1   Let x be a discrete random variable denoting the number of 'successes' in a series of n independent trials. If the outcome of each trial can either be designated a 'success' or a 'failure' and if the probability of a 'success', p, is constant from trial to trial then the random variable x has a binomial distribution with probability function $p(x)$ given by $p(x = r) = \binom{n}{r}p^r q^{n-r}$ for $r = 0, 1, 2, \ldots, n$.

101

Marine Statistics

'Success' and 'failure' have been shown in inverted commas because it is arbitrary which of two events has which name. In the previous example a 'success' was defined to be a mistake since we were interested in the number of mistakes. However, the number of 'successes' = the number of trials − the number of 'failures', so the distribution of one is the reverse of the distribution of the other.

6.5.2 **The mean number of 'successes' in a binomial distribution** $E(x) = np$.

6.5.3 **The variance of the number of 'successes' in a binomial distribution** $V(x) = npq$.

These two results may be proved algebraically using the formula developed earlier. Thus in the navigator mistakes example, the mean number of mistakes in 6 calculations with $p = 1/20$ is $6/20 = 0.3$. The variance of the number of mistakes $V(x) = 6 \times 1/20 \times 19/20 = 57/200 = 0.285$. The probability histogram for this same example is given in Fig. 6.2.

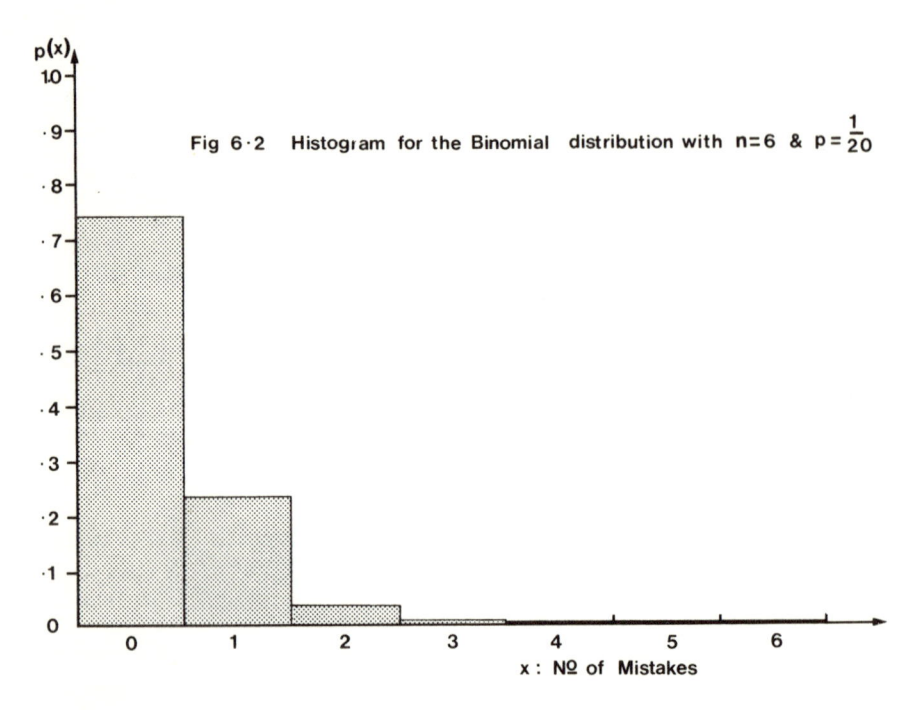

Fig 6·2 Histogram for the Binomial distribution with $n=6$ & $p=\frac{1}{20}$

There is no characteristic shape for a binomial distribution because it depends on the values of p and q. If p and q are equal ($\frac{1}{2}$) then the histogram is symmetrical as shown in Fig. 6.3.

As a second example of a binomial distribution consider the following situation. A growler paints on a radar screen with the same brilliance as wave

Discrete Distributions

returns, but can be distinguished by the observer if it paints in the same position on two or more sweeps out of the five sweeps for which he watches the screen. If the probability that it will paint on a particular sweep is $\frac{3}{4}$, what is the probability that he will pick out the growler?

For each sweep, either the growler paints on the screen or not. The sweeps may be considered independent with respect to picking up the growler and the probability of painting is constant on each sweep. Thus x, the number of paints in 5 sweeps, has a binomial distribution with probability of a paint $p = \frac{3}{4}$. Hence

$$p(x = r) = \binom{5}{r}\left(\frac{3}{4}\right)^r\left(\frac{1}{4}\right)^{5-r} \quad r = 0, 1, 2, 3, 4, 5$$

If $x = 0$

$$p(0) = \frac{5!}{5!\,0!}\left(\frac{3}{4}\right)^0\left(\frac{1}{4}\right)^5 = \frac{1}{1{,}024}$$

If $x = 1$

$$p(1) = \frac{5!}{4!\,1!}\left(\frac{3}{4}\right)^1\left(\frac{1}{4}\right)^4 = \frac{15}{1{,}024}$$

$p(0) + p(1) = 16/1{,}024 = 1/64$ is the probability that the growler paints once or not at all in 5 sweeps. Hence the required probability which is the probability that it paints 2, 3, 4 or 5 times in 5 sweeps is the complement of this, i.e. $1 - 1/64 = 0.9844$. Hence the probability of detection of the growler is 63/64. This is another case where it is easier to calculate the complementary probability than the required one.

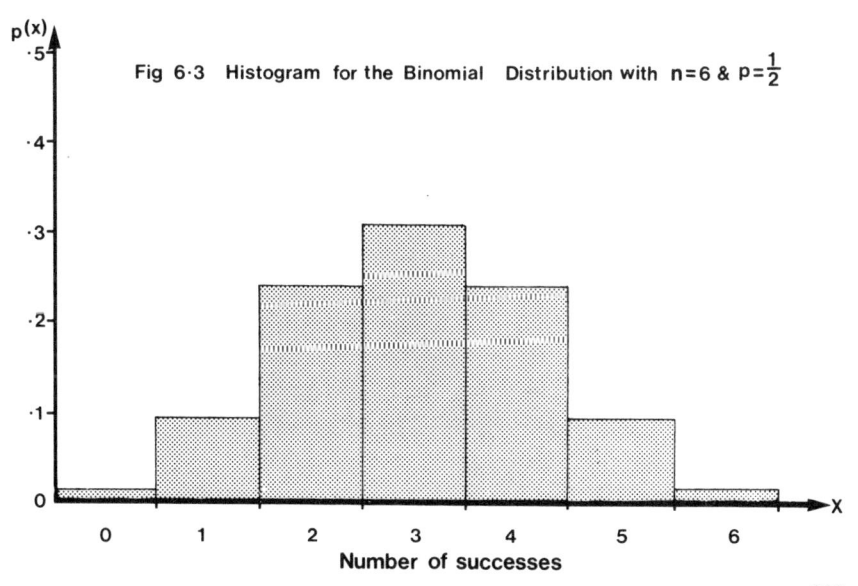

Fig 6·3 Histogram for the Binomial Distribution with $n=6$ & $p=\frac{1}{2}$

Marine Statistics

If x denotes the number of successes in a binomial distribution with p the probability of success and n the number of trials then this may be written as $x \sim b(n, p)$. Thus in the first example on navigator mistakes x, the number of mistakes, $\sim b(6, 1/20)$.

Tables on probabilities associated with binomial distributions exist but obviously it takes a lot of space to cover all combinations of n and p. It is usually quicker to calculate the probabilities from first principles than to search for a set of tables covering the appropriate values of n and p.

6.6 Poisson Distribution

Consider the following situation. It is known that a certain area of the Caribbean is hit, on average, by 6 hurricanes a year. Find the probability that in a given year fewer than 4 hurricanes hit this area. The arrival of a hurricane in a particular area in a region which is prone to them may be considered as a random event. Just because one hurricane hits a particular area in the region does not mean that the next few, or even the next one, will necessarily hit it at all—but on the other hand they might. However, there is a certain constancy about the process in that the mean level of hurricanes per year stays the same in this area over a long period of years. There is also no limit, within reason, to the number of hurricanes which could hit the particular area in any one year.

The probability distribution of a random variable x which describes the occurrence of an event, random in time or space, such as the situation above, is called a *Poisson distribution*. In the above example we are concerned with a random process in time, the number of hurricanes per year. An example of a random process in space might be the number of ships per square mile in the shipping lanes of the Dover Strait. If the distribution were uniform then we would expect more or less the same number of ships in every square mile. However, what occurs in practice is that in some square miles there will be considerably more ships than the usual number and in others considerably less.

This situation has also been shown to have a Poisson distribution with respect to time so that for any one square mile the number of ships per hour varies according to this random process. In any situation where there is an observed amount of clustering at different times or in different places it is often useful to examine, to see if there is an underlying Poisson distribution or not. If there is then the clustering can be said to be due to random causes but if not then it is worth delving further to try to establish any reasons for the clustering. The mathematical proof of the derivation of the formula for the Poisson probability function is rather complicated and is not necessary for an understanding of how the distribution works so the results will be quoted on their own.

Discrete Distributions

6.6.1 A situation whereby the *Poisson distribution* can be used to give the probability of different values of a discrete random variable occurring must possess the following properties:

(a) The mean number of occurrences, μ, in a given time interval or specified region is known.

(b) The probability of a single occurrence in a very short time interval or in a very small region is proportional only to the length of the interval or the size of the region and does not depend on the number of occurrences at other times or in other places.

(c) The probability of more than one occurrence in the very short time interval or very small region is negligible.

6.6.2 The probability function $p(x)$ for the number of occurrences, x, per given time interval or specified region of a certain event is

$$p(x) = \frac{e^{-\mu}\mu^x}{x!} \quad x = 0, 1, 2, \ldots$$

where μ is the mean number of occurrences per time interval or per region and e is the mathematical exponential function. Tables of values of e^y, where y can be positive or negative, are published and the value of e itself is 2·71828, ... (see Table 9)

If x denotes the number of occurrences per given time interval or specified region of a certain event and x has a Poisson distribution with mean, μ, then we can write $x \sim p(\mu)$. This is not used very often because of the confusion with the usual probability function.

The given time interval can obviously be of any suitable length such as a year, a week, an hour etc. and similarly the specified region of any suitable size. We can now answer our hurricane problem. The mean number of hurricanes per year is 6 and the conditions suggest that we may use the Poisson distribution as a model of probability distribution. Thus

$$p(x) = \frac{e^{-6}6^x}{x!} \quad x = 0, 1, 2, \ldots$$

Thus

$$p(x < 4) = p(0) + p(1) + p(2) + p(3)$$

$$p(0) = \frac{e^{-6}6^0}{0!} = e^{-6} = 0 \cdot 0025$$

$$p(1) = \frac{e^{-6}6^1}{1!} = 6e^{-6} = 0 \cdot 0149$$

$$p(2) = \frac{e^{-6}6^2}{2!} = 18e^{-6} = 0 \cdot 0446$$

$$p(3) = \frac{e^{-6}6^3}{3!} = 36e^{-6} = 0 \cdot 0892$$

Marine Statistics

Thus

$$p(x < 4) = 0.1512$$

The probability histogram for this example is given in Fig. 6.4.

As with the binomial distribution there is no standard recognisable form for the probability histogram of the Poisson distribution but its shape depends on the particular value of μ. Situations where this distribution is considered appropriate are therefore best recognised through the conditions for the distribution rather than the shape of the histogram.

The Poisson distribution can also be used as an approximation to the

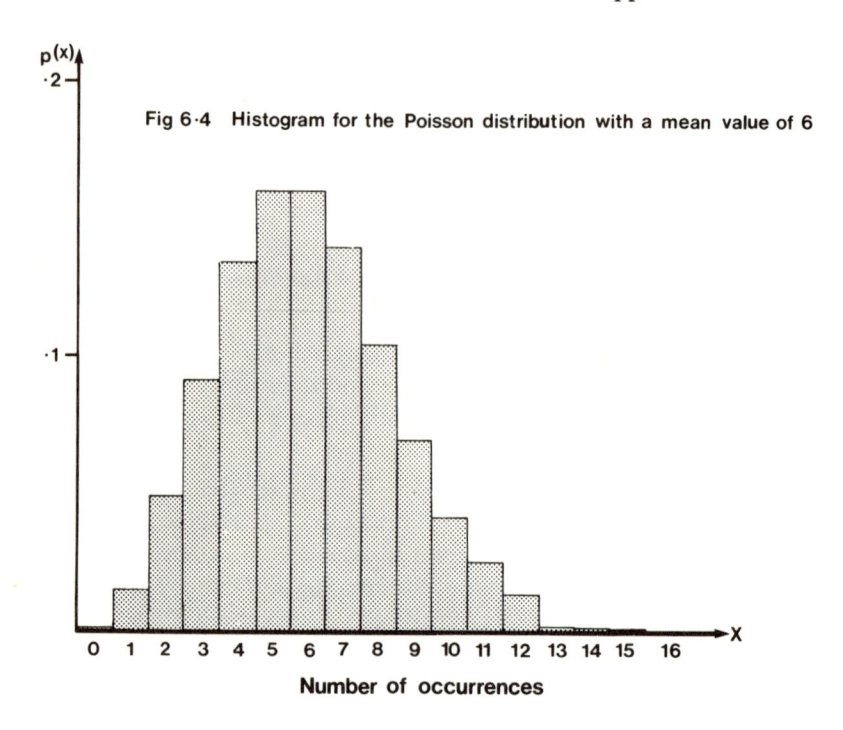

Fig 6·4 Histogram for the Poisson distribution with a mean value of 6

Number of occurrences

binomial distribution in certain cases. Suppose n, the number of independent trials in a binomial distribution, is very large and p, the probability of success on any one trial, is very small, then the occurrence of a success is a rare event among all the numerous failures which will occur. However, the mean number of successes, np, is a constant for constant n and p. This gives similar conditions as for the occurrence of a Poisson distribution. Obviously the practical question is how large must n be and how small p before the approximation can be used, and the answer is not really cut and dried but depends on how accurate the required probabilities should be. As a rough guide n should be larger than 30 and p small, such that the mean np is not bigger than 10. In the next chapter we will consider other approximations to both the binomial and

Discrete Distributions

the Poisson to cover larger means. If p is very large the result still applies but counting failures rather than successes, as it is q that is very small then.

As an example of this approximation method, we can consider the problem of the number of ships grounding which we started the chapter with. The port authority's statistician would have to make some initial assumptions. It would be assumed that firstly all ships were independent with respect to their individual chance of touching the ground, and that the probability of grounding was the same for all ships. These assumptions are obviously open to debate but at least we will get some idea about the situation. If we assume that the buoyage changes have no effect on navigation then the probability of a ship touching ground should stay constant. Thus, if we call a ship touching the ground a success, then x, the number of successes in 2,000 ships, has a binomial distribution with probability of success $p = 1/1,000 = 0.001$ from past records, and mean number of successes $2,000 \times 1/1,000 = 2$. Since 2,000 is large and 0.001 is small we may use the Poisson approximation to the binomial distribution. We want to know $p(x \geq 5) = 1 - p(x \leq 4)$, where

$$p(x) = \frac{e^{-2}2^x}{x!} \quad x = 0, 1, 2, \ldots$$

$$p(0) = \frac{e^{-2}2^0}{0!} = e^{-2} = 0.135$$

$$p(1) = \frac{e^{-2}2^1}{1!} = 2e^{-2} = 0.270$$

$$p(2) = \frac{e^{-2}2^2}{2!} = 2e^{-2} = 0.270$$

$$p(3) = \frac{e^{-2}2^3}{3!} = \tfrac{4}{3}e^{-2} = 0.180$$

$$p(4) = \frac{e^{-2}2^4}{4!} = \tfrac{2}{3}e^{-2} = 0.090$$

$$p(x < 4) = p(0) + p(1) + p(2) + p(3) + p(4)$$
$$= e^{-2} + 2e^{-2} + 2e^{-2} + \tfrac{4}{3}e^{-2} + \tfrac{2}{3}e^{-2} = 7e^{-2} = 0.95$$

Thus under the same conditions as before there is an 0.95 chance of getting fewer than 5 accidents in 2,000 ships and only an 0.05 or 1/20 chance of getting 5 accidents or more. It is up to the port authority to decide how they view a 1-in-20 chance, which is a topic we will be spending time on in later chapters but at least the situation has been quantified.

Marine Statistics

6.6.3 The *mean* of a Poisson distribution is given by $E(x) = \mu$ as defined previously, and the *variance* of a Poisson distribution is given by $V(x) = \mu$. Thus the Poisson distribution has the interesting property that its mean and variance are equal.

Another useful property is that if we have a random variable x which is made up of the sum of two or more other random variables each with Poisson distributions, then x will also have a Poisson distribution, with mean the sum of the means of the individual distributions. Both of these properties may be proved mathematically but as with all the results on the Poisson distribution the algebra is rather complex.

6.7 Negative Binomial Distribution

The three distributions which we have discussed so far are among the most common of the standard discrete probability distributions but there are three others which will be mentioned briefly as they too can arise on a number of situations. The first of these is the negative binomial distribution. It gives the probability that the kth success of an ordinary binomial distribution occurs on the xth trial.

6.7.1 If repeated independent trials can result in a success with probability p and a failure with probability $q = 1 - p$, then the probability distribution of the random variable, x, the number of the trial on which the kth success occurs is the negative binomial distribution with $p(x)$ given by

$$p(x) = \binom{x-1}{k-1} p^k q^{x-k} \quad x = k, k+1, k+2, \ldots$$

6.7.2 The mean of this distribution is given by $E(x) = \mu = k/p$ and the variance is given by $V(x) = kq/p^2$.

We can illustrate the use of this probability distribution with the following example. A target on a radar screen can be distinguished from the background noise if it paints three times in the same position during the period of observation. If the probability that a particular target will paint during one sweep is 0·6, what is the probability of an observer identifying the target on the fifth sweep for which he watches the screen? Each sweep is an independent trial with constant probability of success, a paint, 0·6 and a probability of failure of 0·4. Thus we have a binomial situation but with the number of trials

Discrete Distributions

unknown. Hence the probability of the third paint being on the xth sweep is given by a negative binomial distribution with

$$p(x) = \binom{x-1}{2}(0{\cdot}6)^3(0{\cdot}4)^{x-3} \quad \text{for } x = 3, 4, \ldots$$

If $x = 5$

$$p(5) = \binom{4}{2}(0{\cdot}6)^3(0{\cdot}4)^2$$

$$= \frac{4!}{2!\,2!}(0{\cdot}6)^3(0{\cdot}4)^2 = 6(0{\cdot}6)^3(0{\cdot}4)^2$$

$$= 0{\cdot}207$$

which is the required probability. The mean number of sweeps required to identify the target is $3/0{\cdot}6 = 5$.

If one considers the frequency distribution for the number of sea collisions per month over a recent 20-year period it may be shown that the Poisson distribution does not provide a good model because factors affecting collisions such as visibility do not remain constant from month to month. In fact the best model is given by the negative binomial distribution. It may be shown mathematically that this is the resulting distribution when the probability of success, or mean level of success, is not constant from trial to trial. The probability function is then often quoted in a slightly different form as

$$p(x) = \binom{\alpha}{x}\left(\frac{\gamma}{\gamma+1}\right)^{\alpha}\left(\frac{1}{\gamma+1}\right)^x \quad x = 0, 1, 2, \ldots$$

where x is the number of occurrences of an event per given time interval or specified region. The mean of x is $E(x) = \alpha/\gamma$, and the variance is $V(x) = \alpha(\gamma+1)/\gamma^2$.

The negative binomial distribution has been found to apply to the distribution of road accidents and air accidents together with marine casualties.

6.8 Geometric Distribution

If instead of considering when the kth success occurs we consider the number of trials needed to get one success, then we have a special case of the negative binomial distribution known as the geometric distribution.

6.8.1 If repeated independent trials can result in a success with constant probability p and a failure with probability $q = 1 - p$, then x, the random variable which gives the number of the trial on which the first success occurs, has a geometric distribution with $p(x)$ given by,

$$p(x) = pq^{x-1} \quad x = 1, 2, \ldots$$

It can be shown that $E(x) = 1/p$ and $V(x) = q/p^2$.

109

Marine Statistics

The distribution gets its name because successive terms form a geometric progression. As an example we can find the probability that a student passes the examinations for his Master's certificate on his third attempt if the probability of passing on one sitting is 0·7.

We have to assume that the probability of passing remains constant and the examinations are independent. If these conditions apply then the required probability distribution is the geometric distribution with $p = 0·7$ and $q = 0·3$. Thus the probability that he passes on his third attempt is given by $p(3) = 0·7(0·3)^2 = 0·063$. It should be pointed out that the probability function for the geometric distribution is very easily obtained from first principles using the multiplication law of probability, since we require the probability of $x - 1$ failures followed by 1 success.

6.9 *Hypergeometric Distribution*

The hypergeometric distribution provides the answer to one case of departure from the binomial distribution when it cannot be assumed that the probability of success stays constant from trial to trial. In particular suppose that it is known that in N items there are K successes and $N - K$ failures. If a sample of n items is taken, then the probability that the first item is a success is K/N but the probability that the second item is a success is $(K - 1)/(N - 1)$ if the first item was a success and $K/(N - 1)$ if it was a failure.

6.9.1 If a population of N items contains K successes and $N - K$ failures then the random variable x describing the number of successes in a sample of size n has a hypergeometric distribution with $p(x)$ given by,

$$p(x) = \frac{\binom{K}{x}\binom{N-K}{n-x}}{\binom{N}{n}} \quad x = 0, 1, 2, \ldots, n$$

$$E(x) = \frac{nK}{N} \quad \text{and} \quad V(x) = \frac{N-n}{N-1} \cdot \frac{nK}{N} \cdot \frac{N-K}{N}$$

Example

A small coastal container vessel can carry one container on the after deck, two on the fore-deck and six below decks. The nine containers loaded for a particular voyage are put aboard in random order, but it is known that there is heavy cargo in four of the containers and light cargo in the remaining five. If two of the heavy containers are carried on deck, the ship has to fill ballast tanks in order to achieve satisfactory stability and if three of the heavy containers are loaded on deck, the ship becomes unstable. What is the probability that:

(a) the ballast tanks will need to be filled?
(b) it will be unsafe for the ship to sail?

Discrete Distributions

We have a population of 9 items, consisting of 4 heavy containers (successes!) and 5 light containers (failures!). The sample of containers on deck is of size 3, and the probability distribution of successes within this sample is given by the hypergeometric distribution with

$$p(x) = \frac{\binom{4}{x}\binom{5}{3-x}}{\binom{9}{3}} \quad x = 0, 1, 2, 3$$

To answer (a) we want to know $p(2)$ the probability of 2 heavy containers on deck:

$$p(2) = \frac{\binom{4}{2}\binom{5}{1}}{\binom{9}{3}} = \frac{\frac{4!}{2!\,2!}\frac{5!}{4!\,1!}}{\frac{9!}{6!\,3!}} = \frac{6 \cdot 5}{84} = \frac{5}{14}$$

To answer (b) we want to know $p(3)$:

$$p(3) = \frac{\binom{4}{3}\binom{5}{0}}{\binom{9}{3}} = \frac{\frac{4!}{3!\,1!}\frac{5!}{5!\,0!}}{\frac{9!}{6!\,3!}} = \frac{4 \cdot 1}{84} = \frac{1}{21}$$

6.10 Summary

In this chapter we have considered the general concept of a discrete probability distribution and its relationship to an observed frequency distribution. Following that, a brief discussion was given on the standard discrete probability models. The chapter has moved fairly quickly over the various models discussed and the reader will probably find it easier to grasp the implications of each particular model when he is faced with a practical situation for which one of these is the appropriate model. The next chapter will consider an extension of the concept to probability distributions involving a continuous random variable.

Exercises

1. Two consignments of crated cargo are loaded simultaneously into a ship's lower hold, using pallets each carrying 4 crates. The number of crates of cargo 'A' per pallet were noted for the first 20 pallets as follows:

Marine Statistics

Number of crates marked 'A'	Frequency
0	2
1	8
2	6
3	3
4	1

Construct a relative frequency table for this information and use the inferred probabilities to calculate the mean and the variance of the distribution.

2. A ship is provided with 6 life-rafts. During a lifeboat drill one of these rafts is chosen at random for testing. If the life-rafts are identified by the letters A to F, construct a histogram showing the probability that each raft will be chosen, and state what distribution it represents.

3. A large consignment of cargo consists of equal numbers of four types of cases weighing 30, 40, 50 and 80 kg respectively. Construct a table showing the probability that a case selected at random will be of each of the four weights. Find the mean and the standard deviation of this distribution.

4. A ship on a regular run to a port can enter a lock if the water level is half tide or above. If the height of tide is less than this, the ship has to wait in the river. If no attempt is made to synchronise the time of arrival of the ship with the tidal cycle, what is the probability that, in six trips, the ship will have to wait in the river before entering the lock:
   (a) At least three times?
   (b) Five times or more?

5. In a large parcel of bagged cargo it is known that 10% of the bags are underweight. In a sling of eight bags, what is the probability that:
   (a) None of the bags are underweight?
   (b) At least three of the bags are underweight?
   Find the mean and the standard deviation of the number of defective bags per sling.

6. Over a period of several years a protection and indemnity club received a number of claims for structural damage to ships where the cause was ascribed to 'freak waves'. The mean number of claims per year in this category was 4. Assuming that conditions remain similar, what is the probability that the club will receive more than 4 claims in the coming year?

7. In cargoes of case oil loaded at a particular port it is found that leaking drums occur at random and that, overall, 0·5% of the total are likely to

Discrete Distributions

be defective in this way. If a part cargo of 500 drums is loaded, what are the probabilities of 0, 1, 2 or 3 drums having leakages?

8. At a particular port, the probability that the visibility will be less than 1 km is 0·2. For a ship on a regular run to that port, what is the probability that, on the 10th trip, the visibility will be less than 1 km for the 4th occasion?

What is the mean number of trips made so that, on the final trip, the third occasion of less than 1 km visibility is experienced?

9. A survivor adrift in a lifeboat estimates the probability of his pyrotechnic signal being seen as 0·2 if he ignites it when he sees a ship hull down on the horizon and as 0·6 if he ignites it when he sees a ship hull up. On the basis of his estimate, what is the probability that his plight will be discovered on his third signal:

(a) If he ignites his signals on first seeing another ship?
(b) If he waits until another ship is hull up before igniting his signals?

Does the lower result for section (b) indicate that the survivor has a smaller chance of being rescued if he waits until a ship is hull up before igniting his signals?

10. Of the 12 fire extinguishers with which a ship is equipped it is known that 4 are faulty. If a sample of 5 extinguishers is chosen at random for testing:

(a) What is the probability that the sample will contain 2 of the faulty extinguishers?
(b) What is the probability that the sample will contain less than 2 of the faulty extinguishers?

Answers

1.

Number of crates marked 'A'	Frequency	Relative frequency
0	2	0·10
1	8	0·40
2	6	0·30
3	3	0·15
4	1	0·15

$$\mu = \sum_{x} x_i p(x_i)$$
$$= 0 \times 0\cdot 1 + 1 \times 0\cdot 4 + 2 \times 0\cdot 3 + 3 \times 0\cdot 15 + 4 \times 0\cdot 05$$
$$= \underline{1\cdot 65}$$

Marine Statistics

$$\sigma^2 = \sum_x x^2 p(x) - \left[\sum_x x p(x)\right]^2$$
$$= 3 \cdot 75 - 2 \cdot 72$$
$$= 1 \cdot 03$$

2. Since each raft has the same probability of being chosen, and it is certain that one of them will be, the probability for each is $\frac{1}{6}$. This is a uniform distribution.

3. Since each type of case has an equal chance of being selected, the probabilities are tabled as follows, giving a uniform distribution:

Weight of case	Probability of selection
30	$\frac{1}{4}$
40	$\frac{1}{4}$
50	$\frac{1}{4}$
80	$\frac{1}{4}$

Mean,

$$\mu = \sum_x x p(x)$$
$$= \frac{30}{4} + \frac{40}{4} + \frac{50}{4} + \frac{80}{4}$$
$$= \underline{50 \text{ kg}}$$

Standard deviation,

$$\sigma = \sqrt{\sum_x x^2 p(x) - \left[\sum_x x p(x)\right]^2}$$
$$= \sqrt{\left(\frac{30^2}{4} + \frac{40^2}{4} + \frac{50^2}{4} + \frac{80^2}{4}\right) - 50^2}$$
$$= \sqrt{350}$$
$$= \underline{18 \cdot 7 \text{ kg}}$$

4. Over a prolonged period of time the tide may be assumed to be above half tide level for half the time and below half tide level for half the time. If we take p as the probability of the ship not having to wait and q as the probability of the ship having to wait to enter the lock, we thus have:

$$p = q = \frac{1}{2}$$

We have a binomial distribution with $n = 6$, $p = \frac{1}{2}$. Putting $x =$ the number of times the ship has to wait in 6 arrivals we calculate the following probabilities:

Discrete Distributions

(a) $p(x = 0) = \binom{6}{0} p^6 q^0 = 0.5^6 \qquad = 0.0156$

$p(x = 1) = \binom{6}{1} p^5 q = 6 \times 0.5^6 \qquad = 0.0938$

$p(x = 2) = \binom{6}{2} p^4 q^2 = 15 \times 0.5^6 = 0.2344$

$p(x < 3) \qquad\qquad\qquad = 0.3438$

$\Rightarrow p(x \geqslant 3) = 1 - 0.3438 \qquad\qquad = 0.6562$

(b) $p(x = 5) = \binom{6}{5} pq^5 = 6 \times 0.5^6 \qquad = 0.0938$

$p(x = 6) = \binom{6}{6} p^0 q^6 = 0.5^6 \qquad = 0.0156$

$p(x \geqslant 5) \qquad\qquad\qquad = 0.1094$

5. Let p represent the probability that a bag is underweight and q the probability that it is not underweight; then:

$p = 0.1$
$q = 0.9$

Assuming the bags are independent we have a binomial distribution with $n = 8$ and $p = 1/10$. Putting $x = $ number of underweight bags in a sling of eight, we calculate the following probabilities:

(a) $p(x = 0) = \binom{8}{0} p^0 q^8 = 0.9^8 \qquad\qquad = 0.430$

(b) $p(x = 0) = \binom{8}{0} p^0 q^8 = 0.9^8 \qquad\qquad = 0.430$

$p(x = 1) = \binom{8}{1} pq^7 = 8 \times 0.1 \times 0.9^7 \qquad = 0.383$

$p(x = 2) = \binom{8}{2} p^2 q^6 = 28 \times 0.1^2 \times 0.9^6 = 0.149$

$p(x \leqslant 2) \qquad\qquad\qquad\qquad = 0.962$

$p(x > 2) = 1 - 0.962 \qquad\qquad\qquad = 0.038$

The mean number of underweight bags per sling, μ or $E(x)$, is given by:

$E(x) = np$
$\qquad = 8 \times 0.1 = 0.8$

115

Marine Statistics

The standard deviation, σ or $\sqrt{V(x)}$, is given by:

$$\sigma = \sqrt{npq}$$
$$= \sqrt{8 \times 0 \cdot 1 \times 0 \cdot 9} = \underline{0 \cdot 85}$$

6. We can assume a Poisson distribution since these claims are independent. Putting x = number of claims:

$$p(0) = \frac{0 \cdot 0183 \times 4^0}{0!} = 0 \cdot 0183$$

$$p(1) = \frac{0 \cdot 0183 \times 4}{1!} = 0 \cdot 0732$$

$$p(2) = \frac{0 \cdot 0183 \times 4^2}{2!} = 0 \cdot 1464$$

$$p(3) = \frac{0 \cdot 0183 \times 4^3}{3!} = 0 \cdot 1952$$

$$p(4) = \frac{0 \cdot 0183 \times 4^4}{4!} = 0 \cdot 1952$$

$$p(x \leqslant 4) \qquad\qquad\qquad = 0 \cdot 6283$$
$$p(x > 4) = 1 - 0 \cdot 6283 = \underline{0 \cdot 3717}$$

7. This is strictly a binomial distribution but the quantities are such that it is convenient and permissible to use the Poisson approximation to the binomial. Thus

$$n = 500 \quad \text{which is greater than 30}$$

and

$$np = 500 \times \frac{0 \cdot 5}{100} = 2 \cdot 5 \quad \text{which is less than 10}$$

Putting x = number of defective drums:

$$p(x) = \frac{e^{-\mu} \mu^x}{x!}$$

where $e^{-\mu} = e^{-2 \cdot 5} = 0 \cdot 082$, we calculate the following probabilities:

$$p(x = 0) = \frac{0 \cdot 082 \times 2 \cdot 5^0}{0!} = 0 \cdot 082$$

$$p(x = 1) = \frac{0 \cdot 082 \times 2 \cdot 5}{1!} = 0 \cdot 205$$

$$p(x = 2) = \frac{0 \cdot 082 \times 2 \cdot 5^2}{2!} = 0 \cdot 256$$

$$p(x = 3) = \frac{0 \cdot 082 \times 2 \cdot 5^3}{3!} = 0 \cdot 214$$

Discrete Distributions

8. We have a negative binomial distribution. Putting

   x = number of trips
   k = number of occasions when visibility <1 km
   p = probability of visibility <1 km
   q = probability of visibility $\geqslant$1 km

   the probability that the 4th occasion of visibility <1 km falls on the 10th trip is given by:

   $$p(x) = \binom{x-1}{k-1} p^k q^{x-k}$$
   $$= \binom{9}{3} 0.4^4 \times 0.8^6$$
   $$= \underline{0.035}$$

   The mean number of trips such that the 3rd occasion of visibility <1 km falls on the final day is given by:

   $$\mu = \frac{k}{p} = \frac{3}{0.2} = \underline{15}$$

9. For this situation, a geometric distribution is appropriate and, if we let x represent the number of the trial on which the first success occurs, we have:

   $$p(x) = pq^{x-1}$$

   where p is the probability of success in each trial and q is the probability of failure in each trial. Substituting appropriate values, we have:

   (a) $p(3) = 0.2 \times 0.8^2$
   $= \underline{0.128}$

   (b) $p(3) = 0.6 \times 0.4^2$
   $= \underline{0.096}$

   (c) No. It simply means that the probability of his signal being first seen on the third attempt is less. The overall probability that at least one of his first three signals will be seen is much greater if he waits until another ship is hull up before he ignites them.

10. For this example, the hypergeometric distribution is applicable since the proportion of faulty extinguishers in the population is altered when the first one is chosen for the sample. If we let x represent the number of faulty extinguishers in the sample, then:

    $$p(x) = \frac{\binom{K}{x}\binom{N-k}{n-x}}{\binom{N}{n}}$$

Marine Statistics

where: the number in the population, $N = 12$
the number in the sample, $n = 5$
the number of faulty extinguishers in the population, $K = 4$

Thus:

(a) $\quad p(2) = \dfrac{\binom{4}{2}\binom{8}{3}}{\binom{12}{5}} = \dfrac{6.56}{792} = \underline{0.424}$

and

(b) $\quad p(0) = \dfrac{\binom{4}{0}\binom{8}{5}}{\binom{12}{5}} = \dfrac{1.56}{792} = 0.071$

$\quad p(1) = \dfrac{\binom{4}{1}\binom{8}{4}}{\binom{12}{5}} = \dfrac{4.70}{792} = 0.354$

$p(x < 2) \qquad\qquad\qquad = \underline{0.425}$

Chapter 7
Continuous Probability Distributions

7.1 *Introduction*

For a certain radio aid it is known that the errors in position lines obtained from it have a mean value of zero nautical miles but a standard deviation of 1·5 miles. It is required to know the probability that the error on any one occasion is likely to exceed 6 nautical miles. The exact value that the error can take at any time is measurable on a continuum so we are now dealing with a continuous variable. However the information that we want is similar to answers we were providing in the last chapter so again we need to know how probability is distributed over all possible values of the error. In this chapter we are concerned with continuous probability distributions and will follow the same pattern, discussing the general concept of continuous probability distributions, followed by an introduction to some of the standard distributions which can be used to model practical behaviour.

7.2 *Continuous Probability Distributions in General*

Imagine we have a clock that has stopped. If we look at that clock to know the time then the error that is made in our estimate of the time is the difference between the actual time when we look and the time at which the clock stopped. Whether we measure this time error in hours, minutes or seconds, there is an equal probability that the error will take any one of the possible values. Suppose first we measure only in hours and the clock stopped at about 3 o'clock and the time now is about 10 o'clock. Then our particular error

Marine Statistics

slow is 7 hr. However, if x denotes the size of error slow in hours then the probability that x is 0, 1, 2, ..., 12 hr is given by

$$p(x) = \frac{1}{12} \qquad x = 1, 2, \ldots, 12$$

which means that x has a discrete uniform probability distribution.

Suppose now that we measure in minutes and the clock stopped at about 3.02 and the time is actually about 9.56. Then our particular error is 6 hr 54 min = 414 min. Again if y denotes the size of error slow in minutes then the probability that y is 0, 1, 2, ..., 720 min is given by

$$p(y) = \frac{1}{720} \qquad x = 1, 2, \ldots, 720$$

Again y has a uniform distribution but imagine how difficult it would be to draw a histogram with 720 subdivisions.

To a further degree of accuracy suppose we measure the error in seconds, and we know that the clock stopped at 3.02.10 and that the actual time is 9.56.12. Now our error is 6 hr 54 min 2 sec = 414 min 2 sec = 24,842 sec. Let z denote the size of error measured in seconds and it can be seen that z can take any value from 0 to 43,200 sec, with

$$p(z) = \frac{1}{43,200} \qquad z = 0, 1, \ldots, 43,200$$

The probability of z being any particular number of seconds is so small that it may to all intents and purposes be treated as 0. However, if we take a range of values of z, say, from 24,810–24,869 sec, then if that time is measured to the nearest minute we have the range of values of z which form the value $y = 414$ min. The probability that $y = 414$ min may be found directly or by adding together the probability that z is one of the contributing seconds, i.e. $p(y) = 60/43,200 = 1/720$. The probability is still small but considerably larger than the previous result and by combining larger ranges of seconds, larger probabilities are obtained, which certainly cannot be considered negligible.

When drawing the histogram of the probability distribution of error measured in seconds it would be well-nigh impossible to show an area subdivided into 43,200 equal parts but it would be possible to simply show the outline of the final figure as shown in Fig. 7.1.

This is equivalent to only showing a frequency polygon as discussed in Chapter 2 and in fact for most continuous distributions the outline will be a smooth curve as again the individual lines cannot be made distinct.

It is usual to term the formula for this smooth curve as $f(x)$ and to call it the probability density function. As with a histogram, the area under this curve over a given length of the baseline is equal to the probability associated with the range of values represented by the given length of baseline.

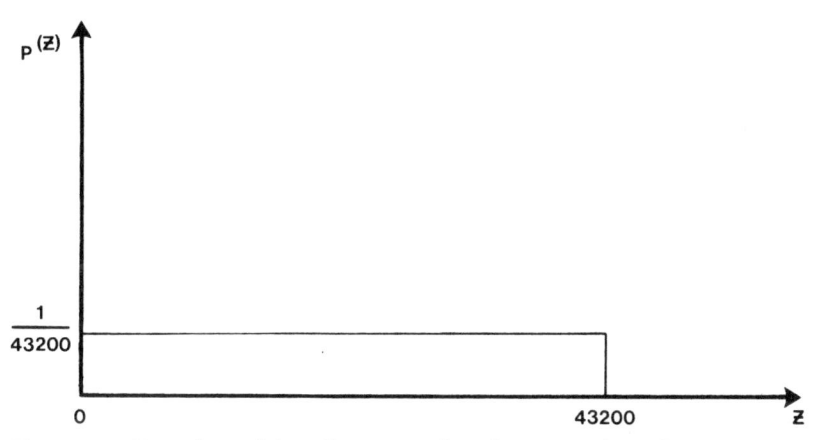

Fig 7.1   Probability Histogram of the Clock Error-Slow Distribution

For a small section of baseline extending from a value of x and with a length of δx the area under the curve is approximately the area of the rectangle whose area is the product of its height $f(x)$ and its width δx. Thus in the clock example, when the error is measured in seconds, we have:

$$f(x) = \frac{1}{43,200}$$

which is the equation of a straight line parallel to the x axis.

The probability of the error falling within the two-second period 24,809·5 to 24,811·5 for example is given by taking $\delta x = 2$, thus:

$$f x\, \delta x = \frac{2}{43,200} = \frac{1}{21,600}$$

Similarly, the probability of the error falling within the sixty-second period 24,809·5 − 24,869·5 is given by:

$$f x\, \delta x = \frac{60}{43,200} = \frac{1}{720}$$

as before. Although $f(x)$ does not, in general, define a straight line curve, provided δx is small, the product $f(x)\, \delta x$ is still a good approximation of the area under the curve and standing on the baseline segment of width δx. The areas standing on longer segments of baseline may thus be found by summing such products between whatever limits are required. The probability that x lies between two values a and b is found by calculating the area standing on the baseline segment from a to b as:

$$p(a \leqslant x \leqslant b) = \sum_{x=a}^{x=b} f(x)\, \delta x$$

Marine Statistics

If we consider this expression for the case where δx becomes vanishingly small, then it is proper to use the calculus notation of integration which deals with summation over a continuous variable. Thus:

$$p(a \leqslant x \leqslant b) = \int_{x=a}^{x=b} f(x)\,dx = \int_a^b f(x)\,dx$$

For readers who have done no calculus, the right-hand side of this equation is read as the integral of $f(x)$ with respect to x from $x = a$ to $x = b$. It will not be necessary in this book for the reader to be able to perform any integration but the notation is useful from time to time and is, in any case, so widely used that the reader is likely to meet it in other statistical situations. Whenever a continuous variable is involved the integration sign will take the place of the Σ sign in discrete cases to denote summation. Thus the mean or expected value of a continuous probability distribution is defined to be:

$$\mu = E(x) = \int_{-\infty}^{\infty} x f(x)\,dx$$

Readers who have studied some integral calculus will recognise the formula $\int_{x=a}^{x=b} f(x)\,dx$ as denoting the area under the curve $y = f(x)$ from $x = a$ to $x = b$. This of course ties in with the idea of the histogram having areas proportional to frequency and hence it makes sense to find an area under a curve between two limits, when calculating the probability of the continuous random variable lying between those limits. If the curve $y = f(x)$ is not defined until $x = c$, say, then the area under the curve, from the smallest value x might possibly have, negative infinity, written as $-\infty$, to the point $x = c$, is 0. Similarly, if the curve is not defined above $x = d$, then the area under the curve from $x = d$ to $x = +\infty$, the largest value x might conceivably have, is also 0. Hence it is reasonable to consider integrals from $-\infty$ to $+\infty$ as in the definition of the mean below.

Having introduced the various concepts we are now in a position to summarise this section in a series of more formal definitions.

7.3 Definitions

7.3.1 A *continuous probability distribution* shows how the probability of an outcome of a situation varies over different ranges of values of a continuous random variable x.

7.3.2 The function $f(x)$ is termed the *probability density function* and is defined so that the probability that x lies in a small interval of length δx is $f(x)\,\delta x$. It has the following properties:

(i) $f(x) \geqslant 0$, and
(ii) $\int_{-\infty}^{+\infty} f(x)\,dx = 1$.

Continuous Distributions

7.3.3 A *cumulative frequency function* $F(x)$ may be defined so that the probability that x takes a value $< a$ is given by

$$F(a) = \int_{-\infty}^{a} f(x)\, dx$$

Thus

$$p(a \leqslant x \leqslant b) = \int_{a}^{b} f(x)\, dx = F(b) - F(a)$$

7.3.4 The *mean* μ of a continuous probability distribution is given by the formula

$$\mu = E(x) = \int_{-\infty}^{\infty} x f(x)\, dx$$

7.3.5 The *variance* σ^2 of a continuous probability distribution is given by the formula:

$$\sigma^2 = V(x) = \int_{-\infty}^{\infty} x^2 f(x)\, dx - \left(\int_{-\infty}^{\infty} x f(x)\, dx\right)^2$$

$$= E(x^2) - [E(x)]^2$$

as before.

7.3.6 In general the expectation of x^r for a continuous probability distribution is written as $E(x^r)$ and is defined mathematically by

$$E(x^r) = \int_{-\infty}^{\infty} x^r f(x)\, dx$$

7.3.7 If a situation is observed N times then the number of times that the range of values $a \leqslant x \leqslant b$ would be likely to be observed is $N[F(b) - F(a)]$.

7.4 Continuous Uniform Distribution

The example on the error-slow when we take the time from a stopped clock, if we measure the error in seconds, provides a good instance of a continuous uniform distribution. The probability density function is constant over the range of values for which the error is defined and 0 for all other values.

7.4.1 In general if x is a continuous variable with $f(x) = 1/(b - a)$ for $a < x < b$ and $f(x) = 0$ elsewhere, then we say that x has a continuous

Marine Statistics

uniform distribution. It can be shown that $E(x) = (b + a)/2$ and $V(x) = (b - a)^2/12$. Thus in the clock example where $a = 0$ sec and $b = 43{,}200$ sec, the mean error = 21,600 sec or 6 hr, with a variance of 1.56×10^8 sec or 12 hr², i.e. a standard deviation of 3.5 hr.

7.5 Exponential Distribution

As a second example of a continuous distribution we will consider an exponential distribution. This distribution may be thought of as the converse to a Poisson distribution. Suppose the number of ships per hour crossing a datum line may be considered to have a Poisson distribution, then the interval between each arrival will be a random variable and have an exponential distribution. It is often referred to as the 'exponential waiting-time distribution' because it gives the probability distribution of the waiting time required before another success occurs.

7.5.1 If x is a continuous random variable with

$$f(x) = \frac{1}{\mu} e^{-x/\mu} \quad \text{for } x \geq 0$$

and

$$f(x) = 0 \quad \text{for } x < 0$$

then x is said to have an exponential distribution with a mean value of μ; μ may be thought of as the mean waiting-time between successes. e^x is the exponential function as used in the Poisson distribution. Figure 7.2 shows the characteristic shape of an exponential distribution.

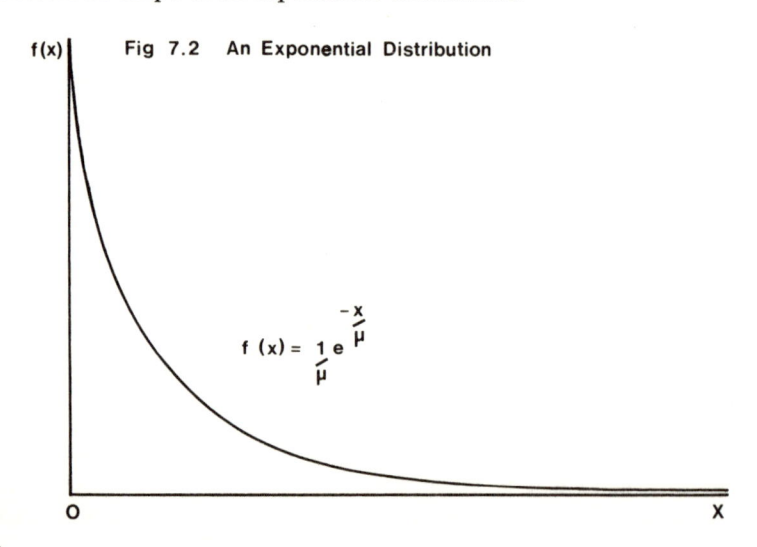

Fig 7.2 An Exponential Distribution

$$f(x) = \frac{1}{\mu} e^{-\frac{x}{\mu}}$$

Continuous Distributions

The two most important features to notice are the behaviour of the curve at either end. As x gets very small and closer and closer to 0 then

$$f(x) = \frac{1}{\mu} e^{-x/\mu} = \frac{1}{\mu} \frac{1}{e^{x/\mu}}$$

tends to $1/\mu$. Similarly as x gets very large then $f(x)$ gets smaller and smaller and closer to the value of 0. However, it is always possible in theory to find a larger x than a previous one observed, hence it is possible for very long waits to occur between successes. These strange mathematical facts make it more difficult to understand than in a situation where there is a closed curve but it can be shown that $\int_0^\infty 1/\mu \, e^{-x/\mu} \, dx = 1$, so the curve does obey the standard requirements of a probability density curve. In practice one looks for a situation where there is a high probability of only having to wait a short time and a low probability of having to wait a very long time, to consider the exponential distribution as a suitable model.

As a practical case we can consider the following example. A ship operating a short sea ferry service has a schedule which is based on its maximum speed under favourable conditions. It is thus never early but is sometimes late in arriving at its destination. The number of minutes that the ship was late during 30 trips was found to be:

```
 1  27  18  92   0  35
 5  14  65   8  23   6
16  12  36  13  17  22
43  32   0  24  51  41
 0   2  54  46   4  15
```

The frequency distribution of these times in 10 min intervals is as follows:

Under 10 min late	9
10 but under 20 min late	7
20 but under 30 min late	4
30 but under 40 min late	3
40 but under 50 min late	3
50 but under 60 min late	2
60 min late and over	2
Total	30

The histogram of this distribution, with in fact full detail shown instead of an open-ended class, given in Fig. 7.3 has the characteristic shape of an exponential distribution.

The mean of the observed distribution is 24 min, so it would seem reasonable to approximate the situation using the exponential distribution with

Marine Statistics

Frequency

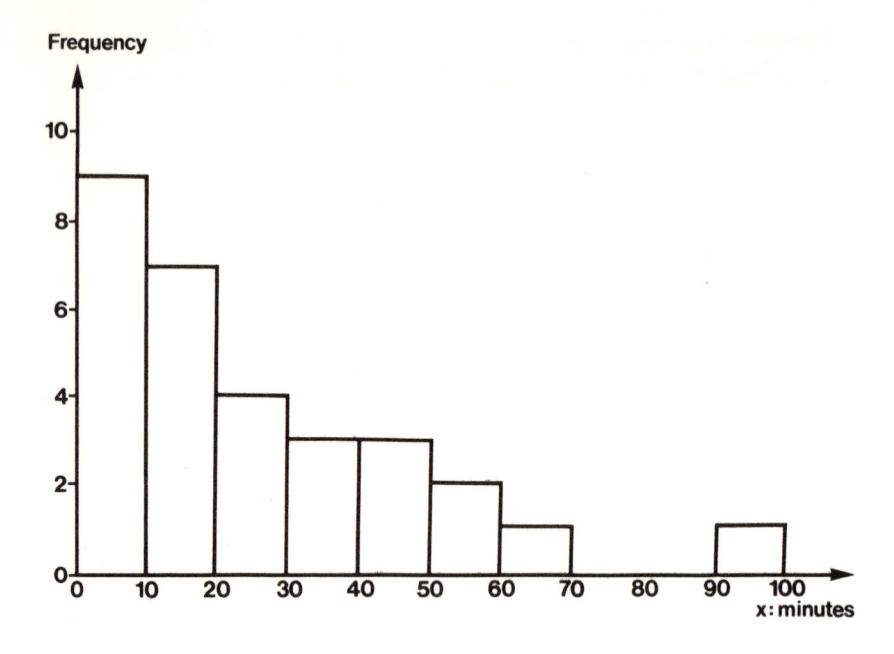

Fig 7·3 Histogram of the distribution for the time late in 30 ferry crossings

$f(x) = (1/24) \, e^{-x/24}$. In a later chapter we will learn how to make an objective decision if this is a reasonable model to take for the situation or not instead of having to rely on eyeball judgement.

7.6 *Normal Distribution*

As a third instance of a continuous probability distribution we will consider the normal distribution, which is probably the most important probability distribution in statistics. Figure 7.4 shows the graph of the normal distribution which is known as the normal curve. Although the mathematical equation of the curve was first developed by De Moivre in 1733, a considerable amount of work was done on it later by Gauss and hence it is often referred to as the Gaussian distribution, particularly in the engineering world. Gauss derived its equation from a study of errors in repeated measurements of the same quantity, but since then it has been found that a large number of types of measurement arising in nature, industry and research are distributed according to this curve. Of great interest to anyone concerned with the sea is the fact that random errors, which are inherent in any measurement situation, can be assumed to be normally distributed. In addition it can be shown

Continuous Distributions

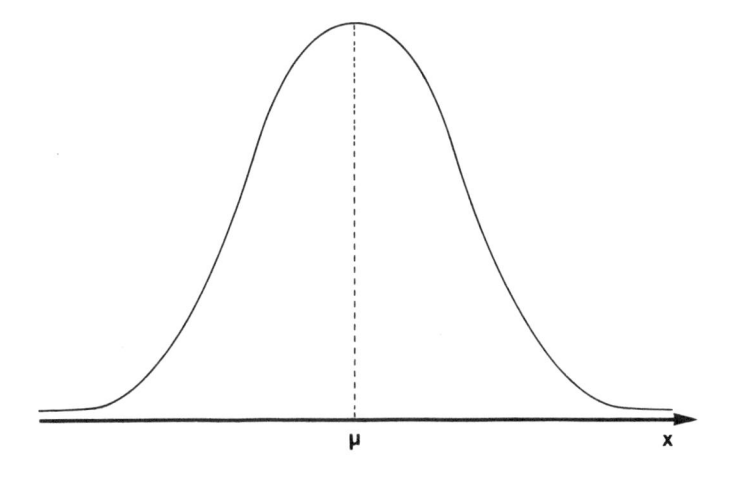

Fig 7·4 The normal distribution

mathematically that the combination of normal distributions is also a normal distribution.

A normal curve is shown in Fig. 7.4 and its main characteristics may be summarised as follows:

(i) It is a bell-shaped curve, suggesting that there is a gradual fall-off in probability from the highest point.

(ii) It is symmetrical about the central point, which is the mean.

(iii) The mean = the median = mode = μ in the diagram.

(iv) The two tails get closer and closer to the base line but never touch or cross it. This is a similar situation to that seen with the exponential distribution, and mathematically we say the tails are *asymptotic* to the x axis. This again implies that there is always a very small chance that something much smaller or much larger than the usual range of values will occur.

Mathematically the equation of a normal curve is given by the formula

$$f(x) = \frac{1}{\sqrt{2\pi}\sigma} e^{-(x-\mu)^2/2\sigma^2}$$

where the mean of the distribution is μ and the variance is σ^2; e is the exponential constant $\simeq 2.72$ we met earlier with the Poisson distribution and the exponential distribution, and π is the mathematical constant $\simeq 3.14$ that occurs in circular measure. There is however no need for the ordinary user of statistics to be able to manipulate this expression as the normal distribution is very well tabulated.

7.6.1 **If x is a continuous random variable with**

$$f(x) = \frac{1}{\sqrt{2\pi}\sigma} e^{-(x-\mu)^2/2\sigma^2}$$

Marine Statistics

for all values of *x*, then we say that *x* has a normal distribution with a mean of μ and a variance of σ² and we write it as $x \sim N(\mu, \sigma^2)$.

The characteristic shape of the curve stays the same whatever the values of the mean and the variance. Two curves with the same variance but different means would look identical but would be centred at different points on the *x* axis, as shown in Fig. 7.5(a).

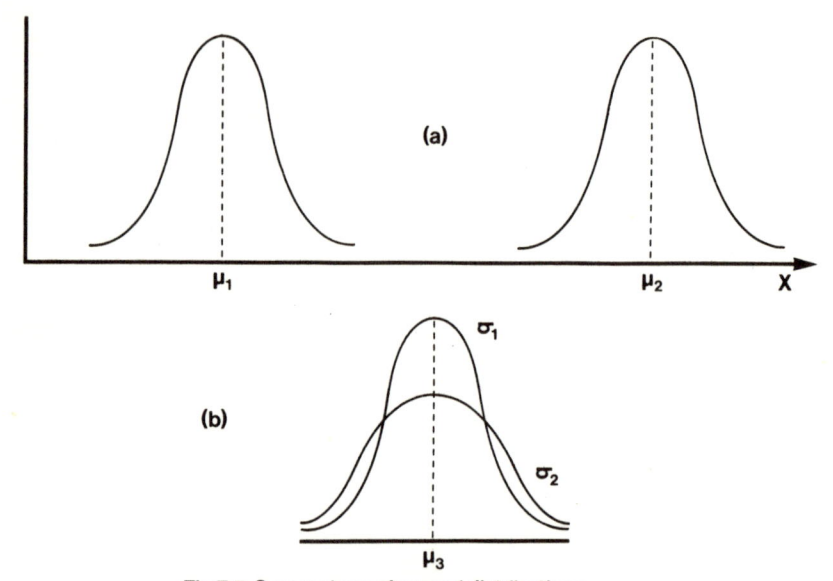

Fig 7.5 Comparison of normal distributions
(a) with the same variance but different means
(b) with the same mean but different variance

Two curves with the same mean but different variances would be centred on the same point but the curve with the smaller variance (σ_1^2 in Fig. 7.5(b)) would not appear to extend so far over the *x* axis as the curve with the larger variance. It is usually very easy to recognise situations where the normal distribution is appropriate from the shape of the frequency curve.

7.6.2 **Any normal distribution which has a mean of 0 and a variance of 1 is known as a** *standard normal distribution*.

7.6.3 **A random variable, *z*, whose distribution is a standard normal distribution is known as a** *standard normal variable*. We write $z \sim N(0, 1)$.

7.6.4 **The equation of the standard normal curve is**

$$f(z) = \frac{1}{\sqrt{2\pi}} e^{-z^2/2}$$

Continuous Distributions

The standard normal distribution has a particularly important role to play because a series of areas under the standard normal distribution have been tabulated. Table 1 in the back of the book gives the values of these areas from $-\infty$ to the point indicated. Thus if we require to know the probability that a standard normal variable, z, takes a value between $-\infty$ and 1·52, then since probability is equal to area under a curve, the required answer is 0·9357. This is equivalent to the shaded area in Fig. 7.6. In mathematical notation we may

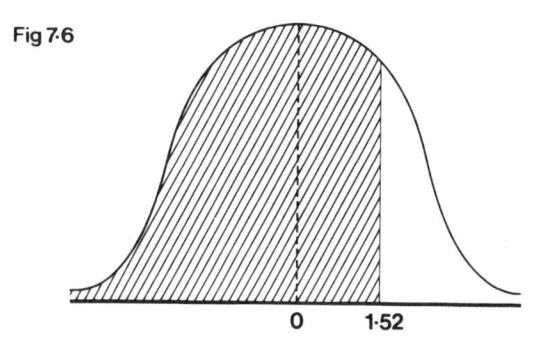

Fig 7·6

write this as $p(z \leqslant 1·52) = F(1·52) = 0·9357$. In these particular tables we are given the area of the standard normal curve lying to the left of a given point but they may be used to answer a variety of different questions.

Suppose now we require the probability that a standard normal variable z takes a value greater than or equal to 0·69. This is equivalent to the shaded area in Fig. 7.7. But $\text{prob}(z \geqslant 0·69) = 1 - \text{prob}(z < 0·69)$ since the total

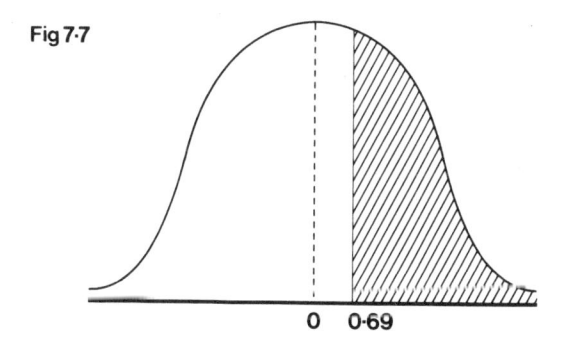

Fig 7·7

area is 1. Thus $F(0·69) = 0·7549$ and required probability is $1 - F(0·69) = 0·2451$.

As a third example suppose we require the probability that a standard normal variable z takes a value less than or equal to $-2·43$. This is equivalent to the shaded area in Fig. 7.8(a). In our particular tables we only have areas for points in the top half of the distribution but we can use the symmetry of the normal distribution to answer questions like this. The required area is

Marine Statistics

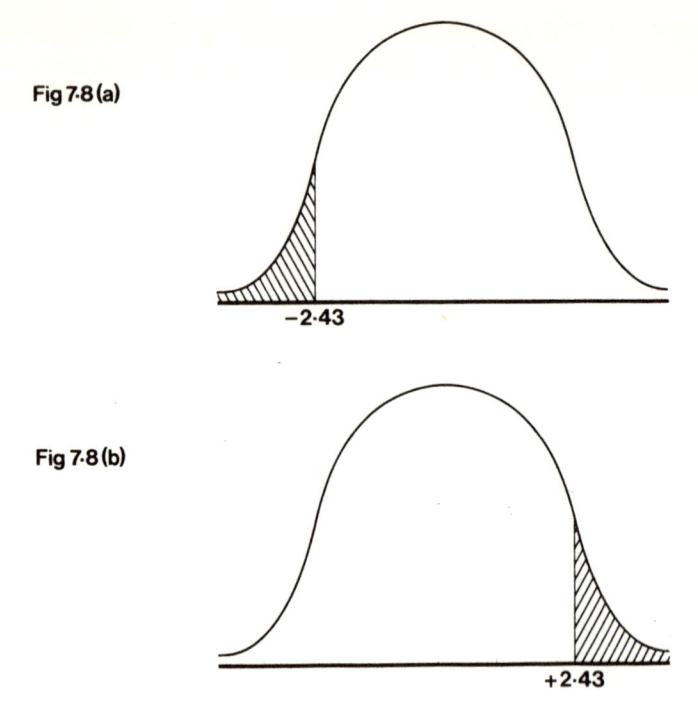

Fig 7·8 (a)

Fig 7·8 (b)

equal in size to the symmetrical area in the top of the distribution shown in Fig. 7.8(b). Thus

$$\text{prob}(z \leqslant -2 \cdot 43) = \text{prob}(z \geqslant +2 \cdot 43)$$

But by the previous case $\text{prob}(z \geqslant +2 \cdot 43) = 1 - F(2 \cdot 43)$. Thus

$$F(-2 \cdot 43) = 1 - 0 \cdot 9925 = 0 \cdot 0075$$

Another type of situation we might have to deal with is finding the probability that the standard normal variable lies between two non-infinite limits, say 0·48 and 1·39 inclusive. The required area is shown in Fig. 7.9. This area

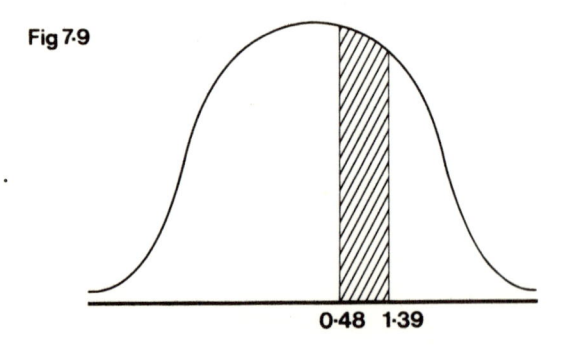

Fig 7·9

is the difference between the area to the left of 1·39 and the area to the left of 0·48. Thus

$$p(0·48 \leqslant z \leqslant 1·39) = F(1·39) - F(0·48)$$
$$= 0·9177 - 0·6844$$
$$= 0·2333$$

The tables may also be used in reverse form. Suppose we wanted to find the point on the standard normal distribution such that there was an 0·85 chance of the random variable lying below it. Thus we want z_0 such that $F(z_0) = 0·8500$. Looking in the body of the table we find the nearest value is 0·8508 under $z = 1·04$. Hence we can conclude that $z_0 = 1·04$.

All our examples so far have been concerned with the standard normal distribution but it is very straight-forward to change from a standard normal distribution to a normal distribution with mean 5, say, and a variance 4, say, and vice versa. In general, suppose x has a normal distribution with mean μ and variance σ^2 then if we put $z = (x - \mu)/\sigma$, which is the standardising transformation, z will have a standard normal distribution. As an example consider the following situation. In a cargo of bagged grain, the mean weight of a bag is 99·2 kg. If the weights of the bags are normally distributed with a standard deviation of 0·5 kg, what proportion of the bags are likely to weigh more than 100 kg each?

If we let x be the weight of any one bag, then we may write that $x \sim N(99·2, 0·25)$ (i.e. x has a normal distribution with mean of 99·2 kg and variance of 0·25 kg²). We require to know the probability that $x > 100$ kg, which is the area shown in Fig. 7.10. Using the standardising transformation

$$z = \frac{x \text{ value} - \text{mean}}{\text{standard deviation}}$$

then an x value of 100 kg corresponds to a z value given by

$$z = \frac{100 - 99·2}{0·5}$$
$$= 1·6$$

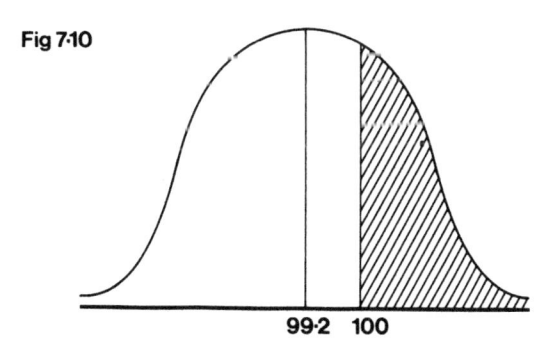

Fig 7·10

Marine Statistics

Thus the required probability for a standard normal variable is $p(z > 1.6)$. This is equal to $1 - F(1.6) = 1 - 0.9452 = 0.0548$. Thus we can say that there is a 5·48% chance that any one bag will weigh more than 100 kg (i.e. we would expect about 55 such bags out of every 1,000 bags loaded).

As a second example we can now answer the problem on errors in the radio aid we started the chapter with, if we assume that the errors have a normal distribution. Let x be the size of error on any one occasion, then $x \sim N(0, 2.25)$. We want $p(x \geq 6)$ so using the transformation $z = (x - \mu)/\sigma = (x - 0)/1.5$ we have that z_0 corresponding to $x = 6$ is given by $z_0 = 6/1.5 = 4.0$, $p(z \geq 4) = 1 - p(z \leq 4) = 1 - F(4)$. But in the tables $F(4)$ is not listed and may be taken as approximately $= 1$. Thus $p(x > 6) = p(z > 4) \simeq 0$. Hence we can be reasonably certain that a value of error larger than 6 nautical miles will not occur. Notice that we must still attach a vague note of uncertainty to the statement rather than a categorical assertion because there is always a slight chance that an extreme value will occur although the mathematical value of the chance is extremely small.

The above examples have dealt with a variety of ways in which we can use the tables of areas under a normal curve and in none of the examples have we had to go back to the original mathematical equation of the curve. There are some areas under the normal curve which are probably worth remembering:

(i) If we take a band of one standard deviation width either side of the mean then $68.26\% \simeq \frac{2}{3}$ of all observations lie within this band.

(ii) If we take a band of two standard deviation width either side of the mean then $95.44\% \simeq 95\%$ of all observations lie within this band.

(iii) If we take a band of three standard deviation width either side of the mean then $99.73\% \simeq 99.7\%$ of all observations lie within this band, i.e.

$$p(\mu - \sigma \leq x \leq \mu + \sigma) \simeq 0.68$$
$$p(\mu - 2\sigma \leq x \leq \mu + 2\sigma) \simeq 0.95$$
$$p(\mu - 3\sigma \leq x \leq \mu + 3\sigma) \simeq 0.997$$

7.7 The Normal Distribution as an Approximation to the Binomial Distribution

At the beginning of the section on the normal distribution it was stated that this distribution provided a good model for many practical situations. One of the reasons for this is that the normal distribution provides a good approximation to the binomial distribution under certain circumstances and even to the Poisson distribution under other circumstances. As normal distribution tables are relatively simple to use, in cases where this approximation may be made, it can save a lot of calculation. Suppose two members of the crew are tossing a fair coin which they toss 50 times and they want to know the probability of getting more than 40 heads. If x denotes the number of heads in 50 tosses then x is a discrete random variable which can take the values

Continuous Distributions

0–50 inclusive. Since each toss is independent and can result in either a head or a tail then we have a binomial distribution for x with $n = 50$ and $p = \frac{1}{2}$. The mean number of heads expected in 50 tosses is $np = 25$ and the variance is $npq = 50.1/2.1/2 = 25/2$, which gives a standard deviation of approximately 3·5. If we were to draw the histogram of the probability distribution, we would find a symmetrical shape as shown in Fig. 7.11. It can be seen

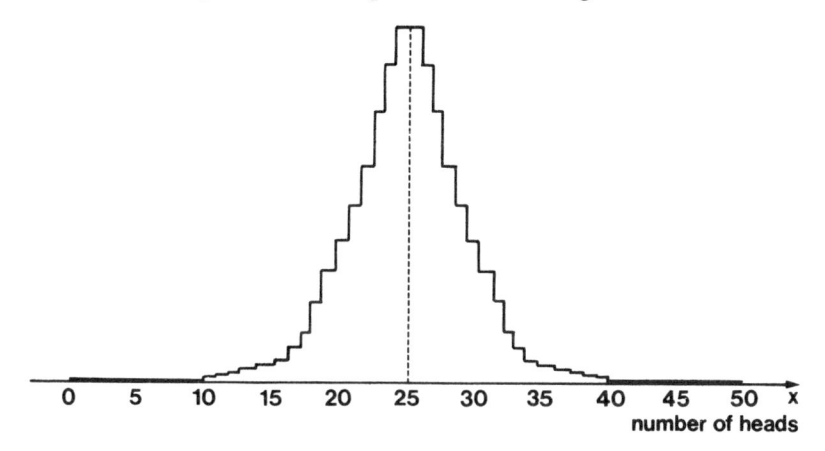

Fig. 7·11 Histogram of the binomial distribution for the number of heads in 50 tosses.

though that the frequency polygon of this distribution could easily be approximated to by a normal curve and this may be shown mathematically.

Thus if x has a binomial distribution, then a normal variable, y, with the same mean, 25, and variance, 25/2, provides a good approximation. We need to find the probability that x is more than 40. The only problem is one we have met earlier with approximating discrete variables by continuous ones. We have to treat the discrete value 40 as though it were measured on a continuum and stretched from $39\frac{1}{2}$ to $40\frac{1}{2}$. Thus $p(x > 40)$ is equivalent to $p(y > 40.5)$. We must now transform y to a standard normal variable z, where $z = (y - \mu)/\sigma$ and $\mu = 25$, $\sigma \simeq 3·5$. Thus if

$$y_0 = 40·5 \qquad z_0 = \frac{40·5 - 25}{3·5} = \frac{15·5}{3·5} = 4·4$$

However

$$p(z > 4·4) = 1 - F(4·4) = 0$$

Thus the probability of throwing more than 40 heads in 50 tosses of a fair coin may be taken as 0.

The situation described was a perfectly symmetrical one with $p = q = \frac{1}{2}$. However, the normal distribution has been shown to be a reasonable approximation to the binomial distribution provided p is not too close to 0 or 1.

Marine Statistics

If p is very small (or very large) then the Poisson distribution may provide a better approximation. As a practical rule the normal approximation to the binomial distribution can be used if n is of a reasonable size, say more than 30 and np, the mean, is greater than or equal to 15.

Thus if we had a binomial distribution with $p = 1/100$ and $n = 300$ then as $np = 3$ we would use a Poisson approximation but if $p = 1/100$ and $n = 2{,}000$, then, as $np = 20$, we would use a normal approximation. In this latter case the normal distribution may be considered as providing an approximation to the Poisson distribution. The main point to remember is what is termed the 'continuity correction' whereby we add or subtract a $\frac{1}{2}$ as appropriate from the discrete value.

7.8 Double Exponential Distribution

As a final example of a continuous probability distribution, a short mention will be given of the double exponential distribution. This distribution, as its name suggests, is made up of two exponential distributions and its characteristic shape is shown in Fig. 7.12. Theoretically, there is a break in the curve

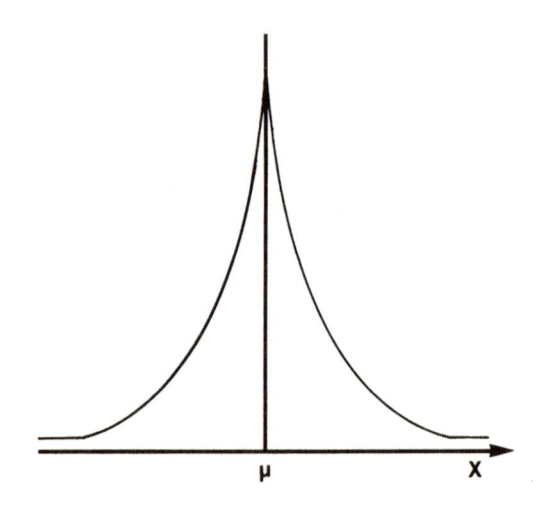

Fig 7·12 A double exponential distribution

at the peak which is the mean of the distribution but in practical situations this may be ignored. The curve may be defined mathematically as

$$f(x) = \frac{1}{2\mu} e^{(x-\mu)/\mu} \quad x < \mu$$

and

$$f(x) = \frac{1}{2\mu} e^{-(x-\mu)/\mu} \quad x > \mu$$

134

From these equations it can be shown that the mean of the curve is μ. This distribution is of interest to navigators because some authors argue that it represents a more realistic distribution for error under certain circumstances than does the normal distribution. The main difference between this and the normal is that the bell-shaped property of the normal distribution is not displayed here and there is a much more abrupt falling away from the central value. This suggests that whereas zero, or minimal errors, are the most frequent occurrences, errors which are made are more likely to be large in magnitude. Some research into this hypothesis has been done on navigational errors in the air but very little has been done in a marine context. For practical purposes the normal distribution has been found to provide a satisfactory model, so most work still makes the assumption that errors are normally distributed.

7.9 Summary

In this chapter the concept of a probability distribution has been extended to cover continuous random variables, and we have seen examples of some of the more common of these distributions. In particular we have seen how the most important continuous probability distribution, the normal distribution, may be applied in practice. We will find in the next chapter that the normal distribution arises in a series of other contexts, when we attempt to generalise from information we have about a small sample, to information on the parent population.

Exercises

1. A fault in a servo-motor causes a gyro compass repeater attached to radio direction finding equipment to rotate slowly but constantly in a clockwise direction.
   (a) Describe the error distribution for radio bearings observed at random times by an operator who is unaware of the defect.
   (b) Sketch the probability density function.
   (c) What is the mean error?
   (d) What is the standard deviation of the error?

2. A ship's log develops a mechanical fault when the reading is 125·0 nautical miles, such that the reading then becomes frozen. The fault is not recognised until the ship has travelled another 20 miles.
   (a) What is the probability density function for the error which would have affected a log reading during the interval between the occurrence of the fault and its discovery?
   (b) Sketch the curve which represents the PDF.
   (c) What is the mean value of the error?
   (d) What is the standard deviation of the error?

Marine Statistics

3. In a ship with an unmanned engine room, the distances travelled by the ship between 25 occasions when the bridge instrumentation indicated a fault requiring attention were recorded in miles as follows:

$$\begin{array}{rrrr}
1{,}020 & 24 & 372 & 1{,}864 \\
235 & 1{,}193 & 5{,}381 & 3{,}270 \\
2{,}176 & 642 & 930 & 361 \\
842 & 135 & 2{,}694 & 2{,}382 \\
1{,}654 & 3{,}593 & 1{,}550 & 240 \\
4{,}327 & 760 & 493 & 1{,}387
\end{array}$$

(a) Sketch a histogram for this distribution, using a class interval of 500 miles.
(b) Suggest a standard distribution which might form a reasonable approximation to the observed distribution.
(c) What is the probability density function for the distribution suggested under (b)?

4. The mean time between failures for a particular make of navigation lamp is quoted by the manufacturer as 3,000 hr. If the failures can be considered to occur at random, suggest an appropriate distribution for the intervals between failures of the lamps. Give the equation of the probability density function and sketch the corresponding curve.

5. Coils of rope supplied by a particular manufacturer are quoted as having a nominal length of 250 m. Tests indicate that the coil lengths are normally distributed with a mean length of 254 m and a standard deviation of 1·5 m. What is the probability that a particular coil will be:
(a) Shorter than 256 m?
(b) Longer than 257 m?
(c) Shorter than the nominal length?

6. A position fixing aid was tested with the ship in a known position. A number of position lines were observed and subsequent analysis gave the mean displacement as 0·8 miles to the south of the true position with a standard deviation of 0·8. If the ship subsequently takes a reading of the position fixing aid in the same general locality and under similar conditions, what is the probability that the observed position line lies:
(a) More than one mile south of the true position?
(b) More than one mile north of the true position?
(c) Between one mile south and one mile north of the true position?

7. A roro ferry across a river has space for 22 vehicles on the upper deck where the headroom is unlimited and 42 vehicles on a lower deck where the headroom is 2 m. Amongst the vehicles carried throughout the season, the proportion of vehicles with a height of 2 m or less is $\frac{3}{4}$ and of vehicles with a height of more than 2 m it is $\frac{1}{4}$. Vehicles are accepted on a first come, first

Continuous Distributions

served basis and the loading procedure is such that no vehicles of 2 m or less in height are put onto the upper deck until the lower deck is full. The ferry always takes a full complement of vehicles on board before sailing.

Estimate the probability that, during a particular loading operation, the upper deck will become full before the ship has taken its full complement of vehicles on board.

8. Over a period of a year, the mean number of ships per day arriving at a pilot cutter to pick up a pilot was 25. Consideration of the data suggests that the Poisson distribution provides a good fit for the number of ship arrivals per day. What number of pilots should be assigned to the cutter each day to give an expectation of there being insufficient pilots to service the ships on only one day in fifty?

Answers

1. (a) Because of the nature of the fault, a particular bearing observed by the operator may take any reading from 0°–360° with equal probability. Since we are dealing with a closed, circular scale, the bearing error can be taken as lying within the range −180° to +180° with an equal probability of taking any value within those limits.

(b) PDF, $p(x) = \dfrac{1}{360}$ for $-180° < x \leq +180°$

(c) It is evident that the mean value, $E(x)$, is zero, the formula giving:

$$E(x) = \frac{b+a}{2} = \frac{180 + (-180)}{2} = \underline{0°}$$

(d) $V(x) = \dfrac{(b-a)^2}{12} = \dfrac{[180-(-180)]^2}{12} = \dfrac{360^2}{12} = 10{,}800$

$\sigma = \sqrt{V(x)} = \underline{104°}$

2. (a) PDF is given by: $p(x) = \dfrac{1}{b-a} = \dfrac{1}{20-0} = \underline{\dfrac{1}{20}}$

(c) $E(x) = \dfrac{b+a}{2} = \dfrac{20+0}{2} = \underline{10 \text{ nautical miles}}$

(d) $V(x) = \dfrac{(b-a)^2}{12} = \dfrac{20^2}{12} = 33.33$

$\sigma = \sqrt{33.33} = \underline{5.77 \text{ nautical miles}}$

Marine Statistics

3. (a) Frequency table:

Interval, nautical miles	Tally	Frequency	Relative frequency
0– < 500	ЖН ‖	7	0·292
500– < 1,000	‖‖‖	4	0·167
1,000– < 1,500	‖‖‖	3	0·125
1,500– < 2,000	‖‖‖	3	0·125
2,000– < 2,500	‖	2	0·083
2,500– < 3,000	ǀ	1	0·042
3,000– < 3,500	ǀ	1	0·042
3,500– < 4,000	ǀ	1	0·042
4,000– < 4,500	ǀ	1	0·042
4,500– < 5,000		0	0·000
5,000– < 5,500	ǀ	1	0·042

(b) An exponential distribution is suggested as a possible approximation to the data given, partly because the distances given may be thought of as the intervals between the occurrence of random events and partly because it is suggested by the shape of the histogram.

(c) We find μ by:

$$\mu = \frac{\Sigma x}{n} = \frac{37,525}{24} = 1,563 \cdot 5$$

PDF for exponential distribution is given by:

$$f(x) = \frac{1}{\mu} e^{-x/\mu} \qquad \text{(for } x \geqslant 0\text{)}$$

$$= \frac{1}{1,564} e^{-x/1,564} \qquad \text{(for } x \geqslant 0\text{)}$$

$$f(x) = 0 \qquad \text{(for } x < 0\text{)}$$

4. Since this is a 'waiting time' situation, the exponential distribution is appropriate, with:

$$\text{PDF}, f(x) = \frac{1}{\mu} e^{-x/\mu} \qquad \text{(for } x \geqslant 0\text{)}$$

$$= \frac{1}{3,000} e^{-x/3,000} \qquad \text{(for } x \geqslant 0\text{)}$$

$$f(x) = 0 \qquad \text{(for } x < 0\text{)}$$

Continuous Distributions

5. (a) We require the probability that the length of a coil will be less than the mean value plus 2 m, i.e. less than $2/1 \cdot 5 = 1 \cdot 33$ standard deviations. Alternatively, putting $z = (x - \mu)/\sigma = (256 - 254)/1 \cdot 5$ gives $z = 1 \cdot 33$. Referring to Table 1, we find that:

$$p(z < 1 \cdot 33) = F(1 \cdot 33) = \underline{0 \cdot 908}$$

(b) Here we need the probability that the length of a coil will be greater than 257 m. Standardising this gives $z > z_0$, where $z_0 = (257 - 254)/1 \cdot 5 = 2 \cdot 0$. Referring to Table 1, we find that:

$$p(z > 2 \cdot 0) = 1 - F(2 \cdot 0) = 1 - 0 \cdot 977 = \underline{0 \cdot 023}$$

(c) Here we require the probability that the length of a coil will be less than the mean value by 4 m, i.e. less than $4/1 \cdot 5 = 2 \cdot 67$ standard deviations. Referring to Table 1, we find that:

$$\begin{aligned} p(z < -2 \cdot 67) &= p(z > 2 \cdot 67) \\ &= 1 - F(2 \cdot 67) \\ &= 1 - 0 \cdot 996 \\ &= \underline{0 \cdot 004} \end{aligned}$$

6. (a) We will assume a normal distribution. We require the probability that the observed position line will lie $1 \cdot 0$ miles south of its true position. Putting $z = (x - \mu)/\sigma$ gives $z = (1 - 0 \cdot 8)/0 \cdot 8 = 0 \cdot 25$. Referring to Table 1:

$$p(z > 0 \cdot 25) = 1 - F(0 \cdot 25) = 1 - 0 \cdot 60 = \underline{0 \cdot 40}$$

(b) We require the probability that the observed position line will lie more than $1 \cdot 0$ miles north of the true position, i.e. $1 \cdot 8$ miles north of its mean position or $1 \cdot 8/0 \cdot 8 = 2 \cdot 25$ standard deviations. Referring to Table 1:

$$p(z > 2 \cdot 25) = 1 - F(2 \cdot 25) = 1 - 0 \cdot 99 = \underline{0 \cdot 01}$$

(c) We require the probability that the observed position line will lie between the limits of $1 \cdot 0$ miles south and $1 \cdot 0$ miles north of the true position. In this case, we have:

$$\begin{aligned} p(0 \cdot 25 < z < 2 \cdot 25) &= F(2 \cdot 25) - F(0 \cdot 25) \\ &= 0 \cdot 99 - 0 \cdot 40 \\ &= \underline{0 \cdot 59} \end{aligned}$$

Note that the three cases above exhaust all the possible limits within which a position line may lie and the sum of the three probabilities is consequently unity.

Marine Statistics

7. In this problem, we are concerned with a binomial distribution but the normal approximation may be used since $n > 30$ and $np > 15$ ($n = 64$ and $p = \frac{1}{4}$).

The mean number of vehicles over 2 m high in a particular load is

$$np = 64 \times \frac{1}{4} = 16$$

The variance

$$\sigma^2 = npq = 64 \times \frac{1}{4} \times \frac{3}{4} = 12$$

The standard deviation

$$\sigma = \sqrt{12} = 3.46$$

We require the probability that more than 22 high vehicles will be included in a batch of 64 vehicles and, to allow for continuity, this is equivalent to the probability that more than 22·5 such vehicles will be included.

Transforming to a standard normal variable, we have:

$$z = \frac{22 \cdot 5 - 16}{3 \cdot 46} = 1 \cdot 88$$

$$\begin{aligned} p(z > 1 \cdot 88) &= 1 - F(1 \cdot 88) \\ &= 1 - 0 \cdot 97 \quad \text{[consulting Table 1]} \\ &= 0 \cdot 03 \end{aligned}$$

Thus we would expect that the upper deck would become full before the ship had taken on its full complement of vehicles on about 3 out of every 100 trips.

8. Since the mean number of ship arrivals per day is greater than 15, we may use a normal approximation to the Poisson.

The mean number of ships per day	$\mu = 25$
The variance	$\sigma^2 = 25$
The standard deviation	$\sigma = 5$

We require the probability that there will be insufficient pilots to be $1/50 = 0 \cdot 02$. So the probability that there will be sufficient pilots is $1 - 0 \cdot 02 = 0 \cdot 98$. Using our normal tables, we find:

$$F(z_0) = 0 \cdot 98$$
$$\Rightarrow \quad z_0 = 2 \cdot 05$$

The number of ships which will not be exceeded with a probability of 0·98 is thus:

$$\mu + 2 \cdot 05 \sigma = 25 + 2 \cdot 05 \times 5$$
$$= 35 \cdot 25$$

Hence we require that 36 pilots should be assigned to the cutter each day so that there will be insufficient pilots to service the arriving ships on less than one day in fifty.

Chapter 8
Sampling Distributions and Estimation

8.1 *Introduction*

After structural alterations a vessel is subjected to an inclining experiment to establish new stability criteria. A standard weight was moved across the deck on 12 separate occasions and the resulting deflections, measured in cm of the pendulum, used to determine the amount of inclination were: 97, 98, 103, 95, 100, 91, 100, 98, 94, 96, 99, 93. Based on this sample of results it is required to estimate the true mean deflection of the pendulum whenever the ship is subject to the movement of a weight such as the standard one.

The second situation to be considered is as follows: over a period of some years similar ships on a particular trade route had a normal distribution for passage time with a mean of 25·5 days and a standard deviation of 1·0 days. During the following year, weather routeing procedures were adopted and the mean passage time for 20 ships on that trade route was 24·0 days, with a standard deviation of 2·0 days. Is there sufficient evidence to conclude that the weather routeing procedures led to a worthwhile reduction in passage time?

In both these examples the common factor is that we have some evidence for a small number of occasions but we want to know what the general situation is likely to be. This problem is one of the most common in statistical analysis, because it is very rarely that information may be obtained from an entire population, so decisions have to be made based on a very small sample of results. This area of work is known as statistical inference because the statistician has to infer results about the general case based only on knowledge of the particular situation. This is a complete contrast to the sort of work tackled by the statistician's colleague, the mathematician, who usually works from the general case and deduces results about a particular situation.

There are various sorts of questions we might like to answer about a population using the information from a sample, such as what its distribution or general shape is or what its mean value (or standard deviation) is and how they compare with the mean and standard deviation of another population about which we are also making inferences. It is these questions concerning the values in the population, such as the mean and standard deviation, which we are going to concentrate on first and at this stage a couple of definitions will be useful.

Marine Statistics

8.1.1 **A numerical quantity in a population such as the mean, standard deviation, median etc. is known as a** *parameter.*

8.1.2 **A numerical quantity in a sample such as the mean, standard deviation, median, etc. is known as a** *statistic.*

It is usual to denote a parameter with a Greek letter and a statistic with a Roman letter. Thus the mean of a population would be μ and the mean of a sample would be m or $\bar{x}$. Similarly the standard deviation of a population would be σ and the standard deviation of a sample would be s. This notation is consistent with what we have used in the previous chapters.

To return to our introductory examples we can see that both are concerned with inference about a population parameter based on a sample statistic. However in the first one we are actually trying to estimate a value for the unknown parameter, the mean, whereas in the second example we are concerned with a hypothesis about the parameter which we wish to investigate. An answer to the second example will be deferred until the following chapter when we will concentrate on hypothesis testing. The first example we will deal with later in this chapter when we consider estimation. However, before we can consider either of these two areas in detail we must investigate what happens when we take samples from a population.

8.2 *Sampling*

Our main consideration here is to investigate what is happening when we take a sample from a population. We are not going to go into any real detail on how to take samples from a population as this is in itself a very wide and complex area known as survey methods. It is however worth pausing a moment or two to consider why we usually only take samples when we want to discover information about a population. The main reasons may be listed as follows. Firstly, it is quicker to go to a smaller number of people or use a smaller number of units and, secondly, it is cheaper. Thirdly, it is easier to follow up anomalies in the data given a small number of readings, so the actual sample data may itself be more accurate. Fourthly, there are many situations whereby the testing procedure involves destruction of the item, e.g. burning a lamp to see how long it lasts. Finally, there are situations where we want to make decisions as soon as possible, e.g. in implementing as general practice the weather routeing of ships, or using a new instrument in navigation.

We are going to assume that any sample which we are dealing with is a random one; in other words a non-biased selection procedure has been used to draw the sample. The selection procedure and hence the sample would be

Sampling and Estimation

said to be biased if parts of the population were excluded in any way when the sample was drawn. For instance, if we only collected results on the weather routeing of ships on a particular trade route during the summer months, then our results would be biased in that they were not applicable to all times of the year. One of the statistician's main tasks is to ensure that any data is truly representative of the situation required; otherwise only very limited inferences may be possible. Slight care has to be taken with the strict interpretation of the term *random sample*. In England the term random sample means 'a sample drawn so that every member of the population had a known non-zero chance of selection' whereas in America this is termed a 'probability sample'. A special case of this is a simple random sample which is defined as follows: 'A simple random sample of size n is one drawn by a method whereby every possible simple random sample of size n in the population had an equal chance of selection'. In American books this is often called just a 'random sample' for short. In this book we will not worry about the more complicated sampling procedures and it is reasonable to assume that in real life most samples in practical situations can be taken as simple random samples or their equivalent, unless the sampling method has been spelled out suggesting otherwise or one doubts whether it is random at all. We will for the most part only use the words 'sample' or 'random sample' even though we will be dealing with simple random samples, for which the basic results which we will be considering are applicable.

To conclude this section let us consider briefly how we would try to get a random sample in one or two situations. If we were collecting the results on the stability of the ship described at the beginning, then the main consideration would be that there were no circumstances in the experiment which would in any way make the behaviour of the ship atypical such as a very shallow depth of water. Another type of situation is where we want to draw a sample from a known population. Suppose, for instance, for a particular shipping company we want to test whether the recruitment policy adopted meant that a higher percentage of officers stayed for more than ten years than in the shipping industry as a whole. We decide to look first at a sample of the records of officers recruited between ten and twenty years ago to see what percentage of them are still at sea. Suppose there were 454 records and we are going to examine a sample of size 10. The easiest method is to assign each record a three digit number from 001 to 454 inclusive and then to pick a sample of three-digit numbers using random number tables such as those given at the back of the book. Taking a start anywhere in the table and reading off the digits in threes, a typical selection of digits might be: 598, 468, 062, 692, 401, 209, 342, 436, 086, 995, 572, 490, 930, 520, 749, 378, 767, 082, 847, 431, 736, 129. Ignoring all the three-digit numbers larger than 454 our sample would consist of records numbered 62, 401, 209, 342, 436, 86, 378, 82, 431 and 129. Random number tables are usually generated by a computer, but in the past this could have been done by shaking a die or other similar lottery method. They have

Marine Statistics

the property that every digit from 0 to 9 inclusive, at any point in the table, has an equal chance of being the next to appear. Thus if a large-sized table were taken each digit would appear roughly an equal number of times.

8.3 Sampling Distributions

Let us imagine we have four deck officers aged 25, 27, 28 and 32. The mean of this small population, μ, is 28 and its standard deviation, σ, is $\sqrt{6.5} = 2.5$. Suppose now we take a sample of size 2 from this population and calculate the mean age of the two men in the sample. We will assume that we pick our sample without replacement which means that once a man is picked, the second man is picked from the remaining three. The alternative is to pick the sample with replacement which means that the man first picked is put back into the pool before the second is chosen and could therefore be picked again. This second method is obviously not very practical but is of important theoretical significance.

Sampling without replacement there are $\binom{4}{2} = 6$ possible samples of two men we could pick which are listed below together with the sample means.

Ages of members of the sample	Mean age: $\bar{x}$
25, 27	26
25, 28	26.5
25, 32	28.5
27, 28	27.5
27, 32	29.5
28, 32	30

The first point to notice in this example is that in fact the mean age in every sample is different from the population mean age. Thus in this case, no matter which sample we had, if we use the mean age of the sample to estimate the mean age of the population we would never get the true estimate of 28. This will not always be the case as the second example illustrates. Suppose we have a population now of five deck officers of ages 21, 23, 25, 27 and 29. The mean, μ_2, here is 25 and the standard deviation $\sigma_2 = 2.8$. Sampling without replacement, the possible number of samples of size 2 is $\binom{5}{2} = 10$ and these together with the sample means are listed below.

Ages of members of the sample	Mean age: $\bar{x}$
21, 23	22
21, 25	23
21, 27	24
21, 29	25
23, 25	24
23, 27	25
23, 29	26
25, 27	26
25, 29	27
27, 29	28

We can rewrite the results for the mean age in a distribution form as follows:

Sample mean age	Frequency	Probability
22	1	$\frac{1}{10}$
23	1	$\frac{1}{10}$
24	2	$\frac{2}{10}$
25	2	$\frac{2}{10}$
26	2	$\frac{2}{10}$
27	1	$\frac{1}{10}$
28	1	$\frac{1}{10}$

In this example there is a 2 in 10 probability that if we have only one sample then the particular sample we have will have the same mean of 25 as the population, since there is an equal chance that any one of the samples will be picked. If we convert all the frequencies to probabilities in this example then we get the probability distribution associated with the different values of the sample mean. Similarly in the first example we can consider our results as a distribution of mean age each value of which occurs with an equal probability of $\frac{1}{6}$. These are both examples of a distribution of the sample means which is known usually as the sampling distribution of the mean.

Marine Statistics

8.3.1 The sampling distribution of the mean for samples of size n is the probability distribution formed by the means of all possible samples of size n from a given population.

Since we have a probability distribution then it is possible to calculate its mean and variance. In the first example we have that the mean of the probability distribution which we will denote as $\mu_{\bar{x}}$ is given by:

$$\mu_{\bar{x}} = E(\bar{x}) = \sum_{\bar{x}} \bar{x}p(\bar{x}) = 26 \times \frac{1}{6} + 26 \cdot 5 \times \frac{1}{6} + 28 \cdot 5 \times \frac{1}{6} + 27 \cdot 5$$

$$\times \frac{1}{6} + 29 \cdot 5 \times \frac{1}{6} + 30 \times \frac{1}{6} = 28$$

Similarly,

$$V(\bar{x}) = \sum_{\bar{x}} \bar{x}^2 p(x) - \left[\sum_{\bar{x}} \bar{x}p(\bar{x})\right]^2$$

$$= \frac{1}{6} \times 26^2 + \frac{1}{6} \times 26 \cdot 5^2 + \frac{1}{6} \times 28 \cdot 5^2 + \frac{1}{6} \times 27 \cdot 5^2$$

$$+ \frac{1}{6} \times 29 \cdot 5^2 + \frac{1}{6} \times 30^2 - 28^2 = 2\frac{1}{6}$$

giving a standard deviation of 1·5 which we call $\sigma_{\bar{x}}$. For the second example, $\mu_{\bar{x}} = 25$ and $\sigma_{\bar{x}} = 1 \cdot 7$. The notation $\mu_{\bar{x}}$ and $\sigma_{\bar{x}}$ is used to indicate that we are talking about the mean and standard deviation of a population comprised of sample means, i.e. the sampling distribution of the mean. Alert readers may well have noticed that in both examples the mean of the sampling distribution of the mean is equal to the mean of the population from which the samples were drawn, i.e. $\mu_{\bar{x}} = \mu$. This result holds for all populations provided the samples were drawn in an unbiased way. Our sampling method would be biased if some part of the population could never be included in a sample. Suppose in example one it was impossible ever to draw the person aged 32 then we would think that the only samples possible were: 25, 27 mean 26 and 25, 28 mean 26·5 and 27, 28 mean 27·5. The mean of these three sample means is 26·67 which is not equal to the population mean of 28 but then we have in effect sampled from a different population.

8.3.2 The mean of the sampling distribution of the mean is equal to the population mean whenever unbiased random samples are drawn, $\mu_{\bar{x}} = \mu$. The standard deviations of the population and the sampling distribution of the mean are also connected but by a slightly more complicated formula.

8.3.3 The standard deviation of the sampling distribution of the mean is called the standard error of the mean and is written as $\sigma_{\bar{x}}$.

Sampling and Estimation

8.3.4 **If random samples of size n are drawn without replacement from a population of size N with standard deviation σ then the standard error of the mean, $\sigma_{\bar{x}}$, is given by the formula**

$$\sigma_{\bar{x}} = \sqrt{\frac{\sigma^2}{n} \cdot \frac{N-n}{N-1}}$$

In example one, $N = 4$, $n = 2$ and $\sigma^2 = 13/2$, giving:

$$\sigma_{\bar{x}} = \sqrt{\frac{13}{2 \cdot 2} \cdot \frac{2}{3}} = \sqrt{\frac{13}{6}} = 1 \cdot 5$$

Similarly in example two, $N = 5$, $n = 2$ and $\sigma^2 = 8$, giving:

$$\sigma_{\bar{x}} = \sqrt{\frac{8}{2} \cdot \frac{3}{4}} = \sqrt{3} = 1 \cdot 7$$

If we sample from a small population with replacement, then the formula for $\sigma_{\bar{x}}$ becomes much simpler and $\sigma_{\bar{x}} = \sigma/\sqrt{n}$. To illustrate this consider the sampling distribution of the mean for samples of size 2 drawn with replacement from the population in example one.

Sample members	Mean $\bar{x}$	Sample members	Mean $\bar{x}$
25, 25	25	28, 25	26·5
25, 27	26	28, 27	27·5
25, 28	26·5	28, 28	28
25, 32	28·5	28, 32	30
27, 25	26	32, 25	28·5
27, 27	27	32, 27	29·5
27, 28	27·5	32, 28	30
27, 32	29·5	32, 32	32

Note that we have to treat the samples consisting of officers aged 25, 27 and 27, 25 as distinct samples to ensure that every possible sample has an equal chance of selection. We can calculate $\mu_{\bar{x}} = 28$ and $\sigma_{\bar{x}} = \sqrt{13/4} = 1 \cdot 8 = \sqrt{\sigma^2/n}$.

8.3.5 **If random samples of size n are drawn with replacement from a population with standard deviation σ then the standard error of the mean $\sigma_{\bar{x}}$ is given by the formula $\sigma_{\bar{x}} = \sigma/\sqrt{n}$.**

We said earlier that in practice we do not like sampling with replacement because we could end up with a sample consisting of only one distinct member. However in most practical situations we are dealing with very large populations where the size, N, is very much greater than the size of the sample, n. In these cases even though we sample without replacement we can get a sufficiently accurate value of $\sigma_{\bar{x}}$, by taking $\sigma_{\bar{x}} = \sigma/\sqrt{n}$, since the additional term $\sqrt{(N-n)/(N-1)}$ is then approximately equal to 1.

Marine Statistics

8.3.6 The term $\sqrt{(N-n)/(N-1)}$ is called the finite population correction or f.p.c. for short.

8.3.7 If random samples of size n are drawn from a large population with standard deviation, σ, then the standard error of the mean, $\sigma_{\bar{x}}$, is given by $\sigma_{\bar{x}} = \sigma/\sqrt{n}$.

As an illustration of these points, suppose we have a population of 400 seamen and we draw without replacement a random sample of size 30 from it. Then the f.p.c. would be

$$\sqrt{\frac{400-30}{400-1}} = \sqrt{\frac{370}{399}} = \sqrt{0.93} = 0.96$$

which may reasonably be approximated to 1. In most of the future work we will ignore the finite population correction unless it is essential.

In our discussion so far we have only considered a sampling distribution of the mean. However, whatever statistic we chose to calculate from a random sample, if we were to calculate the same statistic from the other random samples of the same size which could be drawn from the same population, then the set of different values of the statistic would form the sampling distribution of the statistic. Thus, for instance, we might talk about the sampling distribution of a proportion and the term 'standard error of a proportion' would refer to the standard deviation of this distribution. At the moment while we are considering general concepts, it is easier if we restrict our attention to sample means, but we will bring in the sampling distributions of other statistics once we want to apply them.

8.4 *Sampling Distribution of the Mean for a Large Sample*

One of the most useful results from theoretical statistics concerns the form of the sampling distribution of the mean when we have large samples. From a practical point of view we may reasonably consider a large sample to be one of size greater than 30, although as we saw in our discussion on approximations in previous chapters, this is simply a rough working guide.

The important result comes from a theorem known as the *Central Limit Theorem* but whose mathematical proof is beyond the scope of this book. It states that the sampling distribution of the mean for random samples of size n, where n is large, drawn from any population of mean μ and variance σ^2, may be approximated to by a normal distribution with mean μ and variance σ^2/n (or $\sigma^2/n[(N-n)/(N-1)]$ if the f.p.c. is appropriate). The point to note carefully is that we will get a normal distribution, provided the sample size is large, no matter what distribution the population has.

Sampling and Estimation

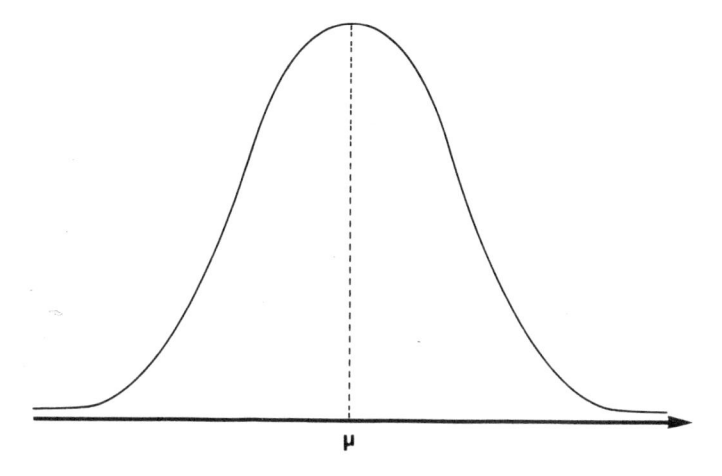

Fig 8·1 The sampling distribution of the mean for large samples of size n

Using the connection between the standard deviation of a normal distribution and the area under the curve, therefore, we know that approximately 67% of the means of all possible random samples of size n from a population lie within one standard error of the population mean, i.e. within a band from $\mu - \sigma_{\bar{x}}$ to $\mu + \sigma_{\bar{x}}$. The relationship between the population mean and the mean of the sampling distribution of the mean we considered earlier. Similarly, approximately 95% of the means of all possible random samples of size n lie within two standard errors of the population mean, i.e. within a band from $\mu - 2\sigma_{\bar{x}}$ to $\mu + 2\sigma_{\bar{x}}$. Further results of this type may be obtained by reference to tables of standard normal curve areas.

Example

An instructor, from past experience, knows that the error in the reading of a certain instrument has a mean value of zero with a standard deviation of 0·5 units. If a class of 40 cadets each take a reading from the instrument independently, what is the probability that the mean of these values will have an error of more than 0·2 units?

The first point we must make is that we assume that the error distribution in the reading is a property of the instrument and does not vary depending on the person taking the reading. Hence we can assume that the mean of the 40 readings, $\bar{x}$, is a member of the sampling distribution for the mean of samples of size 40, of readings from the machine and hence has a mean value of 0 and a standard deviation of $0\cdot5/\sqrt{40} = 0\cdot08$ since the finite population correction may be ignored. By the Central Limit Theorem, since the sample is large, we may assume that the sampling distribution has the form of a normal distribution without having to know the distribution of errors. Hence we require to know the probability that our value of $\bar{x}$ is larger than 0·2 or −0·2, which is

Marine Statistics

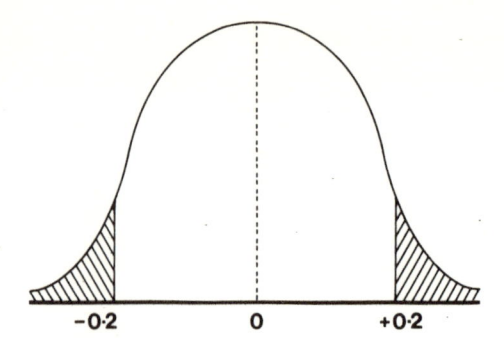

Fig 8·2 The region where the mean error is more than $|0.2|$ for 40 readings from the instrument

given by the shaded area in Fig. 8.2. Using the standardising transformation $z = (x - \mu)/\sigma$, the standard value corresponding to 0·2 is $(0·2 - 0)/0·08 = 2·5$. Thus we require the probability

$$(-2·5 \leqslant z \text{ and } z \geqslant 2·5) = 2(\text{probability } (z \geqslant 2·5))$$
$$= 2(1 - F(2·5)) = 2 \times 0·0062 = 0·0124$$

In this example we can see that there is only a 1% chance of getting an error of more than 0·2 units whereas if the original error distribution was assumed to be normal the probability based on only one reading would be 0·69. Hence we can see that statistically the error will be less if we take the mean of a few readings instead of relying on one observation of an instrument alone. In practice we obviously cannot take a large sample of readings so we will discuss the exact form of our results when considering the sampling distributions for small samples later.

8.5 Estimation in General

It was said at the start of the chapter that the main purpose when taking a sample was either to estimate a population parameter or to test a hypothesis about a population parameter (which we will consider in the next chapter). Turning to the first aim, that of estimation, there are some general definitions which we must first note. An estimate of a population parameter may be given as a point estimate or an interval estimate.

8.5.1 A *point estimate* of a parameter θ is a single value of a statistic *t*, often written as $\hat{\theta}$.

8.5.2 An *interval estimate* of a parameter θ is a range of values for θ, based usually on the sampling distribution of a statistic *t*.

8.5.3 A *confidence interval* for a parameter θ is an interval estimate for a given probability that θ lies within the interval.

Sampling and Estimation

8.5.4 **The extreme points of a confidence interval for a parameter θ are termed confidence limits for θ.**

8.5.5 **The statistic that is used to obtain a point estimate of a parameter is called an *estimator*.**

Thus the mean of a random sample would be an estimator for the mean of the population, and so also would the median of the random sample. One essential property that we require of an estimator is that it is unbiased.

8.5.6 **An estimator is said to be *unbiased* if the mean of the sampling distribution of the estimator is the required population parameter.** Both the sample mean and the sample median are unbiased estimators of the population mean.

8.5.7 **The most efficient estimator of a parameter is the one out of all the unbiased estimators whose sampling distribution has the least variance.**

It is only at this point that we find that the sample mean of a sample of size n is in fact preferable to the sample median for estimating the population mean. For example in a normal population with variance σ^2 the standard error of the mean is $\sigma/\sqrt{n}$ and the standard error of the median is $1 \cdot 25\sigma/\sqrt{n}$, ignoring the f.p.c. in both cases. The smaller the standard error, the smaller the confidence interval for a given probability as we shall see in a minute.

8.6 Confidence Intervals

Consider now the following example. A consignment of bags are nominally supposed to have a mean weight of 100 kg. However, it is known from past experience with similar consignments that the mean is often slightly different from this, although the standard deviation remains roughly constant at 8 kg. A random sample of 50 bags is taken from the latest consignment and its mean is found to be 97 kg. Find a 95% confidence interval for the mean of the latest consignment.

In other words, what we require is an interval estimate for the unknown mean of the latest consignment so that we are 95% certain that the unknown mean lies within it. By the Central Limit Theorem, since we have a large sample, we know that the sampling distribution of the mean is a normal distribution, i.e. symbolically $\bar{x} \sim N(\mu, \sigma^2/n)$. In this case our $\bar{x}$ is the sample mean = 97, μ is the unknown population mean, σ^2 is the population variance = 64, and n is the sample size = 50. Thus

$$\bar{x} \sim N(\mu, 1 \cdot 28)$$

Since we have a normal distribution, we know that 95% of all sample means lie within the range $\mu \pm 1 \cdot 96\sigma/\sqrt{n}$. Hence our particular sample mean of 97

Marine Statistics

is 95% certain to lie within that range. This can be stated alternatively that we are 95% certain that our $\bar{x}$ is not more than $1 \cdot 96\sigma/\sqrt{n}$ kg away from μ. This therefore implies that we are 95% certain that μ is not more than $1 \cdot 96\sigma/\sqrt{n}$ kg away from our $\bar{x}$. Hence we are 95% certain that μ is to be found within an interval given by:

$$\bar{x} \pm \frac{1 \cdot 96\sigma}{\sqrt{n}}, \quad \text{i.e.} \quad 97 \pm \frac{1 \cdot 96 \times 8}{\sqrt{50}}, \quad \text{i.e.} \quad 97 \pm 2 \cdot 2$$

Thus a 95% confidence interval for μ is $94 \cdot 8 \text{ kg} \leqslant \mu \leqslant 99 \cdot 2 \text{ kg}$. There is a 5% chance that our particular sample mean lies in one of the two extreme tails of the distribution and if this were the case, then repeating all the arguments again there is a 5% chance that μ lies outside the stated interval. However as we do not know what μ actually is, this is the best we can do. A higher level of confidence inevitably means a larger interval. We are 99% certain that our value of $\bar{x}$ lies within the range $\mu \pm 2 \cdot 58\sigma/\sqrt{n}$, where $+2 \cdot 58$ is the value on a standard normal curve such that only $\tfrac{1}{2}\%$ lies above it and similarly $\tfrac{1}{2}\%$ of the curve lies below $-2 \cdot 58$. Thus a 99% confidence interval for μ is given by:

$$\bar{x} \pm \frac{2 \cdot 58\sigma}{\sqrt{n}}, \quad \text{i.e.} \quad 97 \pm 2 \cdot 58 \times \frac{8}{\sqrt{50}}, \quad \text{i.e.} \quad 97 \pm 2 \cdot 9$$

which is $94 \cdot 1 \text{ kg} \leqslant \mu \leqslant 99 \cdot 9 \text{ kg}$. Our estimate for μ has, therefore, more flexibility than when we took a 95% confidence interval.

8.6.1 An α% confidence interval for the population mean μ based on the mean $\bar{x}$ of a large random sample of size n drawn from the population whose standard deviation is σ is given by:

$$\bar{x} - z_{[(100-\alpha)/2]\%} \frac{\sigma}{\sqrt{n}} \leqslant \mu \leqslant \bar{x} + z_{[(100-\alpha)/2]\%} \frac{\sigma}{\sqrt{n}}$$

where $z_{[(100-\alpha)/2]\%}$ is the standard normal variable such that $[(100 - \alpha)/2]\%$ of the curve lies above it.

If we had used the sample median $\tilde{x}$ as estimator, then although we can assume that for large samples its sampling distribution is normal and centred on the population mean, its standard error is $1 \cdot 25\sigma/\sqrt{n}$. Hence a 95% confidence interval for μ would be $\tilde{x} \pm 1 \cdot 96 \times (1 \cdot 25\sigma/\sqrt{n})$ which is larger than the result using the sample mean. In any situation we aim to use the most efficient estimator. The size of the confidence interval for the population mean using the sample mean is also dependent on σ, the standard deviation of the original population and inversely related to the sample size. There is not much we can do to alter σ as this is a property of the original population but this is possible if we can afford to take a larger sample to get a smaller interval. Suppose we had taken 100 bags instead of 50 then our 95% interval would be:

$$97 \pm 1 \cdot 96 \times \frac{8}{10} \text{ kg}, \quad \text{i.e.} \quad 95 \cdot 4 \text{ kg} \leqslant \mu \leqslant 98 \cdot 6 \text{ kg}$$

Sampling and Estimation

Since there is inherent variability between the mean of one random sample and that of another from the same population, it can be misleading to use a single point estimate of the population mean. As a matter of course it is always a good idea, when trying to estimate a population mean, to base one's assessment of a situation on the range of values given by the sample mean ± 2 standard errors, which gives roughly the 95% confidence interval, rather than on the sample mean alone. A similar approach should be used for estimating any population parameter.

We have now considered the most important concepts involved in sampling distributions and estimation, and the final sections of this chapter are less complicated as our concern is a factual one of noting the sampling distributions of some statistics under different circumstances. These results will be used again in the next chapter on hypothesis testing so examples will not be given in all cases here.

8.7 Sampling Distribution of the Mean for Large Samples When the Population Standard Deviation is Unknown

In practice we often know nothing about a population at all, so we have to estimate the standard deviation, σ, before we can get a confidence interval for the mean. Our estimate for the population standard deviation is the standard deviation of the sample of size n, s, calculated from the formula

$$s = \sqrt{\frac{1}{n-1}\left[\sum_i (x_i - \bar{x})^2\right]}$$

or in its easier computational form,

$$s = \sqrt{\frac{1}{n-1}\left[\sum_i x_i^2 - \frac{\left(\sum_i x_i\right)^2}{n}\right]}$$

This, as we saw in an earlier chapter, is different from the usual formula for standard deviation since the divisor is $n - 1$ rather than n. Unfortunately if we divide by n, the resulting statistic is a biased estimator for σ, but if we divide by $n - 1$ then the result, s, is an unbiased estimator for σ. As a general rule it can be shown mathematically that for every population parameter in a formula that is unknown we must take off one from the denominator. In this case the population mean is unknown in the formula and $\bar{x}$, the sample mean, has been substituted so the divisor is $n - 1$.

The standard error of the mean is then given by $s/\sqrt{n}$ (ignoring the f.p.c.) and if the sample is large the Central Limit Theorem applies as before.

Marine Statistics

8.7.1 If a large random sample of size n, with mean $\bar{x}$ and standard deviation s, is taken from a population of unknown mean, μ, and unknown standard deviation, then the sampling distribution of the mean is normal with mean, μ, and standard deviation given by $s/\sqrt{n}$.

Example

A shipping company conducts trials of ropes from a manufacturer and finds that of 64 ropes purchased the mean life was 4·8 yr with a standard deviation of 0·5 yr. Find a 99% confidence interval for the mean life of all similar ropes from the manufacturer.

We are told that in the sample, the mean $\bar{x} = 4\cdot 8$ yr, the standard deviation $s = 0\cdot 5$ yr and the sample size n is 64. Since we have a large sample we may assume that the sampling distribution of the mean is normal with mean, μ, the unknown population mean, and standard deviation

$$\frac{s}{\sqrt{n}} = \frac{0\cdot 5}{\sqrt{64}} = \frac{0\cdot 5}{8} = 0\cdot 0625$$

Thus 99% confidence limits for μ are given by $\bar{x} \pm 2\cdot 58(s/\sqrt{n}) = 4\cdot 8 \pm 2\cdot 58 \times 0\cdot 0625 = 4\cdot 8 \pm 0\cdot 16$ yr. Hence we are 99% certain that the true mean life of the ropes is between 4·64 and 4·96 yr.

8.8 *Sampling Distribution of the Mean for Small Samples*

When we are considering small samples whose size roughly is under 30, then the situation is more complicated and we have to be careful about the distribution of the population being sampled. If this population does not have a normal distribution then the methods of this type fall down completely. The only sort of work we can do are some hypothesis tests and for this we need special methods known as non-parametric tests which will be considered in a later chapter. However, if the population can be assumed to have a normal distribution then the sampling distribution of the mean is readily defined. We do however have to distinguish between situations where the population standard deviation is known and situations where it has also to be estimated from the sample.

8.8.1 If a random sample of size n, where n is small, is taken from a normal population with unknown mean, μ, and known standard deviation σ, then the sampling distribution of the mean is normal with mean, μ, and standard deviation $\sigma/\sqrt{n}$.

Thus if the population standard deviation is known we have essentially the same as in previous results and it is applied in the same way. However if the population standard deviation is unknown, then we have to use a new distribution known as the 't' distribution.

Sampling and Estimation

8.8.2 If we have a random sample of size n, where n is small, with mean, $\bar{x}$, and standard deviation, s, from a normal population of unknown mean, μ, and unknown standard deviation, σ, then the statistic $t = (\bar{x} - \mu)/(s/\sqrt{n})$ is said to have a 't' distribution with $n - 1$ degrees of freedom.

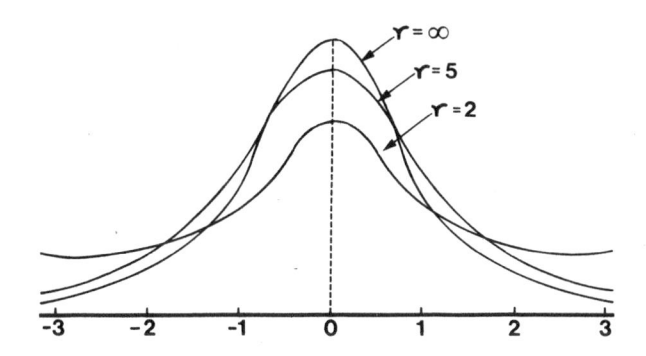

Fig. 8·3 't' distribution curves on different degrees of freedom

Figure 8.3 shows the curves for a series of t distributions on different degrees of freedom, ν (pronounced 'nu'), where ν is one less than the sample size. When the sample size is large, then we can treat the degrees of freedom as being infinite and the distribution becomes the standard normal distribution. It has probably occurred to some readers that the formula given for the 't' statistic is completely analogous to the formula for the standard normal transformation. For smaller number of degrees of freedom, the curves become less bell-shaped but still preserve the properties of symmetry and zero mean. The main difference, essentially, is that the tails of the distributions become larger and larger as the degrees of freedom decrease as can be seen numerically in the table of t values, Table 2 at the end of the book. As the distribution is different for each number of degrees of freedom, it is completely impractical to give areas under each curve. Instead for each number of degrees of freedom, read vertically, the value of t is given so that a proportion α of the curve, read horizontally, lies to the right of it. Only the most frequently used values of α are given such as 0·1, 0·05, 0·025, etc.

In the formula for t, given as $t = (\bar{x} - \mu)/(s/\sqrt{n})$, s is calculated as usual, using the definition

$$s = \sqrt{\frac{1}{n-1} \sum_{i=1}^{n} (x_i - \bar{x})^2}$$

The divisor in s, $n - 1$, is the number of degrees of freedom for t, since it is the number of independent values in the sample algebraically in the formula.

Marine Statistics

This arises because we also use the value of the sample mean in the formula for s, and hence the nth sample value may be written as a combination of the sample mean and the other $n - 1$ values.

We use the 't' distribution to calculate confidence limits for the unknown population mean, μ, in the same way as we use the standard normal distribution when the sample is large. The example at the beginning of the chapter helps to illustrate this point. We must assume that the deflections of the pendulum measuring the stability of the ship have a normal distribution and we will calculate a 95% confidence interval for the true mean deflection of the pendulum based on the 12 results.

The first step is to calculate the mean, $\bar{x}$, and standard deviation, s, of the 12 results:

$$\bar{x} = \frac{1}{n}\sum_{i=1}^{n} x_i = \frac{1}{12}(97 + 98 + 103 + 95 + 100 + 91 + 100 + 98 + 94 + 96 + 99 + 93)$$

$$= 97$$

$$s = \sqrt{\frac{1}{n-1}\left[\sum_i x_i^2 - \frac{\left(\sum_i x_i\right)^2}{n}\right]}$$

$$= \sqrt{\frac{1}{11}\left[97^2 + 98^2 + \cdots + 93^2 - \frac{(1,164)^2}{12}\right]} = 3\cdot 38$$

Hence

$$t = \frac{\bar{x} - \mu}{s/\sqrt{n}} = \frac{97 - \mu}{3\cdot 38/3\cdot 46}$$

As the sample size is 12, this t statistic is from a t distribution on 11 degrees of freedom (d.o.f.) 95% of all values of t_{11}, the t distribution on 11 d.o.f., lie between $-2\cdot 20$ and $+2\cdot 20$. Hence one extreme value for μ at the 95% confidence level is given by $97 + 2\cdot 2 \times (3\cdot 38/3\cdot 46) = 99\cdot 1$ and the other is given by $97 - 2\cdot 2 \times (3\cdot 38/3\cdot 46) = 94\cdot 9$. We are therefore 95% certain that the mean incline of the pendulum lies between 94·9 cm and 99·1 cm.

8.8.3 An $\alpha\%$ confidence interval for the population mean μ, of a normally distributed population whose standard deviation is unknown, is given by

$$\bar{x} - t_{\nu,[(100-\alpha)/2]\%}\, s/\sqrt{n} \leqslant \mu \leqslant \bar{x} + t_{\nu,[(100-\alpha)/2]\%}\, s/\sqrt{n}$$

where $\bar{x}$ is the mean and s is the standard deviation of a small random sample of size n drawn from the population and $t_{\nu,[(100-\alpha)/2]\%}$ is the point of the 't' distribution on $\nu = n - 1$ degrees of freedom such that $[(100 - \alpha)/2]\%$ of the curve lies above it.

8.9 Sampling Distribution of the Difference Between Two Means

Another common practical situation that arises is when we take a sample from each of two populations and wish to compare the two means.

We will assume that the first population has mean μ_1 and standard deviation σ_1, and we take a sample of size n from it having mean $\bar{x}_1$ and standard deviation s_1. The corresponding values for the second population are μ_2, σ_2, n_2, $\bar{x}_2$ and s_2. We form the statistic of the difference between the two sample means, $\bar{x}_1 - \bar{x}_2$.

8.9.1 If we take a large sample from each of two populations then the sampling distribution of the difference between the sample means $\bar{x}_1 - \bar{x}_2$ is normal with mean $\mu_1 - \mu_2$ and variance $(\sigma_1^2/n_1) + (\sigma_2^2/n_2)$ using the notation given above.

This result is analogous to the large-sized single sample case, and similarly if we do not know the population standard deviations we can estimate them using the sample standard deviations and simply replace σ_1 by s_1 and σ_2 by s_2. If the sample sizes are both larger than 30 then these results hold very well, but if not they only apply when both populations have normal distributions and known standard deviations. If the populations do not have normal distributions then non-parametric methods have to be applied for hypothesis testing and confidence intervals cannot easily be found. The other remaining case is for small samples from normally distributed populations when the standard deviations are unknown. This subdivides again into situations where it may be assumed that the standard deviations are equal and those for which this is not true.

If we can assume that $\sigma_1 = \sigma_2 = \sigma$, say, then we can estimate σ from the two samples by using the formula:

$$\hat{\sigma}^2 = \frac{1}{n_1 + n_2 - 2} \left[\sum_i (x_{1i} - \bar{x}_1)^2 + \sum_i (x_{2i} - \bar{x}_2)^2 \right]$$

The first summation is over the values in the first sample alone and the second over the values in the second sample alone. Our divisor is $n_1 + n_2 - 2$ since we have two unknowns in the formula which are replaced by $\bar{x}_1$ and $\bar{x}_2$ respectively.

8.9.2 If we take a small sample from each of two normally distributed populations whose variances can be assumed to be equal but are unknown, then

Marine Statistics

the statistic,

$$t = \frac{(\bar{x}_1 - \bar{x}_2) - (\mu_1 - \mu_2)}{\sqrt{\dfrac{\hat{\sigma}^2}{n_1} + \dfrac{\hat{\sigma}^2}{n_2}}}$$

has a t distribution with $n_1 + n_2 - 2$ d.o.f., using the notation above.

Hence confidence limits may be calculated as before. If we cannot assume that the population standard deviations are equal then no exact result is known. For a practical situation the result above could be used but great care would be necessary in using the answers. An example to illustrate the general principle of estimating the difference between means will be helpful now.

Example

A shipping company conducts trials on the breaking strength of ropes from two manufacturers. A sample of 40 ropes from manufacturer A is found to have a mean breaking strength of 2,400 kg with a standard deviation of 100 kg. A sample of 60 ropes from manufacturer B is found to have a mean breaking strength of 2,800 kg with a standard deviation of 200 kg. Give 95% confidence limits for the true difference between the mean breaking strengths.

As we have large samples we can assume a normal distribution for the sampling distribution of the difference between the means. Hence 95% confidence limits for $\mu_1 - \mu_2$, the difference between the two population means, are given by $\bar{x}_1 - \bar{x}_2 \pm 1.96$ standard errors, i.e.

$$\bar{x}_1 - \bar{x}_2 \pm 1.96\sqrt{\frac{s_1^2}{n_1} + \frac{s_2^2}{n_2}} \quad \text{(since the population standard deviations are unknown)}$$

$\bar{x}_1 = 2,400 \quad \bar{x}_2 = 2,800 \quad s_1 = 100 \quad s_2 = 200 \quad n_1 = 40 \quad n_2 = 60$

Thus we get

$$-400 \pm 1.96\sqrt{\frac{100^2}{40} + \frac{200^2}{60}}$$

which gives

$$-400 \pm 59.3$$

i.e.

$$-460 \leqslant \mu_1 - \mu_2 \leqslant -340 \text{ kg}$$

Hence we can be 95% certain that the ropes of manufacturer B have a breaking strength between 340 and 460 kg better than the ropes of manufacturer A.

8.10 Sampling Distributions of Proportions

Another population parameter which we are often interested in estimating is the proportion of a population having a particular attribute. We usually refer to this with the Greek letter π, and use the Roman letter p to indicate the corresponding sample proportion.

The standard error of a proportion, σ_p, for samples of size n is given theoretically by $\sqrt{[\pi(1-\pi)/n]}$, a result derived from the binomial distribution. However if our aim is to estimate π, the population proportion, then automatically the standard error is undefined as well and must be estimated. It is usual to estimate σ_p by $\hat{\sigma}_p = \sqrt{pq/n}$, where $q = 1 - p$.

8.10.1 **If a large random sample of size n is taken from a population and p is the sample proportion having a particular attribute, then the sampling distribution of p is normal with mean π, the population proportion, and standard deviation given by $\sqrt{pq/n}$ where $q = 1 - p$.**

From this result it can be seen that a confidence interval for π may be calculated in a similar manner to those for a mean.

Example

In a survey of deck officers over the age of 30, it was found that 34 out of 50 were married. Find 95% confidence limits for the proportion of all deck officers over the age of 30 who are married.

Since we have a large sample ($n = 50$) we may assume a normal distribution for the sampling distribution of the proportion. In the sample p, the proportion of married deck officers $= 34/50$. Let this proportion be π in the population:

$$p \sim N\left(\pi, \frac{p(1-p)}{n}\right)$$

Thus 95% confidence limits for π are given by:

$$p \pm 1\cdot 96\sqrt{\frac{p(1-p)}{n}} = \frac{34}{50} \pm 1\cdot 96\sqrt{\frac{34}{50} \cdot \frac{16}{50} \cdot \frac{1}{50}}$$

$$= 0\cdot 68 \pm 0\cdot 129$$

Hence $0\cdot 551 \leqslant \pi \leqslant 0\cdot 809$. This can be stated alternatively that we are 95% certain that between 55% and 81% of deck officers over the age of 30 are married.

If we are dealing with the difference between the proportions in two populations based on the evidence of a large sample from each then similar results apply. Let π_1 be the proportion in the first population having a certain

Marine Statistics

attribute, with p_1 the corresponding proportion in a large sample of size n_1 from that population. Let π_2, p_2 and n_2 be the corresponding values for the second population.

Result

8.10.2 **If a large sample is taken from each of two populations then the sampling distribution of the difference between the proportion in each of the two samples possessing a certain attribute is normal with mean $\pi_1 - \pi_2$ and standard deviation given by**

$$\sqrt{\frac{p_1(1 - p_1)}{n_1} + \frac{p_2(1 - p_2)}{n_2}}$$

using the notation above.

This result enables confidence intervals to be calculated as before. If we only have small samples then it is not easy to calculate confidence intervals for proportions. However it is possible to do hypothesis tests using the small samples as we shall see later.

8.11 Sample Sizes

A practical problem which often arises concerns the size of sample which should be taken if we want to estimate a result to a certain degree of accuracy. It was mentioned earlier that the size of confidence interval depends on the standard error of the statistic and the level of confidence we are willing to use.

If we reconsider an earlier example of the instructor taking readings from an instrument with a known error, standard deviation of 0·5 units but assuming now an unknown mean error, then we might ask how large a sample of readings we should take so that we are at least 95% certain our sample mean is not more than 0·1 unit from the true mean. Hence we require $0·1 = 1·96 \times (0·5/\sqrt{n})$ when n is sample size. Then $\sqrt{n} = 1·95 \times 5$, giving $n = 96·08$. We conclude that n must be 97 to ensure that we are at least 95% certain. If we are dealing with a proportion then we meet the problem again that the standard error of a proportion is dependent on the actual population proportion. Suppose we wanted to know how large a sample of deck officers over the age of 30 we needed to take to be 95% certain of estimating the proportion who were married to within 0·05. Hence

$$0·05 = 1·96 \sqrt{\frac{p(1 - p)}{n}}$$

where p is the unknown proportion and n the sample size. However, the largest that $\sqrt{p(1-p)}$ can be is when $p = \frac{1}{2}$. Hence

$$0{\cdot}05 = 1{\cdot}96 \times \frac{1}{2} \times \frac{1}{\sqrt{n}}$$

$$\therefore\ n = 385$$

8.12 Summary

In this chapter we have considered one of the most basic concepts in statistical inference, that of a sampling distribution. We then considered how we could utilise this knowledge to provide an estimate of a population parameter which took into account the variability in any sampling situation. As a general rule we found that when dealing with means and proportions the confidence interval estimate which we arrived at could be summarised as:

**sample statistic ± probability factor
× standard error of the statistic**

In the next chapter we will use the sampling distribution concept to test hypotheses concerning population parameters.

Exercises

1. The weights of cases of fruit loaded at a certain port are known to be distributed normally such that the mean weight of a case is 20 kg and the standard deviation is 1·5 kg. The cases are loaded 36 to a sling. What is the probability that the weight of a sling will exceed 725 kg?

2. During the loading of a consignment of bagged grain, 50 bags were taken at random and the total weight of this sample was found to be 2,400 kg. It is known that the weights of bags have a normal distribution and that they consistently have a standard deviation of 2 kg although the mean weight of a bag tends to decrease as the season advances. If the consignment consists of 2,000 bags, estimate the total weight of the consignment and give the 95% confidence limits for this estimate.

3. In a survey, the mean speeds of a sample of 60 ships were recorded as they passed through a traffic separation scheme. The results are as shown in the frequency table below:

Marine Statistics

Speed (knots)	Number of ships
0–4	6
5–9	14
10–14	24
15–19	10
20–24	4
25–29	2

(a) Assuming that the sample is representative of the general traffic, suggest the confidence limits within which the mean speed of the general traffic can be expected to fall on 95% of occasions.

(b) Why is the value of s in this example slightly larger than the standard deviation calculated from the same numerical information given in question 2 of the exercises in Chapter 4?

4. To investigate the service life of a wire rope used in a crane, 32 test pieces were subjected to simulated working conditions until they failed. The lengths of times recorded for these test pieces are summarised in the following table:

Number of hours of test (thousands)	Number of samples tested
Less than 2	0
2 but less than 3	2
3 but less than 4	6
4 but less than 5	12
5 but less than 6	8
6 but less than 7	3
7 but less than 8	1
8 and over	0

It is required to estimate the limits within which the mean of similar wire ropes can be expected to lie with 99% certainty.

5. The turn-round times for 10 container ships at a particular berth were noted in hours as follows:

24·5   16·3   19·7   27·4   18·6   21·8   22·3   20·4   23·5   21·2

Given that these are typical cases from a normally distributed population, estimate 95% limits for the mean of the population.

Sampling and Estimation

6. A radio navigation aid was used to determine a position line on eight occasions while the ship was in a fixed position. The position lines were found to run to the eastwards of a navigation mark with the following displacements in nautical miles:

   5·3   6·9   7·8   6·4   8·8   7·1   6·0   7·5

   Assuming that these results are a random sample from normally distributed quantities, estimate the mean displacement of similar position lines from the navigational aid and give the 99% confidence limits for this estimate.

7. A port traffic centre wishes to compare the transit times for ships entering the port by day and by night. A sample of 50 ships are timed entering the port during daylight, giving a mean transit time of 3·5 hr with a standard deviation of 0·45 hr. A sample of 45 ships are timed entering the port at night, giving a mean transit time of 4·3 hr with a standard deviation of 0·55 hr.

   Find the 95% confidence limits for the true difference between the transit times.

8. Two consignments of bale goods, nominally of the same unit weights, were loaded at a particular port. As a result of storage conditions, it was suspected that the first consignment may have absorbed an unusual amount of moisture. Ten bales were taken at random from the first consignment and were found to have a mean weight of 105·4 kg with a standard deviation of 1·5 kg. Twelve bales were taken at random from the second consignment and were found to have a mean weight of 101·2 kg with a standard deviation of 1·4 kg.

   Find the 99% limits for the true difference between the mean weight of the bales in the two consignments. Mention any assumptions you make.

9. A consignment of cases consists of units supplied by two manufacturers with appropriate identification marks. In order to estimate the proportion of each type, 60 cases were chosen at random during the loading process and it was discovered that 24 were supplied from manufacturer A and 36 from manufacturer B. Estimate the 95% limits for the proportion corresponding to each manufacturer for the whole consignment.

10. For a ship trading across the North Atlantic it was found that, during the 62 days of the months of July and August, the weather on 32 days was good enough for painting and similar maintenance work. During the 59 days of January and February, it was found that the weather on 15 days was suitable for this work. What limits can be suggested for the difference between the proportions expected in future years, assuming the seasons investigated to have been reasonably representative? We are looking for a 90% confidence level.

Marine Statistics

Answers

1. The population mean, μ, is given as 20 kg. The standard deviation, σ, is 1·5 kg. The standard error for a sample of 36 cases is:

$$\frac{\sigma}{\sqrt{n}} = \frac{1 \cdot 5}{6} = \underline{0 \cdot 25}$$

The probability that the total weight of a sling will exceed 725 kg is the probability that the mean weight of the sample of 36 cases making up the sling will exceed:

$$\frac{725}{36} = 20 \cdot 14 \text{ kg}$$

Since we have a large sample we may assume the sampling distribution of the mean is normal. Hence using

$$z = \frac{\bar{x} - \mu}{\sigma/\sqrt{n}} \quad \text{with} \quad \bar{x} = 20 \cdot 14 \quad \text{gives} \quad z = \frac{0 \cdot 14}{0 \cdot 25} = 0 \cdot 56$$

Referring to the table of normal distribution, we find that:

$$p(z > 0 \cdot 56) = 1 - F(0 \cdot 56) = 1 - 0 \cdot 7123$$
$$= 0 \cdot 2877$$

Thus we would expect about 29 slings out of every hundred to exceed 725 kg in weight.

2. The mean weight of a bag in the sample of 50 is given by:

$$\bar{x} = \frac{2{,}400}{50} = 48 \text{ kg}$$

The population standard deviation, σ, is given as 2 kg. The standard error of the sample mean is given by:

$$\frac{\sigma}{\sqrt{n}} = \frac{2}{\sqrt{50}} = 0 \cdot 28$$

Since we have a large sample we may assume a normal distribution and we can say that on 95% of occasions the true population mean, μ, will lie within $\pm 1 \cdot 96$ standard errors of the sample mean $\bar{x}$, i.e. within:

$$\pm 1 \cdot 96 \times 0 \cdot 28 = \pm 0 \cdot 55$$

Thus:

$$47 \cdot 45 < \mu < 48 \cdot 55$$

Hence the total weight of 2,000 bags lies within the range:

$$47 \cdot 45 \times 2{,}000 \text{ to } 48 \cdot 55 \times 2{,}000 \quad \text{or} \quad 94{,}900 \text{ to } 97{,}100 \text{ kg}$$

Sampling and Estimation

3. (a)

Speed (knots)	Mid mark	f	fx	fx²
0–4	2	6	12	24
5–9	7	14	98	686
10–14	12	24	288	3,456
15–19	17	10	170	2,890
20–24	22	4	88	1,936
25–29	27	2	54	1,458
		60	710	10,450

$$\bar{x} = \frac{\Sigma x}{n} = \frac{710}{60} = \underline{11 \cdot 83 \text{ knots}}$$

$$s = \sqrt{\frac{1}{n-1}\left[\Sigma fx^2 - \frac{(\Sigma fx)^2}{n}\right]}$$

$$= \sqrt{\frac{1}{59}\left[10{,}450 - \frac{710^2}{60}\right]}$$

$$= \underline{5 \cdot 89 \text{ knots}}$$

The standard error of the mean is given by:

$$\frac{s}{\sqrt{n}} = \frac{5 \cdot 89}{\sqrt{60}} = \underline{0 \cdot 76}$$

Since we have a large sample the normal distribution is appropriate.

The 95% confidence limits are: $\bar{x} \pm 1 \cdot 96 \dfrac{s}{\sqrt{n}}$

$$= 11 \cdot 83 + 1 \cdot 96 \times 0 \cdot 76$$
$$= \underline{11 \cdot 83 \pm 1 \cdot 49 \text{ knots}}$$

(b) The value of s is larger because we are now treating the data as a sample and using this to estimate the standard deviation of the population from which it is drawn. In Ex. 2 of Chapter 4 we were treating the data as the population and thus deriving the standard deviation directly. Numerically, our divisor in the calculation is $n - 1$ instead of n.

Marine Statistics

4.

Number of hours of test (thousands)	Class mid mark (x)	Number of samples tested (f)	fx	fx^2
Less than 2	—	0	0	0
2 but less than 3	2·5	2	5	12·5
3 but less than 4	3·5	6	21	73·5
4 but less than 5	4·5	12	54	243·0
5 but less than 6	5·5	8	44	242·0
6 but less than 7	6·5	3	19·5	126·75
7 but less than 8	7·5	1	7·5	56·25
8 and over	—	0	0	0

$$n = \Sigma f = 32 \quad \Sigma fx = 151\cdot0 \quad \Sigma fx^2 = 754\cdot0$$

The mean value of this sample is given by:

$$\bar{x} = \frac{\Sigma fx}{n} = \frac{151}{32} = 4\cdot719$$

The standard deviation is given by:

$$s = \sqrt{\frac{1}{n-1}\left[\Sigma (fx^2) - \frac{\Sigma (fx)^2}{n}\right]}$$

$$= \sqrt{\frac{1}{31}\left[754 - \frac{151^2}{32}\right]}$$

$$= 1\cdot157$$

We require confidence limits within which we can expect the true mean to lie on 99% of occasions. By consulting the normal distribution tables, since we have a large sample, we find appropriate limits lie ±2·58 standard errors on either side of the mean. Thus we expect the true mean for the population of similar wire ropes to lie within the limits:

$$\bar{x} \pm 2\cdot58 \frac{s}{\sqrt{n}}$$

$$= 4\cdot719 \pm \frac{2\cdot58 \times 1\cdot157}{\sqrt{32}}$$

$$= 4\cdot719 \pm 0\cdot528$$

Thus we are 99% certain that the mean life of similar wire ropes lies between 4,191 and 5,247 hr.

5.

x	x^2
24·5	600·25
16·3	265·69
19·7	388·09
27·4	750·76
18·6	345·96
21·8	475·24
22·3	497·29
20·4	416·16
23·5	552·25
21·2	449·44
215·7	4741·13

$$\bar{x} = \frac{\sum x}{n} = \frac{215 \cdot 7}{10} = 21 \cdot 57$$

$$s = \sqrt{\frac{1}{n-1}\left[\sum x^2 - \frac{(\sum x)^2}{n}\right]}$$

$$= \sqrt{\frac{1}{9}\left[4{,}741 \cdot 13 - \frac{215 \cdot 7^2}{10}\right]}$$

$$= 3 \cdot 135$$

The 't' distribution may be used as we have a small sample from a normally distributed population with unknown variance.

Consulting the table for the percentage points of the t distribution with 9 degrees of freedom, we find that the limits within which t will lie on 95% of occasions are $\pm 2 \cdot 26$.

To find the corresponding limits for μ, we use:

$$t = \frac{\bar{x} - \mu}{s/\sqrt{n}}$$

$$\pm 2 \cdot 26 = \frac{21 \cdot 57 - \mu}{3 \cdot 135/3 \cdot 162}$$

Thus:

$$\mu = 21 \cdot 57 \pm 2 \cdot 24 \quad \text{or} \quad 19 \cdot 33 < \mu < 23 \cdot 81$$

Hence we are 95% certain that the population mean for container ship turn-rounds at the berth in question lies between 19·33 and 23·81 hr, with the sample mean of 21·57 hr being our best estimate.

Marine Statistics

6.

x	x^2
5·3	28·09
6·9	47·61
7·8	60·84
6·4	40·96
8·8	77·44
7·1	50·41
6·0	36·00
7·5	56·25
$\Sigma x = 55\cdot 8$	$\Sigma x^2 = 397\cdot 60$

$$\bar{x} = \frac{\Sigma x}{n} = \frac{55\cdot 8}{8} = \underline{6\cdot 975}$$

$$s = \sqrt{\frac{1}{n-1}\left[\Sigma x^2 - \frac{(\Sigma x)^2}{n}\right]}$$

$$= \sqrt{\frac{1}{7}\left[397\cdot 6 - \frac{55\cdot 8^2}{8}\right]}$$

$$= 1\cdot 097$$

The 't' distribution is appropriate here so consulting the table for percentage points of the t distribution with 7 degrees of freedom, we find that the limits within which t will lie on 99% of occasions are $\pm 3\cdot 5$.

Thus we have for the limits of μ:

$$t = \frac{\bar{x} - \mu}{s/\sqrt{n}}$$

$$3\cdot 5 = \frac{6\cdot 975 - \mu}{1\cdot 095/2\cdot 828} \quad \text{or} \quad -3\cdot 5 = \frac{6\cdot 975 - \mu}{1\cdot 095/2\cdot 828}$$

Thus:

$$\mu = 6\cdot 975 - 1\cdot 355 \quad \text{or} \quad \mu = 6\cdot 975 + 1\cdot 355$$
$$\mu = 5\cdot 62 \quad \text{or} \quad \mu = 8\cdot 33$$

Hence the best estimate of the displacement of the mean position line is $\bar{x}$ which has the value 6·975 and we are 99% certain that the mean displacement will lie between 5·62 and 8·33 miles from the navigation mark.

Sampling and Estimation

7. Since we have large samples, we can assume that the difference between the means is normally distributed. Hence the 95% confidence limits for $\mu_1 - \mu_2$ are given by:

$$\bar{x}_1 - \bar{x}_2 \pm 1.96 \text{ standard errors}$$

i.e.

$$\bar{x}_1 - \bar{x}_2 \pm 1.96\sqrt{\frac{s_1^2}{n_1} + \frac{s_2^2}{n_2}} = 4.3 - 3.5 \pm 1.96\sqrt{\frac{0.55^2}{45} + \frac{0.45^2}{50}}$$
$$= 0.80 \pm 0.20$$

We can thus be 95% certain that night-time transits have a mean value which is between 0·60 and 1·00 hr longer than daytime transits.

8. Since we have small samples, we have to assume that each of the parent populations are normally distributed with equal variances so that we can make use of the t distribution.

The appropriate number of degrees of freedom is:

$$n_1 + n_2 - 2 = 10 + 12 - 2 = 20$$

From the t table we find that the limits within which t will lie on 99% of occasions is ± 2.85, but

$$t = \frac{(\bar{x}_1 - \bar{x}_2) - (\mu_1 - \mu_2)}{\sqrt{\frac{s_1^2}{n_1} + \frac{s_2^2}{n_2}}}$$

Thus to find the limits of $(\mu_1 - \mu_2)$, we put:

$$\pm 2.85 = \frac{(105.4 - 101.2) - (\mu_1 - \mu_2)}{\sqrt{\frac{1.5^2}{10} + \frac{1.4^2}{12}}} \qquad \mu_1 - \mu_2 = 4.2 \pm 1.78$$

We can thus be 99% certain that the mean weight of the bales in the first consignment is between 2·42 and 6·0 kg heavier than those in the second consignment.

9. Since we have a large sample, we may assume that the sampling distribution of the proportion is normal.

We designate the sample proportion for manufacturer A as p, so that:

$$p = \frac{24}{60} = \underline{0.4}$$

We estimate the standard error of the sample proportion, $\hat{\sigma}_p$, by

$$\hat{\sigma}_p = \sqrt{\frac{pq}{n}} = \sqrt{\frac{0.4 \times 0.6}{60}} = \underline{0.063}$$

169

Marine Statistics

The 95% confidence limits for π, the population proportion for manufacturer A, are given by:

$$p \pm 1.96\hat{\sigma}_p = 0.4 \pm 1.96 \times 0.063$$
$$= 0.4 \pm 0.12$$

Hence we are 95% certain that the proportion of manufacturer A's cases in the complete consignment lies between 0.28 and 0.52 and the corresponding figures for manufacturer B's cases are 0.48 and 0.72.

10. Since we have two large samples, we may assume that the sampling distribution of the difference between the proportions of working days is normal.

    We designate the proportion of working days observed in July/August as p_1 and the proportion of working days in January/February as p_2. Thus:

    $$p_1 = \frac{32}{62} = 0.516 \qquad p_2 = \frac{15}{59} = 0.254$$

    We estimate the standard error of the difference between the proportions as:

    $$\hat{\sigma}_{(p_1-p_2)} = \sqrt{\frac{p_1(1-p_1)}{n_1} + \frac{p_2(1-p_2)}{n_2}}$$
    $$= \sqrt{\frac{0.516 \times 0.484}{62} + \frac{0.254 \times 0.746}{59}}$$
    $$= 0.085$$

    Using normal distribution tables, we find that the 90% confidence limits for $\pi_1 - \pi_2$, the difference between the working day proportions in summer and winter, are given by:

    $$(p_1 - p_2) \pm 1.65\hat{\sigma}_{(p_1-p_2)} = 0.516 - 0.254 \pm 1.65 \times 0.085$$
    $$= 0.262 \pm 0.140$$

    Thus we are 90% certain that the proportion of working days expected in July/August is between 0.122 and 0.402 higher than the proportion of working days in January/February.

Chapter 9
Hypothesis Testing

9.1 Introduction

A rope manufacturer claims that one type of his ropes have a nominal breaking load of 2,500 kg. A sample of 49 of these ropes were tested to destruction and found to have breaking loads with a mean of 2,400 kg and a standard deviation of 350 kg. Do we have sufficient evidence to conclude that the manufacturer's claim is false?

In this example, as in the second example at the start of the previous chapter, we are not interested in knowing what actual value a parameter has but simply how the unknown parameter compares to a known case. We do not need to know what the actual mean breaking load of the ropes is, but simply how it compares with a value of 2,500 kg. We attempt to answer questions of this sort by statistical hypothesis testing. As already said, most of the fundamental information needed for this was covered in the previous chapter when we discussed sampling distributions but we must first consider some of the ideas specific to statistical hypothesis testing.

9.2 Null and Alternative Hypotheses

9.2.1 **A *statistical hypothesis* is an assumption or statement about one or more populations, which may or may not be true.**

Our usual problem in statistical inference is that as we cannot examine the whole population for one or more of the reasons previously discussed and we must base our conclusions about the population on evidence provided from a small sample of it. Hence we will never be able to state categorically whether a hypothesis is right or wrong, but simply whether the evidence suggests that it is more likely to be right or wrong. A hypothesis may concern the nature of the distribution of the population, e.g. the population has a normal distribution or the population does not have a Poisson distribution. A hypothesis on the other hand may only concern a parameter in the population, such as stating the mean is 15 units or the mean is not 0 units. There can be several populations involved in one hypothesis, e.g. the means of all these populations are equal or the proportions of two populations having an

Marine Statistics

attribute are not equal. In this chapter we are going to concentrate on hypotheses concerning parameters in one or more populations.

A hypothesis concerning a parameter may be termed simple or composite. If it is simple then it assigns a single value to the parameter, e.g. $\mu = 15$, whereas if it is composite then it assigns a range of values, e.g. $\mu \neq 15$, whereby μ could be 14, 13, 16, 17, etc. We must now distinguish between null and alternative hypotheses. The *null hypothesis* is a hypothesis which we make which enables us to consider our sample result based on the assumption that our null hypothesis is true. For instance, if we take as null hypothesis that the population mean is 15 then we may consider how likely it is that a sample mean of 18, say, in fact belongs to this population. The *alternative hypothesis* is the hypothesis that we will accept if we reject, i.e. 'do not like' the null hypothesis using the sample evidence. Our alternative hypothesis may be also a simple hypothesis such as $\mu = 20$ and we would then accept the one that seemed most likely to define the population that a sample mean came from. However this can be a long and tedious mathematical process, especially if there are a whole variety of alternative hypotheses to be tested, so we are going to confine our discussion to the most usual type of hypothesis test, where we are interested in knowing if we should or should not assume a particular value for our unknown parameter. Thus in this situation if we were to test a null hypothesis of $\mu = 15$ we would test it against one of three composite alternative hypotheses, viz.: $H_1: \mu \neq 15$ or $H_1': \mu > 15$ or $H_1'': \mu < 15$.

The most important point about a null hypothesis is that it must set up a yardstick by which we can make a decision. If we take as null hypothesis $\mu = 15$, then we can define the sampling distribution of the mean under this hypothesis. However if we were to take as null hypothesis $\mu < 15$, say, then we have not defined a unique sampling distribution to use as a test. The term null hypothesis arose because in many cases we want to test the no difference situation, e.g. the mean of the new population is no different from the known mean of the old population.

9.2.2 The *null hypothesis* is the hypothesis which provides a basis on which to test our sample result. It is usually written as H_0.

9.2.3 The *alternative hypothesis* is the hypothesis which we will accept if our evidence suggests that we should reject the null hypothesis. It is usually written as H_1.

9.3 One- and Two-tailed Tests

We mentioned in the previous section that our alternative hypothesis to accompany a single valued null hypothesis could be one of three types, viz.: $\mu \neq 15$, $\mu < 15$, $\mu > 15$. If we adopt the first of these then we are allowing

Hypothesis Testing

for a two-tailed alternative, and our interest is in distinguishing between a situation where the population mean can be taken to be 15 as opposed to one where we must take the population mean to be anything about 15 without any predetermined knowledge about whether it is more than or less than 15. For instance, if we were testing an instrument for error, we might well be interested in knowing if the mean error level was 0 or not. Hence we could test the following pair of hypotheses $H_0: \mu = 0$ against $H_1: \mu \neq 0$. As we have a two-tailed alternative we call this a two-tailed test. Under our testing procedure we will reject H_0 if our sample mean $\bar{x}$ seems very large and positive compared to 0 or similarly very large and negative.

Adoption of either of the other two alternative hypotheses, e.g. $\mu < 15$ and $\mu > 15$, means that we are only allowing for a one-tailed alternative and hence we call it a one-tailed test. For instance, a company might be interested in buying new lamps for lighting the bridges of their ships to replace the old ones which past experience has shown to have a mean life of 1,400 hr. A sample of the new lamps is to be tested and hence the correct hypotheses must be set up. If we were only interested in the new lamps if they appear to have a mean life that is greater than 1,400 hr then we would set up the hypotheses; $H_0: \mu = 1,400$ against the alternative hypothesis $H_1: \mu > 1,400$. Thus we would only accept the alternative hypothesis and hence the new lamps if our sample mean was very much greater than 1,400 hr. However we might be interested in the new lamps provided only that they were not actually worse than the existing lamps. In this situation we would set up the hypotheses; $H_0: \mu = 1,400$ against the alternative hypothesis $H_1: \mu < 1,400$. Thus we would only accept the alternative hypothesis and in this case reject the new lamps if our sample mean was very much less than 1,400 hr. Which of these one-tailed alternatives was preferable would obviously depend on a variety of factors such as cost, availability, etc., but it is important that the actual hypotheses should be decided upon before the test is carried out.

9.4 Type One and Type Two Errors

To consider further the actual procedure of carrying out a hypothesis test let us consider the initial example in this chapter concerned with the breaking stresses of a particular type of rope. We have a sample of size 49 taken from a population whose mean, μ, is actually unknown but which we want to compare to a standard of 2,500 kg. The sample mean, $\bar{x}$, is known to be 2,400 kg and the sample standard deviation $s = 350$ kg. If the manufacturer's claim is false then our main interest is whether the true mean breaking stress is actually less than 2,500 kg; hence we would test the pair of hypotheses: $H_0: \mu = 2,500$, $H_1: \mu < 2,500$, giving a one-tailed test. Under H_0 we assume that the population mean is 2,500 kg and hence using the results on sampling distributions of the mean, given a large sample, from the previous chapter we

Marine Statistics

know that the sample mean, $\bar{x}$, comes from a normal distribution with mean 2,500 kg and standard deviation given by $s/\sqrt{n} = 350/\sqrt{49} = 50$ kg, i.e.

$$\bar{x} \sim N(2{,}500, 2{,}500)$$

Figure 9.1 shows the sampling distribution of the mean for this particular situation.

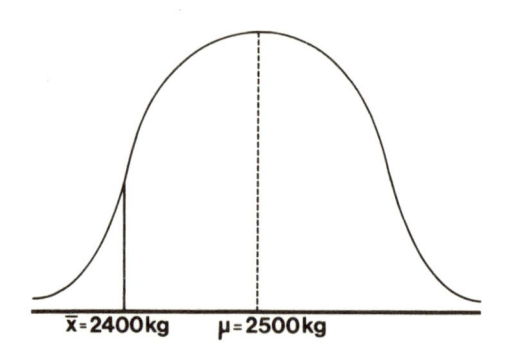

[Normal distribution curve with $\bar{x} = 2400\,kg$ marked on left and $\mu = 2500\,kg$ marked at center]

Fig. 9·1 Sampling distribution of the mean for samples of size 49 from the population whose mean given H_0 is true is 2500 kg

Our particular value of $\bar{x}$ is 2,400 kg and what we have to decide is whether this is sufficiently less than 2,500 kg to justify rejection of the null hypothesis and acceptance of the alternative hypothesis. To make this decision we set up an arbitrary criterion based on how likely it is that our sample mean belongs to the main part of the sampling distribution as defined by H_0. Thus if the sample mean lies in the extreme $100\alpha\%$ of the distribution we would decide to reject H_0 and accept H_1. The most usual values of α that are taken are 0·05, 0·01 and 0·001, but the decision as to which one depends on the subject of the test as errors of different natures are involved. In this situation we took α to be 0·05, thus stating that we would reject the null hypothesis if our value of 2,400 kg lay in the extreme left-hand 5% of the sampling distribution as shown in Fig. 9.2. This extreme 5% is known as the critical region or region of significance for the test.

The unshaded region in the diagram, showing the main 95% of the curve, is termed the acceptance region or region of non-significance for the test.

However, it must be borne in mind that in cutting off the extreme $100\alpha\%$ of the curve as defined by the null hypothesis, we are cutting off $100\alpha\%$ of values which genuinely belong to this curve. Thus if we reject H_0 when our sample value lies in this region we are running a $100\alpha\%$ chance of making the wrong decision. This is known as a Type I error, the error we commit if we reject the null hypothesis when it is true. If this were the only consideration then obviously we would go for α as small as possible, but in doing this we would

Hypothesis Testing

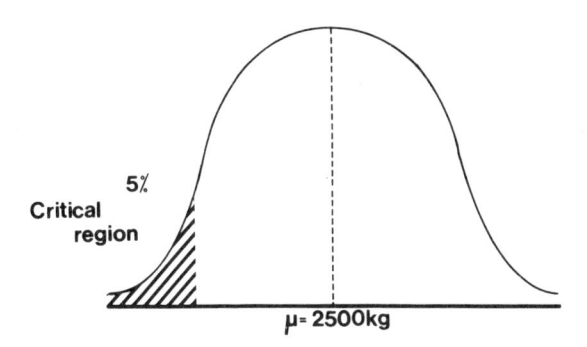

[Figure: Normal distribution curve with shaded 5% critical region on the left tail, centered at μ = 2500kg]

Fig. 9·2 The Critical region given by the extreme 5% of the sampling distribution of the mean given H_0 is true

lessen also the probability of picking out a difference which did genuinely exist. Hence if we choose too small an α, the probability of committing a Type II error is increased. A Type II error is the error that would be committed in accepting the null hypothesis when it is not true. A suitable value for α, which is also known as the level of significance of the test, may be and in fact should be predetermined before the test takes place. Beta, the probability of committing a Type II error, is never known as the true value of the unknown parameter, and so its value is never known. However we can demonstrate diagrammatically in Fig. 9.3 for a general case the effect upon β for an assumed value of the population mean for different values of α.

Suppose we were doing a hypothesis test with a null hypothesis of $\mu = \mu_0$ when in fact unbeknown to us the true value of μ is μ_1. The sampling distributions of the mean for each of these situations are shown in Fig. 9.3(a). If we set a level of significance at 5% with an alternative hypothesis of $H_1: \mu < \mu_0$ then we accept all sample means lying to the right of the critical value in Fig. 9.3(a). This means that the proportion of the curve under $\mu = \mu_1$ lying to the right of the critical value, the shaded region, is wrongly classified. Figure 9.3(b) shows the same situation but with α at 1% and it can be immediately seen that the shaded region whose size is β has increased considerably. This relationship between α and β must always be borne in mind when settling on a level for α, and as in the choice of alternative hypothesis, the decision made bearing in mind the consequences of different actions. There are likely to be costs involved in wrong decisions of either Type I or Type II but if we look only at the Type I error, we can illustrate this point as follows. If the decision concerns new lamps, the amount of money involved is probably such to justify an α of 5%, i.e. a 1 in 20 chance of switching to a new manufacturer, say, when in fact his lamps are no better than the existing ones. If however the decision concerned new engines for a fleet of ships then the amount of money involved might lead to an adoption of $\alpha = 1\%$ as a criterion, i.e. only a 1 in

Marine Statistics

(a)

μ=μ₁   μ=μ₀

Critical value when α=·05

(b)

μ=μ₁   μ=μ₀

Critical value when α=·01

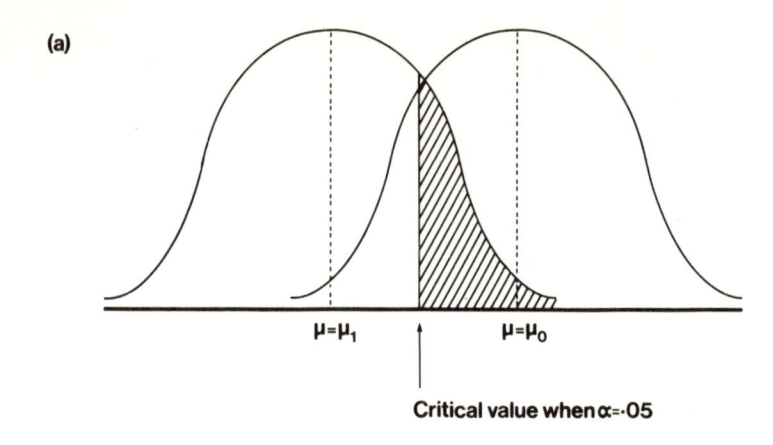

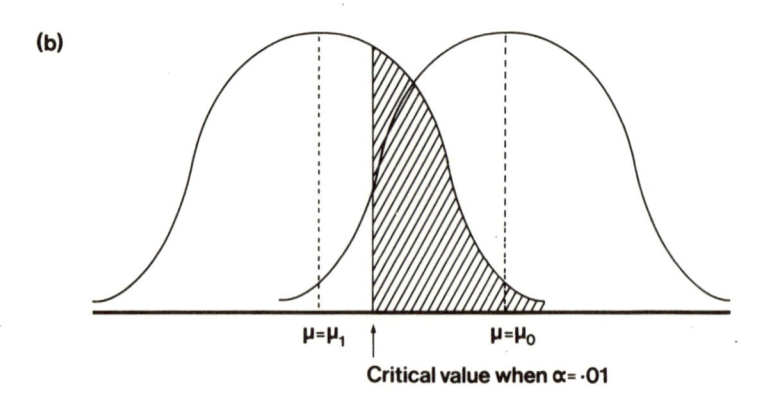

Fig 9·3 Comparison of the probabilities of committing a type II error for different levels of significance

100 chance of making the wrong decision in deciding some new engines were better than the existing ones. Even in these last sentences we have used a slightly obscure word 'better', since the definition of better might itself help to make the decision of α-level clearer. This is why we said the value of α should be predetermined before the test starts as it is bound up with the motivation for performing the test in any case.

In the discussion so far we have concentrated on where the critical region lies under a one-tailed alternative hypothesis of the form $H_1: \mu < \mu_0$. If we had chosen as alternative hypothesis the other one-tailed form $H_1: \mu > \mu_0$, then we would be looking for values of the sample mean which were significantly larger than the majority of values in the distribution. Thus our region of significance would form the upper $100\alpha\%$ of the distribution under the null hypothesis as shown in Fig. 9.4.

Hypothesis Testing

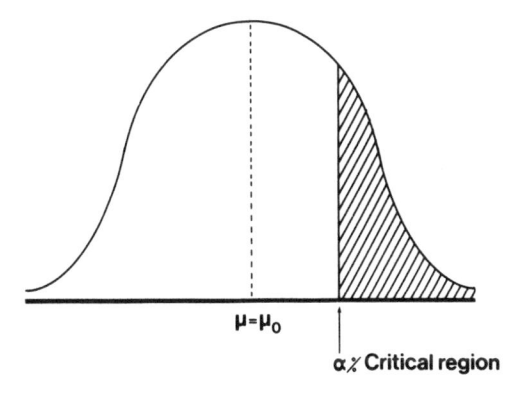

Fig 9·4 Critical region for the of alternative hypothesis $H_1: \mu > \mu_0$

If we are performing a two-tailed test then our alternative hypothesis would be of the form $H_1: \mu \neq \mu_0$ and we want to pick out values which are significantly less than μ_0 and values which are significantly more than μ_0. We thus put critical regions at both ends of the distribution under the null hypothesis but if we are working to an overall level of $100\alpha\%$ then we make the size of each one of them $100\alpha/2\%$. Figure 9.5 gives an illustration of this.

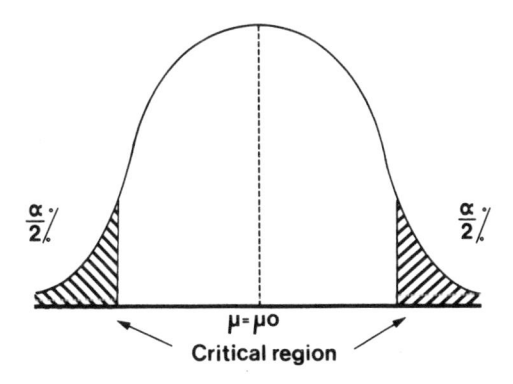

Fig. 0·5 Critical region for the test of alternative hypothesis $H_1: \mu \neq \mu_0$

There have been several important definitions in this section, so to emphasise them they are summarised now.

9.4.1 A *type one error* is the error committed if the null hypothesis is **rejected when it is true.**

9.4.2 A *type two error* is the error committed if the null hypothesis is **accepted when it is false.**

177

Marine Statistics

9.4.3 The *level of significance* of a test is the probability of committing a type one error. It is usually expressed as α.

9.4.4 The *critical region* or *region of significance* is the area lying in the extreme 100α% of the sampling distribution of the test statistic, defined under the null hypothesis. This region may be at one extreme of the curve or split between the two extremes depending on the nature of the alternative hypothesis.

9.4.5 The *acceptance region* or *region of non-significance* is the area lying in the main $100(1 - \alpha)\%$ of the sampling distribution of the test statistic defined under the null hypothesis.

9.4.6 The *critical value* is the value dividing the critical region from the acceptance region.

9.4.7 The probability of committing a type two error is usually expressed as β.

9.4.8 The *power* of a test against a single-valued alternative hypothesis is defined as $100(1 - \beta)\%$.

Since our test procedure involves determining how likely it is that our value of a test statistic belongs to the sampling distribution defined under the null hypothesis, these tests are also known as 'tests of significance.'

9.5 Steps in a Test of Significance

Returning once again to the rope manufacturing problem, we have decided firstly to test the pair of hypotheses $H_0: \mu = 2,500$, $H_1: \mu < 2,500$.

Let us take a level of significance $\alpha = 0.05$. Our sample evidence is

$$\bar{x} = 2,400 \quad n = 49 \quad s = 350$$

and

$$\bar{x} \sim N\left(2,500, \frac{350^2}{49}\right)$$

It is easier if we take as test statistic the standard normal variable, z; hence our test value, corresponding to $\bar{x} = 2,400$, is given by:

$$z = \frac{\bar{x} - \mu}{s/\sqrt{n}} = \frac{2,400 - 2,500}{350/7} = -2.0$$

From standard normal tables the value of z such that 5% of the curve lies to the left of it is $z = -1.65$, which is the critical value for this alternative hypothesis and level of significance. Thus we will reject the null hypothesis if our value of z is less than -1.65 and accept it otherwise; $-2.0 < -1.65$ so as our value lies in the critical region we say we have a significant result and reject H_0. Our conclusion is that we have no evidence to suggest that the mean

Hypothesis Testing

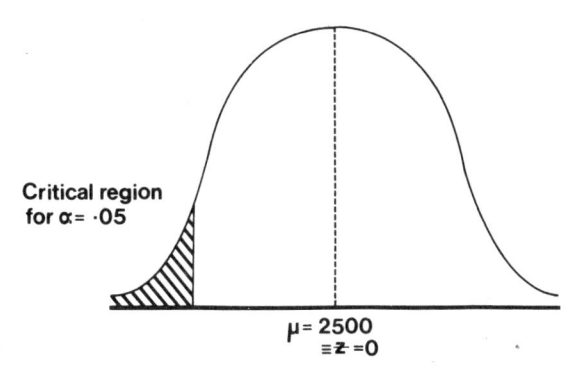

Fig. 9·6 Critical region for $\alpha = \cdot 05$ and alternative hypothesis $H_1: \mu < 2500$

breaking load of the ropes is 2,500 kg and it seems likely that it is in fact lower than this, indicating that there is doubt as to the truth of the manufacturer's claim. It is important to note that we are still cautious in our final managerial conclusion because we know that statistically we cannot *prove* anything conclusively.

We are now in a position to summarise the steps involved in carrying out a statistical significance test.

Step 1. Decide on the appropriate null hypothesis H_0 and alternative hypothesis H_1.

Step 2. Decide on the appropriate significance level.

Step 3. Knowing the relevant sampling distribution for the given sample statistic, form the appropriate test statistic.

Step 4. Determine the critical region for the test.

Step 5. Decide whether the test statistic is in the critical region or the acceptance region and hence draw the statistical conclusion as to whether to accept H_0 or not.

Step 6. Relate the statistical conclusion back to the original problem and so draw the managerial conclusion.

The remainder of this chapter is concerned mainly with considering how to apply these rules to a variety of different situations.

9.6 Test on Means

The example we have just looked at considers the situation of testing the mean when we have a large sample from a population about which nothing is known. The only difference if we had known the population standard deviation would have been to have used that when standardising our sample mean instead of using the sample standard deviation. If we only have a small sample

Marine Statistics

then we saw in the last chapter that we only know something about the sampling distribution of the mean if our original population has a normal distribution. In this situation we can use a standard normal test statistic as before if we know the population standard deviation but if not we have to use a 't' statistic. To illustrate the procedure in a situation of this type consider the second example given in the introduction to the previous chapter. We have a small sample of 20 results on passage time for a certain route under new weather routeing procedures. We know that the old mean passage time was 25·5 days and we want to know if the new mean passage time of 24 days is significantly lower than this. Therefore we take

$$H_0: \mu = 25\cdot 5 \qquad H_1: \mu < 25\cdot 5$$

and

$$\alpha = 0\cdot 05$$

It is reasonable to assume that the new passage times are still normally distributed but as the times are more variable it is probably better to use the new sample standard deviation and not assume that it has stayed constant. In fact, using a result not covered yet, this argument can be upheld statistically. Thus our test statistic is $t = [(\bar{x} - \mu)/(s/\sqrt{n})] \sim t$ on $n - 1$ d.o.f. $\bar{x} = 24\cdot 0$, $s = 2\cdot 0$, $n = 20$.

Under H_0 our value of

$$t = \frac{24 - 25\cdot 5}{2/\sqrt{20}}$$

$$t = -3\cdot 35$$

which belongs to a 't' distribution on 19 degrees of freedom. Critical region is $t < -1\cdot 73$, where $-1\cdot 73$ is the value of the t distribution on 19 degrees of freedom such that 95% of the distribution lies above it; $-3\cdot 35 < -1\cdot 73$. Hence our 't' value lies in the critical region, which implies that we have a

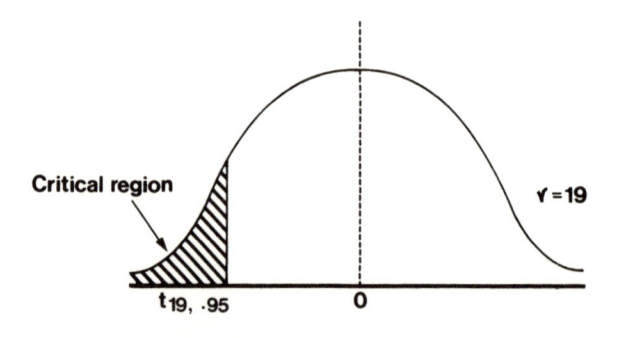

Fig. 9·7 Critical region for the test on passage times

Hypothesis Testing

significant result and therefore we accept H_1. Hence our conclusion is that we have evidence to suggest that the mean passage time under weather routeing has been reduced.

9.7 Tests on Differences Between Means

Let us consider further the problem of buying the light bulbs for all the ships in a company. A random sample of 80 light bulbs manufactured by firm A had a mean lifetime of 1,536 hr with a standard deviation of 94 hr, while a random sample of 60 light bulbs manufactured by firm B had a mean lifetime of 1,307 hr with a standard deviation of 68 hr. However the bulbs from A are so much more expensive than those from B that we have calculated that it is only worth purchasing the bulbs from firm A if we think that they will last over 200 hr more than those from firm B. We want to determine from whom the bulbs should be purchased assuming that we require a 1% level of significance.

Let μ_A and σ_A be the mean and standard deviation of the lifetime of all bulbs produced by firm A and μ_B and σ_B be the mean and standard deviation lifetime of all of the bulbs produced from firm B. Let $\bar{x}_A$ be the mean lifetime for the sample of size n_A from firm A and s_A be the standard deviation; $\bar{x}_B$, n_B and s_B are similarly defined. Then our unknown population parameter is the difference between the means $\mu_A - \mu_B$ and our corresponding sample statistic is $\bar{x}_A - \bar{x}_B$. Then

$$H_0: \mu_A - \mu_B = 200$$
$$H_1: \mu_A - \mu_B > 200$$
$$\alpha = 0.01$$

Since we have large samples from each population,

$$\bar{x}_A - \bar{x}_B \sim N\left(\mu_A - \mu_B, \frac{\sigma_A^2}{n_A} + \frac{\sigma_B^2}{n_B}\right)$$

but since we only know s_A and s_B instead of σ_A, σ_B we may use these as an approximation. Hence our standard normal test statistic is

$$z = \frac{(\bar{x}_A - \bar{x}_B) - (\mu_A - \mu_B)}{\sqrt{\frac{s_A^2}{n_A} + \frac{s_B^2}{n_B}}}$$

$\bar{x}_A = 1,536$, $\bar{x}_B = 1,307$, $s_A = 94$, $n_A = 80$, $s_B = 68$, $n_B = 60$
and $\mu_A - \mu_B = 200$ under H_0

$$\bar{x}_A - \bar{x}_B = 1,536 - 1,307 = 229$$

Thus
$$z = \frac{229 - 200}{\sqrt{\frac{94^2}{80} + \frac{68^2}{60}}} = 2.12$$

Marine Statistics

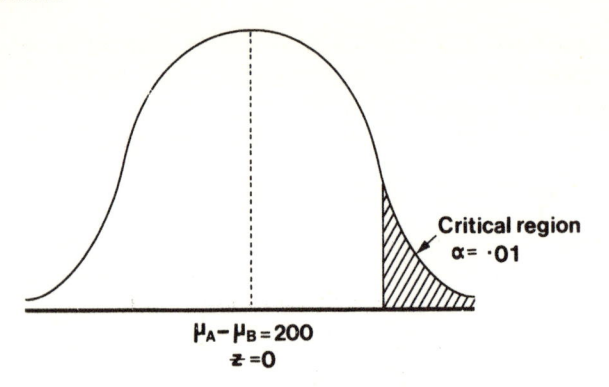

Fig 9·8 Critical region for the light bulbs test

The critical region, since we have a one-tailed test, is given by $z > 2·33$; $2·12 < 2·33$. Hence we have a non-significant result and we accept H_0. Our conclusion for the manager is that we have no evidence to suggest that the difference in lifetimes is more than 200 hr and hence it will be better policy to buy from firm *B*.

As we discussed in the previous chapter, if we have small samples from each of two populations then we only have a precise result of this nature for the sampling distribution of the difference between two means, if both of the populations have normal distributions and equal variances. Let us consider an example of this type.

From a ship in a known position one observer took a series of 20 astro observations. The displacement of his position lines from the true position are tabulated below, displacements to the eastwards being designated positive and displacements to the westwards being designated negative.

+0·1	−0·1	+0·1	+1·1	+1·8	+0·2	+2·8
+0·3	−1·9	+0·3	+1·2	−3·2	−1·3	+0·8
−0·6	+0·9	+0·2	+2·7	+0·7	+0·0	

A second observer took a second series of 20 observations and his displacements are also tabulated below.

−1·2	−0·2	−0·1	−2·7	+0·6	+1·6	+1·1
+0·6	−0·4	−0·8	−1·9	+1·7	+0·1	−1·0
−0·4	−1·4	−0·7	+2·3	−3·1	+0·1	

Is there any significant difference between the mean level of error for the two observers?

Let μ_1 and σ_1 be the population mean and standard deviation of error and $\bar{x}_1$ and s_1 be the sample mean and standard deviation based on a sample of size n_1 for the first observer. Let μ_2, σ_2, $\bar{x}_2$, s_2 and n_2 be the corresponding quantities for the second observer. Then $\bar{x}_1 = 0·31$, $\bar{x}_2 = -0·29$. We must assume that the populations are both normally distributed and have equal variance σ^2. Hence

$$\hat{\sigma} = \sqrt{\frac{1}{n_1 + n_2 - 2}\left[\sum_{\substack{\text{both} \\ \text{samples}}} x_i^2 - [\sum_{\substack{\text{both} \\ \text{samples}}} x_i]^2/n_1 + n_2\right]} = 1.44$$

$H_0: \mu_1 - \mu_2 = 0$ (this is a two-tailed test because we are not
$H_1: \mu_1 - \mu_2 \neq 0$ looking for one mean to be better than the
$\alpha = 0.05$ other in any sense)

Since we have small samples we must use a 't' statistic given by:

$$t = \frac{(\bar{x}_1 - \bar{x}_2) - (\mu_1 - \mu_2)}{\sqrt{\hat{\sigma}^2\left[\frac{1}{n_1} + \frac{1}{n_2}\right]}}$$

$\bar{x}_1 - \bar{x}_2 = 0.31 - (-0.29) = 0.60 \qquad \mu_1 - \mu_2 = 0$ under H_0
$n_1 = n_2 = 20 \qquad \hat{\sigma} = 1.44$

$$\therefore t = \frac{0.60}{1.44\sqrt{\frac{1}{20} + \frac{1}{20}}} \sim t \text{ on 38 d.o.f.}$$

$= 1.32$

Critical region is given by $t < -1.96$ and $t > +1.96$; $-1.96 < 1.32 < 1.96$. Hence we have a non-significant result and we accept H_0.

Our conclusion is that the two observers do not appear to differ significantly in their mean level of error.

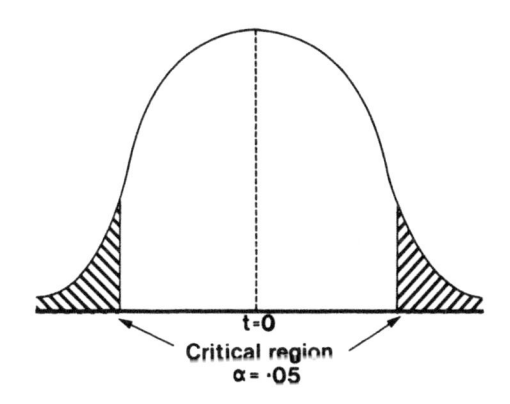

Fig. 9·9 Critical region for the astro observations error test

If we have small samples from two populations and we cannot assume equal variances then we could use a test like this to give an approximate result or we could use a non-parametric test as we do if we cannot assume that the populations are normally distributed.

Another situation which often arises in practice is where we have two sets

Marine Statistics

of results which are paired. For instance, suppose the four deck officers on a ship all decided to give up smoking together and they weighed themselves at that time, and then five weeks after they had stopped, to see if they had gained weight. The results, measured in pounds, were as follows:

	Before	After
Captain	148	154
1st officer	176	179
2nd officer	153	151
3rd officer	116	121

They wanted to see if their experience was evidence to support the claim that the result of giving up smoking is weight gain.

The difference between the two samples in this case is not that they have different numbers but only that the readings were taken at different times. Hence it is sensible to work on the variable defined by taking the difference between the two sets of readings.

Let μ_d be the mean difference in weight for all people before stopping smoking and five weeks after. Let $\bar{x}_d$ be the sample mean, s_d the sample standard deviation and n_d be the number of differences in the sample. If we assume that the weight difference has a normal distribution then we can use as test statistic

$$t = \frac{\bar{x}_d - \mu_d}{s_d/\sqrt{n_d}} \sim t \text{ on } n_d - 1 \text{ d.o.f.}$$

$H_0: \mu_d = 0$ (there is no weight difference before and after giving up smoking)

$H_1: \mu_d > 0$ (there is a weight gain after giving up smoking)

$\alpha = 0.05$

The weight differences are as follows: 6, 3, −2, 5. Hence $\bar{x}_d = 3$, $s_d = 3.56$ gives

$$t = \frac{3 - 0}{3.56/\sqrt{4}} \sim t \text{ on 3 d.o.f. under } H_0$$

$$= 1.69$$

Critical region is given by $t \geq 2.35$; $1.69 < 2.35$. Hence we have a non-significant result and we accept H_0. Thus we can conclude that on the evidence of these four officers there appears to be no support for the claim that giving up smoking can lead to a weight gain.

Hypothesis Testing

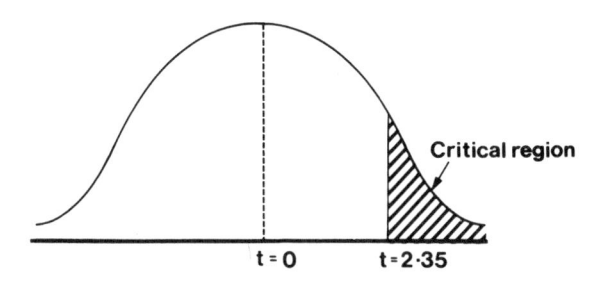

Fig. 9·10 Critical region for the smoking test

9.8 Tests on Proportions

In a particular trade it is assumed that on about 30% of the ships the cargo carried will suffer some amount of damage from absorption of water. To test if this figure is about right random checks are made on cargo from 36 ships and some water damage is found in 15 of them. Should the original assumption be modified as a result of this evidence?

Let π be the true proportion of all ships whose cargo will suffer water damage on this particular route. Let n be the sample size and let p be the corresponding sample proportion. Since we have a large sample, then, we may assume that the sampling distribution of a proportion is normal with mean $\mu_p = \pi$ and variance $\sigma_p^2 = \pi(1-\pi)/n$, i.e. $p \sim N(\pi, \pi(1-\pi)/n)$.

Hence our test statistic is

$$z = \frac{p - \pi}{\sqrt{\frac{\pi(1-\pi)}{n}}} \sim N(0, 1)$$

$H_0: \pi = 0.3$ (We take this as alternative hypothesis because we had
$H_1: \pi \neq 0.3$ no idea before the test was conducted whether the
$\alpha = 0.05$ true proportion was likely to be below or above 0·3.)

$n = 36 \qquad p = \frac{15}{36} = 0.42$

Hence under H_0

$$z = \frac{0.42 - 0.30}{\sqrt{\frac{0.30 \times 0.70}{36}}} = 1.57$$

Critical region is $z < -1.96$ and $z > 1.96$; $-1.96 < 1.57 < 1.96$. Hence our value of z lies in the acceptance region and we have a non-significant result

185

Marine Statistics

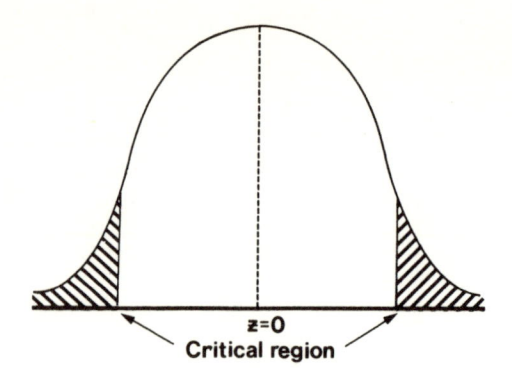

Fig. 9·11 Critical region for the cargo damage test

leading to the acceptance of H_0. Thus we have no evidence to suggest that the assumption of 30% as the rough percentage of ships whose cargo suffers some water damage should be changed.

If we have a small sample then we may work out the results exactly using a binomial distribution.

A naval gunner claims that he hits 80% of the targets he fires at. Would you agree with this claim if on a given day he hits 9 out of the 15 targets he fires at?

Again let π be the true proportion of hits that he achieves and p be the proportion in a sample of size n.

$$H_0: \pi = 0.8$$
$$H_1: \pi \neq 0.8$$
$$\alpha = 0.05$$

In this case $p = 9/15$ and $n = 15$.

Using the binomial distribution we want to know what the probability is of scoring 9 or less hits out of 15, if the probability of success per shot is 0·8. We look at the probability of scoring 9 or less hits, because we want to know how this probability compares with the value of 2·5%, which is the extreme part of the distribution in this direction we have decided to consider for significance at the 5% level on a two-tail test.

Required probability is

$$\binom{15}{0} 0.8^0 0.2^{15} + \binom{15}{1} 0.8^1 0.2^{14} \cdots \binom{15}{9} 0.8^9 0.2^6$$

$$= 0.0611$$

Hypothesis Testing

Thus there is 6·1% chance of scoring 9 or less hits if his claim is true, so we have a non-significant result and conclude that there appears to be no reason to doubt the gunner's claim.

9.9 Tests on Differences Between Proportions

These methods are appropriate for large samples only and we have to distinguish between situations where we are testing if the proportions are equal and situations where we test to see if they differ by a certain amount.

Let us consider an example of the first type of situation. The examination results of some of the navigating cadets from shipping company A showed that 37 out of 62 passed, while from shipping company B 27 out of 49 passed. Does this mean that there is likely to be a difference in the overall pass rates between the two companies?

Let π_A be the proportion passing of all cadets in company A and π_B be the proportion passing of all cadets in company B. We must assume that our results refer to a random sample of cadets from each company. Let p_A be the proportion passing out of a sample of size n_A from company A and p_B and n_B are similarly defined for company B.

$$H_0: \pi_A - \pi_B = 0 \Rightarrow \pi_A = \pi_B = \pi, \text{ say}$$
$$H_1: \pi_A \neq \pi_B$$
$$\alpha = 0.05$$

Since under H_0 we are stating that the proportions are equal, we can estimate the common proportion π treating both samples together. As additionally the sample sizes are large we may take as test statistic

$$z = \frac{p_A - p_B}{\sqrt{\hat{\pi}(1 - \hat{\pi})\left(\dfrac{1}{n_A} + \dfrac{1}{n_B}\right)}} \sim N(0, 1)$$

In this situation

$$p_A = \frac{37}{62} = 0.60 \qquad p_B = \frac{27}{49} = 0.55$$

$$\hat{\pi} = \frac{37 + 27}{62 + 49} = \frac{64}{111} = 0.58 \qquad n_A = 62 \qquad n_B = 49$$

Thus

$$z = \frac{0.60 - 0.55}{\sqrt{0.58 \times 0.42 \left[\dfrac{1}{62} + \dfrac{1}{49}\right]}} = 0.53$$

Marine Statistics

Critical region is $z < -1.96$ and $z > 1.96$; $-1.96 < 0.53 < +1.96 \Rightarrow$ we have a non-significant result. Hence, based on this sample, there seems to be no difference between the pass rates of the two companies.

The second type of situation is illustrated by the following example: In an investigation into the reasons why many qualified officers leave the sea for shore jobs, it was found that in a sample of 56 serving officers under the age of 30, 60% of them were unhappy about the long periods of time at sea, whereas in a sample of 44 serving officers over the age of 30 only 45% of them were unhappy about this fact. Do we have evidence to suggest that the true difference in proportions is more than 10%?

Let π_A be the proportion of all serving officers under 30 who were unhappy about the long period of time at sea and π_B be the corresponding proportion for those serving officers aged over 30. Let p_A be the corresponding sample proportion for the first group based on a sample of size n_A and similarly p_B and n_B can be defined.

$$H_0: \pi_A - \pi_B = 0.10$$
$$H_1: \pi_A - \pi_B > 0.10$$
$$\alpha = 0.05$$

Since we have large samples our test statistic is

$$z = \frac{(p_A - p_B) - (\pi_A - \pi_B)}{\sqrt{\frac{p_A(1-p_A)}{n_A} + \frac{p_B(1-p_B)}{n_B}}} \sim N(0, 1)$$

In this case $p_A = 0.6$, $p_B = 0.45$, $n_A = 56$, $n_B = 44$. Thus

$$z = \frac{0.6 - 0.45 - 0.10}{\sqrt{\frac{0.6 \times 0.4}{56} + \frac{0.45 \times 0.55}{44}}} = 0.51$$

Critical region is $z > 1.65$; $0.51 < 1.65 \Rightarrow$ non-significant result. Hence we have no evidence to suggest that the true difference in proportions of the two age-groups not liking the long periods of time at sea is more than 10%.

In the next chapter we will consider an alternative method of deciding whether there are significant differences between proportions from two or even more populations.

9.10 *Summary*

In this chapter we have considered some of the fundamental concepts of hypothesis testing and then their practical applications in the context of tests concerning means and proportions. In the following two chapters we will consider the extension of these ideas to cover different situations.

Hypothesis Testing

Exercises

1. Forty bags of potatoes, each of nominal weight 25 kg, were taken aboard ship as part of the stores. The mean weight of these bags was found to be 24·8 kg with a standard deviation of 1·2 kg. Is this sufficient evidence to suggest that bags from the supplier in question are generally under weight? Use a 0·05 level of significance as a basis for your answer.

2. A manufacturer claims that the burners he supplies have a mean life of 500 hr. Engine-room log-book records show that, over past years, 36 burners of this type were replaced after the following service times:

    520 648 490 725 593 827
    774 860 621 432 570 681
    701 632 586 859 910 734
    523 559 648 763 663 802
    619 654 388 474 730 623
    812 569 614 668 594 710

   Is this sufficient evidence to conclude that the mean life of the burners is different from that claimed by the manufacturer? A significance level of 0·01 is required.

3. A radio position fixing aid is used to obtain a series of eight position lines while the ship is at a known position. The position line errors for this aid are known to be normally distributed with a standard deviation of 0·6 nautical miles. The mean of the displacements of the eight position lines was found to be 0·25 miles to the eastwards of the ship's actual position. Is this sufficient evidence to suggest, at a 0·05 significance level, that the radio aid is giving position lines with a systematic bias?

4. A ballast pump has a rated capacity of 275 tonnes per hr. It was suspected that the ballast system was not capable of this throughput and, as a test, the pump was used during acceptance trials to fill ten tanks in succession with ballast. The rate for each tank was calculated as follows:

    225 234 255 246 282 252 249 265 258 271

   Is this sufficient evidence to claim, at the 1% level, that the mean performance of the pump is less than its rated capacity?

5. The mean speed of 120 ships which transit a channel during a 48-hour traffic survey in good visibility was observed to be 12·98 knots with a standard deviation of 2·45 knots. The mean speed of 85 ships which transit the same channel during a 48-hour traffic survey when visibility was less than 1 km was observed to be 12·48 knots with a standard deviation of 2·3 knots. Is this sufficient evidence to conclude, at the 5% level of significance, that the mean speed of ships which transit the channel is lower in poor visibility than in good visibility?

Marine Statistics

6. The master of a ship allocates two independent crews, A and B, to operate the accident boat. In the course of a voyage he exercises both crews in recovering a life-buoy dropped without warning. The times for 8 exercises by crew A, in minutes, are:

   10·4   12·9   11·6   15·0   13·2   11·8   12·5   14·2

   The times for 9 exercises by crew B, in minutes, are:

   10·8   11·4   9·6   10·4   12·1   8·4   9·2   10·7   9·8

   A purpose of the trials was to ascertain whether there was a significant difference between the mean times achieved by the two crews. Can the master conclude that such a difference is established at the 5% level on the basis of the trial results?

7. In order to compare the performance of navigating officers using compass stabilised and unstabilised radar displays, a series of tests was devised whereby each of 16 subjects were given exercises to resolve using each type of display on a radar simulator. The closest point of approach (CPA) to other target ships in an exercise was used as the measure of performance and the results of the tests were as tabulated below:

Subject No.	CPA achieved using stabilised display n.m.	CPA achieved using unstabilised display n.m.
1	0·8	1·1
2	1·2	0·6
3	0·8	1·1
4	0·6	0·7
5	1·1	0·0
6	0·7	1·1
7	1·0	1·0
8	0·9	0·3
9	1·7	0·7
10	0·9	0·1
11	1·6	0·8
12	0·7	1·0
13	0·4	0·6
14	0·9	0·6
15	1·2	0·7
16	0·9	0·6

Is there a difference between the CPAs achieved using the two display types at the 0·05 significance level?

8. On a particular trade, a shipping company found that the mix of containers carried by their ships was 25% 30 ft containers and 75% 20 ft containers. In the interests of economy the freight policy was adjusted to attempt to encourage even more shippers to use 20 ft containers rather than 30 ft containers. Subsequently, of 300 shipped, it was found that 60 were 30 ft containers and the remainder were 20 ft containers. Is it possible to say that the new policy had some effect using a 5% level of significance?

9. A newspaper report suggested that on a third of the ships of a certain national fleet the lifesaving equipment was inadequately maintained. To investigate this report, government surveyors carried out random spot checks on 16 ships and found that the lifesaving equipment was substandard on two of them. Is this sufficient evidence to refute the newspaper report at the 5% level?

10. The records of a shipping company showed that, out of 80 consignments of steel plates carried in their ships, there were 48 cases where a claim for rust damage was lodged. Subsequently the company policy on ventilating such cargoes was changed and, out of the following 50 consignments, there were 20 claims for rust damage. Can it be concluded, using a 0·05 level of significance, that the new policy improved the percentage of such claims by more than 5%?

Answers

1. We set up:

$$H_0: \mu = 25 \cdot 0 \text{ kg}$$
$$H_1: \mu < 25 \cdot 0 \text{ kg}$$

Level of significance, $\alpha = 0 \cdot 05$.

Since we have a large sample, we can assume that, under H_0:

$$\bar{x} \sim N\left(25 \cdot 0, \frac{1 \cdot 2^2}{40}\right)$$

We calculate:

$$z = \frac{\bar{x} - \mu}{s/\sqrt{n}}$$

$$= \frac{24 \cdot 8 - 25 \cdot 0}{1 \cdot 2/6 \cdot 32}$$

$$= \underline{-1 \cdot 05}$$

Since we are only testing for a difference in the sense that the true mean is less than 25·0 kg, we use a one-tailed test.

Marine Statistics

From the normal distribution tables we find that the critical value of z, below which 5% of the distribution lies, is: $z = 1\cdot65$.

Thus our value of $1\cdot05$ lies within the acceptance region and we have insufficient evidence to reject H_0. We accept H_0 and conclude that the supplier's bags of potatoes may well have a mean weight of 25 kg although the mean weight of the sample taken as stores was rather less.

2. We calculate the sample mean, $\bar{x}$, and the standard deviation, s:

$$n = 36 \qquad \sum x = 23{,}576 \qquad \sum x^2 = 15{,}971{,}410$$

$$\bar{x} = \frac{\sum x}{n} = \frac{23{,}576}{36} = \underline{654\cdot89}$$

$$s = \sqrt{\frac{1}{n-1}\left[\sum (x^2) - \frac{(\sum x)^2}{n}\right]} = \sqrt{\frac{1}{35}\left[15{,}971{,}410 - \frac{23{,}576^2}{36}\right]}$$
$$= \underline{123\cdot26}$$

We set up:

$$H_0: \mu = 500 \text{ hr}$$
$$H_1: \mu \neq 500 \text{ hr}$$

Level of significance, $\alpha = 0\cdot01$.

We have a large sample and thus assume that, under H_0:

$$\bar{x} \sim N\left(500, \frac{123\cdot26^2}{36}\right)$$

We calculate:

$$z = \frac{\bar{x} - \mu}{s/\sqrt{n}}$$
$$= \frac{654\cdot89 - 500}{123\cdot26/6}$$
$$= \underline{7\cdot54}$$

Since we are testing that μ is not equal to 500 without specifying direction, we use a two-tailed test. From the standard normal table we find that the critical values of z, such that only 1% of the distribution lies outside, are $\pm 2\cdot58$. Thus our observed value of $7\cdot54$ lies well in the rejection region. Hence we reject the null hypothesis H_0 and accept the alternative hypothesis H_1. We conclude that the mean life of the burners is significantly in excess of the manufacturer's claim.

Hypothesis Testing

3. $\bar{x} = 0{\cdot}25$ n.m.
   $n = 8$
   $\sigma = 0{\cdot}6$ n.m.
   We set up:

   $H_0: \mu = 0$ (i.e. the mean error is zero and there is no systematic
   $H_1: \mu \neq 0$ bias)

   Level of significance, $\alpha = 0{\cdot}05$.

   In this case, we are dealing with a small sample but from a normal population of known variance and, under H_0:

   $$\bar{x} \sim N\left(0, \frac{0{\cdot}6^2}{8}\right)$$

   $$z = \frac{\bar{x} - \mu}{\sigma/\sqrt{n}} = \frac{0{\cdot}25}{0{\cdot}6/2{\cdot}83} = \underline{1{\cdot}18}$$

   Since we are testing for a systematic bias without specifying the sense of the bias, we use a two-tailed test.

   From the normal distribution tables we find that the critical value of z, such that 5% of the distribution lies outside those limits, is 1·96.

   Thus our value of 1·18 is within the acceptance region and we conclude that we do not have sufficient evidence to reject the null hypothesis. It is quite possible that the radio aid position lines observed were drawn from a population with a zero mean and thus no systematic bias.

4. We calculate:

   $$\sum x = 2{,}537$$

   $$\bar{x} = \frac{\sum x}{n} = \frac{2{,}537}{10} = \underline{253{\cdot}7}$$

   $$\sum x^2 = 646{,}181$$

   $$s = \sqrt{\frac{1}{n-1}\left[\sum x^2 - \frac{(\sum x)^2}{n}\right]} = \sqrt{\frac{1}{9}\left[646{,}181 - \frac{2{,}537^2}{10}\right]}$$

   $$= \underline{16{\cdot}81}$$

   We set up:

   $H_0: \mu = 275$ tonnes per hr
   $H_1: \mu < 275$ tonnes per hr

   Level of significance, $\alpha = 0{\cdot}01$.

   Since this is a small sample, we must assume that the population has a normal distribution. We can then use as test statistic:

   $$t = \frac{\bar{x} - \mu}{s/\sqrt{n}} = \frac{253{\cdot}7 - 275}{16{\cdot}81/3{\cdot}16} = -4{\cdot}08 \quad \sim t \text{ on 9 d.o.f. under } H_0$$

Marine Statistics

Since we are testing for a difference in the sense that the true mean is less than 275, we use a one-tailed test.

From t distribution tables we find that the critical value of t, below which only 1% of the distribution will lie, is $-2\cdot82$ on 9 d.o.f. Hence we conclude that H_0 should be rejected and that it seems most likely that the pump does not deliver at its rated capacity.

5. Let μ_G be the mean speed in good visibility and μ_P be the mean speed in poor visibility. We set up:

$$H_0: \mu_G - \mu_P = 0$$
$$H_1: \mu_G - \mu_P > 0$$

Level of significance, $\alpha = 0\cdot05$.

Since we have large samples from each population:

$$(\bar{x}_G - \bar{x}_P) \sim N\left(\mu_G - \mu_P, \frac{\hat{\sigma}_G^2}{n_G} + \frac{\hat{\sigma}_P^2}{n_P}\right)$$

Hence:

$$z = \frac{(\bar{x}_G - \bar{x}_P) - (\mu_G - \mu_P)}{\sqrt{\frac{s_G^2}{n_G} + \frac{s_P^2}{n_P}}} = \frac{12\cdot98 - 12\cdot48}{\sqrt{\frac{2\cdot45^2}{120} + \frac{2\cdot30^2}{85}}} = \frac{0\cdot50}{0\cdot33}$$

$$= 1\cdot52 \quad \sim N(0, 1) \text{ under } H_0$$

For this example, a one-tailed test is appropriate and, consulting the standard normal distribution table, we find that $1\cdot65$ is the critical value above which only 5% of the distribution lies.

It follows that the value obtained for z lies within the acceptance region and we are not justified in rejecting H_0.

We conclude that the evidence is not sufficient to state that the mean speed of ships which transit the channel is less when visibility is under 1 km than when visibility is good.

6. For crew A, we calculate:

$$\sum x_A = 101\cdot6 \qquad \bar{x}_A = \frac{\sum x}{n} = \frac{101\cdot6}{8} = \underline{12\cdot70}$$

For crew B, we calculate:

$$\sum x_B = 92\cdot4 \qquad \bar{x}_B = \frac{\sum x}{n} = \frac{92\cdot4}{9} = \underline{10\cdot27}$$

We have to assume that the two populations are normally distributed with equal standard deviations and we estimate this common standard

Hypothesis Testing

deviation, $\hat{\sigma}$, by:

$$\bar{x}_{A+B} = \frac{101 \cdot 6 + 92 \cdot 4}{17} = \underline{11 \cdot 41}$$

$$\hat{\sigma} = \sqrt{\frac{1}{n_A + n_B - 2}\left[\sum x^2 - \frac{(\sum x)^2}{n_A + n_B}\right]}$$

$$= \sqrt{\frac{1}{15}\left[2{,}264 \cdot 56 - \frac{194^2}{17}\right]} = \underline{1 \cdot 84}$$

$H_0: \mu_A = \mu_B$
$H_1: \mu_A \neq \mu_B$
$\alpha = 0 \cdot 05$

In this case we have small samples, and we calculate:

$$t = \frac{(\bar{x}_A - \bar{x}_B) - (\mu_A - \mu_B)}{\hat{\sigma}\sqrt{\frac{1}{n_A} + \frac{1}{n_B}}} = \frac{12 \cdot 70 - 10 \cdot 27}{1 \cdot 84\sqrt{\frac{1}{8} + \frac{1}{9}}}$$

$$= \underline{2 \cdot 72}$$

Here the test is to ascertain whether there is a difference between the mean times for the two crews, irrespective of direction, and hence a two-tailed test is appropriate.

The limits for the value of t outside which only 5% of the distribution will lie are found from the tables to be $\pm 2 \cdot 13$ on 15 degrees of freedom. Hence the calculated value of t lies in the rejection region and we reject H_0 in favour of H_1. The conclusion is that there is a significant difference between the mean times taken for the two crews with crew B turning in the better times.

7. Tabulating the results:

Subject No.	CPA Stab.	CPA Unstab.	Stab. − Unstab. x	x^2
1	0·8	1·1	−0·3	0·09
2	1·2	0·6	0·6	0·36
3	0·8	1·1	−0·3	0·09
4	0·6	0·7	−0·1	0·01
5	1·1	0·0	1·1	1·21
6	0·7	1·1	−0·4	0·16
7	1·0	1·0	0·0	0·00
8	0·9	0·3	0·6	0·36
9	1·7	0·7	1·0	1·00
10	0·9	0·1	0·8	0·64
11	1·6	0·8	0·8	0·64
12	0·7	1·0	−0·3	0·09
13	0·4	0·6	−0·2	0·04
14	0·9	0·6	0·3	0·09
15	1·2	0·7	0·5	0·25
16	0·9	0·6	0·3	0·09

$\sum x = 4 \cdot 4 \quad \sum x^2 = 5 \cdot 12$

Marine Statistics

Mean difference between CPAs, $\bar{x}_d$, is thus:

$$\bar{x}_d = \frac{\sum x}{n} = \frac{4 \cdot 4}{16} = 0 \cdot 275$$

The standard deviation of the difference, s_d, is:

$$s_d = \sqrt{\frac{1}{n-1}\left[\sum x^2 - \frac{(\sum x)^2}{n}\right]} = \sqrt{\frac{1}{15}\left[5 \cdot 12 - \frac{4 \cdot 4^2}{16}\right]}$$
$$= \underline{0 \cdot 51}$$

The appropriate null hypothesis is that the population mean difference, μ_d, of such results is zero and we thus set up:

$H_0: \mu_d = 0$
$H_1: \mu_d \neq 0$

Since we are dealing with a small sample from a population with unknown variance we must assume the population of differences is normally distributed. We then calculate:

$$t = \frac{\bar{x}_d - \mu_d}{s_d/\sqrt{n}} = \frac{0 \cdot 275}{0 \cdot 51/4}$$
$$= \underline{2 \cdot 16}$$

For this example, we are simply looking for a difference without specifying direction and a two-tailed test is clearly appropriate.

From t distribution tables we find that the limits outside which 5% of the distribution falls are $\pm 2 \cdot 13$ on 15 degrees of freedom. Hence our calculated value of t falls just in the rejection region.

Rejecting H_0, we may conclude that there is a significant difference between the mean performance in terms of achieved CPAs for the two types of radar display. However as the result obtained is so near the critical value it might be sensible to gather more evidence.

8. We let π be the true proportion of 30 ft containers after the policy change.

n is the sample size, $= 300$

p is the sample proportion, $= \frac{60}{300} = 0 \cdot 20$

Since we have a large sample, we may assume that the sampling distribution of the proportion is normal with mean,

$$\mu_p = \pi$$

and variance,

$$\sigma_p^2 = \frac{\pi(1-\pi)}{n}$$

Hypothesis Testing

We set up:

$H_0: \pi = 0{\cdot}25$
$H_1: \pi \neq 0{\cdot}25$
$\alpha = 0{\cdot}05$

Hence under H_0:

$\pi = 0{\cdot}25$

and

$\sigma_p = \sqrt{\dfrac{0{\cdot}25 \times 0{\cdot}75}{300}}$

$= \underline{0{\cdot}025}$

Also:

$z = \dfrac{p - \pi}{\sigma_p} = \dfrac{0{\cdot}20 - 0{\cdot}25}{0{\cdot}025}$

$= \underline{-2{\cdot}0}$

For this example, we are only interested in a difference in proportion in one predicted direction and so a one-tailed test is appropriate.

Consulting the standard normal distribution table, we find that the critical value, below which only 5% of the distribution will fall, is $-1{\cdot}65$. The calculated value of 2 is less than this figure and hence in the rejection region.

We thus reject H_0 and conclude that it is likely that company pricing policy is having an effect on the proportion of 30 ft containers used by shippers.

9. In this example we have a small sample and therefore use the binomial distribution to test our null hypothesis. We set up:

$H_0: \pi = 0{\cdot}33$
$H_1: \pi < 0{\cdot}33$
$\alpha = 0{\cdot}05$

We calculate the probability of finding two substandard ships out of 16 if the probability of 'success' is $0{\cdot}33$.

Using the binomial distribution we calculate the probability of finding 0, 1 and 2 such ships as:

$p = \binom{16}{0}\left(\dfrac{1}{3}\right)^0 \cdot \left(\dfrac{2}{3}\right)^{16} + \binom{16}{1}\left(\dfrac{1}{3}\right)^1 \cdot \left(\dfrac{2}{3}\right)^{15} + \binom{16}{2}\left(\dfrac{1}{3}\right)^2 \cdot \left(\dfrac{2}{3}\right)^{14}$

$= 0{\cdot}0015 + 0{\cdot}0122 + 0{\cdot}0457$
$= 0{\cdot}0594$

Since the test is made to find whether the newspaper report may be refuted, we are only concerned with the probability that the proportion of ships with defective equipment is less than one third. We thus compare

Marine Statistics

the result with our significance level of 0·05 and find that we have a non-significant result and cannot therefore reject H_0. The spot checks thus give insufficient evidence to refute the newspaper report.

10. In this case we are dealing with large samples and we can use the normal approximation to the binomial distribution. We let π_A be the proportion of claims initially and π_B the proportion after the policy change. Similarly, $p_A = 48/60 = 0\cdot6$ and $p_B = 20/50 = 0\cdot4$. We set up:

$$H_0: \pi_A - \pi_B = 0\cdot05$$
$$H_1: \pi_A - \pi_B > 0\cdot05$$
$$\alpha = 0\cdot05$$

We calculate:

$$z = \frac{(p_A - p_B) - (\pi_A - \pi_B)}{\sqrt{\frac{p_A(1-p_A)}{n_A} + \frac{p_B(1-p_B)}{n_B}}} = \frac{(0\cdot6 - 0\cdot4) - (0\cdot05)}{\sqrt{\frac{0\cdot6 \times 0\cdot4}{80} + \frac{0\cdot4 \times 0\cdot6}{50}}}$$
$$= 1\cdot70$$

Since a one-tailed test is appropriate, the critical region is for $z > 1\cdot65$. Hence we reject the null hypothesis, H_0, and accept H_1.

We conclude that the new ventilation policy appears to reduce the percentage of claims by more than 5%.

Chapter 10
Tests Based on the χ^2 Distribution

10.1 Introduction

In an investigation into marine traffic flow in a certain area it was found that over a 500 hr observation period the number of ships per hour passing a lightvessel there was given by the following distribution:

Number of ships per hour passing the lightvessel	0	1	2	3	4	5	6	7	8	9	10	11 or more	Total
Number of hours	4	17	42	68	81	98	76	50	32	19	11	2	500

It is now required to know if in fact the arrivals of ships at the lightvessel follow a Poisson distribution and hence may be treated as random events or if there is some sort of bunching effect in their arrival pattern. In other words we are faced with the problem of deciding objectively whether we can use a certain distribution for a population based on some sample results. Similarly, we might want to know if we can use a normal distribution to describe the error distribution for readings from a certain instrument or whether we ought to use a double exponential distribution, say, instead or whether neither of these is appropriate. In this chapter we are concerned with tests to enable us to make this sort of decision and a variety of other types of decision as well which are all based on one particular distribution, the χ^2 distribution (pronounced chi-squared distribution).

10.2 Some Properties of the Chi-Squared Distribution

The probability density function for a χ^2 statistic has a very complicated form and as we will not need to manipulate it there is no need to quote it here. The first most useful property of χ^2 is that if we take a sample of size n from a population with a normal distribution and with variance σ^2, then if s^2 is the variance of the sample the statistic $[(n-1)s^2]/\sigma^2$ has a χ^2 distribution on $n-1$ degrees of freedom. Thus we have another situation analogous to a 't' distribution whereby the shape of the curve actually changes depending on the degrees of freedom. A typical χ^2 curve having 6 d.o.f. is shown in Fig. 10.1.

Since $\chi^2 = [(n-1)s^2]/\sigma^2$ and $n-1$, s^2 and σ^2 must all be positive

Marine Statistics

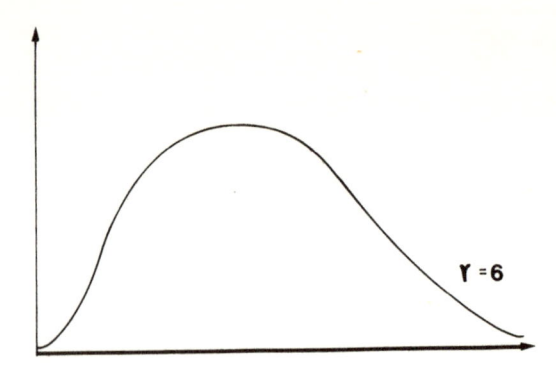

Fig. 10·1 Chi-squared distribution with six degrees of freedom

quantities then it is clear that χ^2 must always take positive values, the smallest value being 0 and the largest value being infinite. The distribution is not symmetrical, so care must be taken when using it. Tables are published showing the value of χ^2 at different percentage points and for different degrees of freedom and a set of these are given at the back of the book, Table 3. Values are given only for the most widely used percentages such as 5%, 1%, etc., since in practical work one very rarely uses any others than these.

Figure 10.2 gives a comparison of these different χ^2 curves to show how the shape can vary depending on the number of degrees of freedom ν, where ν is one less than the sample size. Obviously we cannot have a value of ν smaller than one, i.e. a sample size of two, but that intuitively makes sense as we are concerned with variances.

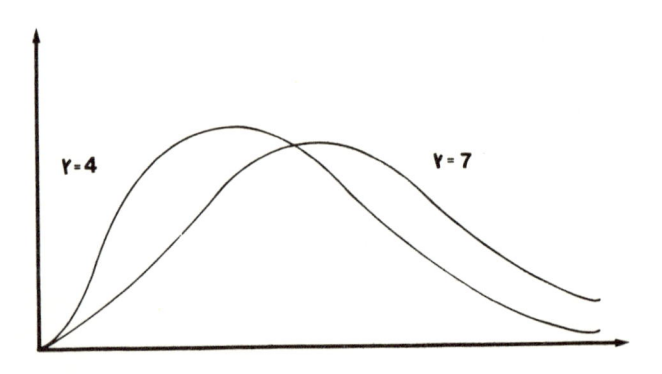

Fig. 10·2 Comparison of chi-squared distributions with different degrees of freedom

χ^2 Distribution Tests

It may be seen that as the sample size increases the curves become less asymmetrical. Hence, for large samples, if no values of χ^2 are tabulated an approximation may be used. This is that the statistic $\sqrt{2\chi^2}$ has a normal distribution with mean $\sqrt{2\nu - 1}$ and variance 1, where ν is the degrees of freedom. Hence if we calculate $z = (\sqrt{2\chi^2} - \sqrt{2\nu - 1})/1$, we have a standard normal variable. An example of this approximation is given later in the chapter.

10.3 Estimation and Tests on Unknown Population Variances

The first type of example of the many uses of the χ^2 statistic is when we want to calculate confidence limits for an unknown population variance, σ^2, or to perform a hypothesis test on it. This is only possible strictly when we are sampling from a normal distribution.

As an example of this consider the situation at the beginning of Chapter 8 on weather routeing of ships. We have a sample of 20 ships whose standard deviation for passage time was 2·0 days under weather routeing. Previously the standard deviation had been 1·0 days and when we discussed the example in Chapter 9 we wanted to know if the new sample standard deviation was significantly different from the old value.

Let σ be the new population standard deviation; then using the usual notation, $n = 20$ and $s = 2·0$:

$$H_0: \sigma = 1·0$$
$$H_1: \sigma \neq 1·0$$
$$\alpha = 0·05$$

We assumed the population was normally distributed; hence the test statistic is

$$\chi^2 = \frac{(n-1)s^2}{\sigma^2} \sim \chi^2 \text{ on 19 d.o.f.}$$

$$= \frac{19·4}{1} = 76·00$$

Since it is a two-tailed test our critical region is in two parts:

$$\chi^2 < \chi^2_{19, 0·975} \quad \text{i.e. } \chi^2 < 8·91$$

and

$$\chi^2 > \chi^2_{19, 0·025} \quad \text{i.e. } \chi^2 > 32·85$$

$76·00 > 32·85$, so we have a significant result and we reject H_0. Hence we conclude that the new population standard deviation is different from the

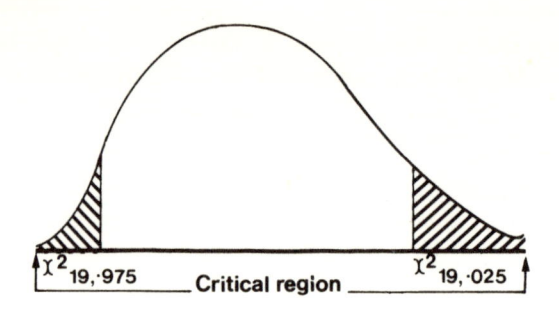

Fig. 10·3 Critical region for the test on the standard deviation of passage time under weather routeing

original one. If we had wanted confidence limits for the new population variance we would have calculated them at the 95% level as follows:

$$8\cdot 91 \leqslant \frac{19 \times s^2}{\sigma^2} \leqslant 32\cdot 85$$

where 8·91 and 32·85 cut off the extreme 2½% of the distribution on 19 d.o.f. in either direction.

From the first inequality we get $\sigma^2 \leqslant 76/8\cdot 91$, i.e. $\sigma^2 \leqslant 8\cdot 53$, and from the second we get $\sigma^2 \geqslant 76/32\cdot 85$, i.e. $\sigma^2 \geqslant 2\cdot 31$. Thus $2\cdot 31 \leqslant \sigma^2 \leqslant 8\cdot 53$ are 95% confidence limits for the new variance and by taking the square roots $1\cdot 52 < \sigma < 2\cdot 92$.

10.4 *Goodness of Fit Tests*

Apart from the connection with the variance of a population and sample, the principal χ^2 tests are concerned with testing frequencies rather than parameters in a distribution. The first of these uses which we will describe, where frequencies are concerned, is in testing if an observed distribution fits an expected theoretical distribution. This is known as a *goodness of fit test*.

As an example of its use consider the following situation. The mix of containers for a particular trade consists of 50% 20 ft containers, 20% 30 ft containers, and 30% 40 ft containers.

During the course of 10 voyages, a ship carried a total of 500 containers of which 300 were 20 ft, 70 were 30 ft and 130 were 40 ft. Is this sufficient evidence to suggest that for some definite reason the cargo delivered to the ship has a different mix to that of the norm for the trade?

In this situation, our null hypothesis H_0 would be that the mix of containers on all voyages on the one ship is the same as the overall mix for the trade and our alternative hypothesis H_1 would be that the mixes were different.

χ^2 Distribution Tests

To test this pair of hypotheses we in fact test a null hypothesis that the distribution of container size for the one ship, as suggested by the sample results, fits the theoretical distribution for the trade against an alternative hypothesis that the observed distribution is a bad fit to the theoretical distribution.

For the different sizes of container we have our observed distribution as follows:

Size of container	20 ft	30 ft	40 ft	Total
Observed distribution	300	70	130	500

In this distribution we have three cells in which the frequencies can be put and we let O_i be the frequency in the ith cell of the observed distribution. If we were to have 500 containers under the theoretical distribution then our expected frequencies in each cell would be as follows:

Size of container	20 ft	30 ft	40 ft	Total
Expected distribution	250	100	150	500

We let E_i be the frequency in the ith cell of the expected distribution. Our test statistic is given by

$$\chi^2 = \sum_i \frac{[O_i - E_i]^2}{E_i}$$

In words this means that we take the difference between the observed and the expected frequency in each cell, square it to avoid all the differences cancelling out and then express it as a fraction of the expected frequency for the cell. With this step we standardise the size of difference for the size of frequency, because obviously a difference of 10 when we only expect 20 in the cell is rather different than a difference of 10 when we expect 200. We then total these values over all the cells. Under the null hypothesis our test value of χ^2 should belong to a χ^2 distribution on ν degrees of freedom. The number of degrees of freedom is determined by the number of parameters the observed and expected distribution have in common. It is equal to the number of cells minus 1 minus the number of parameters in common. The 1 is subtracted because in all cases the two distributions have the same total, so there are only in fact $n - 1$ independent cells to be filled where n is the apparent number. In this example our degrees of freedom would be given by $\nu = 3 - 1 = 2$. Under the null and alternative hypotheses which we are using, we will say that we have a bad fit if our value of χ^2 lies in the upper extreme $100\alpha\%$ of the distribution since we would then have a very large value of χ^2 implying a large difference between the observed and expected frequencies. Goodness of fit tests are usually one-tailed tests of this sort and perhaps ought to be named more aptly badness of fit tests. Very occasionally we suspect an observed distribution as being too good a fit to an expected distribution and then we would look at the lower $100\alpha\%$ of the χ^2 distribution

Marine Statistics

and do a one-tailed test in this direction. However this occurs very seldom and on most occasions we do the one-tailed test previously described.

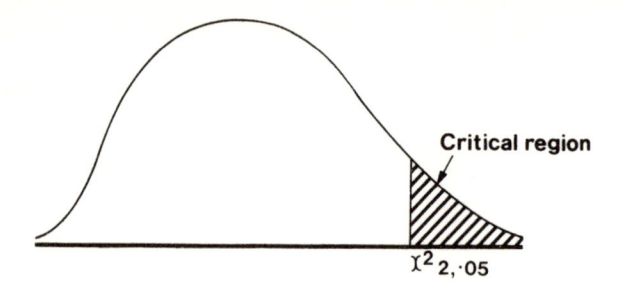

Fig. 10·4 Critical region for the container mix goodness of fit test

Continuing with the example let us set a 5% significance level. Then our critical region is given us by $\chi^2 > \chi^2_{2, 5\%}$, i.e. $\chi^2 > 5.99$. Our test value of χ^2 is given by:

$$\chi^2 = \frac{(300-250)^2}{250} + \frac{(70-100)^2}{100} + \frac{(130-150)^2}{150}$$
$$= 1.00 + 9.00 + 2.67$$
$$= 12.67$$

$12.67 > 5.99 \Rightarrow$ significant result $\Rightarrow$ reject H_0

Our conclusion therefore is that the container mix on the one ship does seem to be different to the mix for the whole trade. Looking at the differences, there are fewer medium and large sized containers than would be expected and hence more small ones.

The goodness of fit test may be used as an alternative to a test on a proportion using the normal distribution. For instance, in an investigation into collisions at sea it was found that for collisions between ships, both over 1,000 grt, when no report of restricted visibility was made there were 178 collisions in darkness and 80 collisions in daylight. If we take $H_0: p$, probability of a collision occurring in darkness under these circumstances $= \frac{1}{2}$, and

$$H_1 : p \neq \frac{1}{2} \quad \text{with} \quad \alpha = 0.05$$

	Darkness	Daylight	Total
Then our observed frequencies are:	178	80	258
and our expected frequencies under H_0 are:	129	129	258

Hence

$$\chi^2 = \sum_{i=1}^{2} \frac{(O_i - E_i)^2}{E_i} = \frac{(178 - 129)^2}{129} + \frac{(80 - 129)^2}{129}$$
$$= 37{\cdot}22$$

Our critical region is $\chi^2 > \chi^2_{1, 5\%} = 3{\cdot}84$ since the degrees of freedom are $\nu = 2 - 1 = 1$. It is also important to note that although our actual alternative hypothesis is two-tailed, the goodness of fit test still demands a one-tailed test since either extreme for p implies a bad fit to the expected distribution:

$$37{\cdot}22 > 3{\cdot}84 \Rightarrow \text{significant result} \Rightarrow \text{reject } H_0$$

Hence our conclusion is that there is strong evidence to suggest a difference between the number of collisions occurring in darkness and daylight under the specified conditions. By inspection it may be further concluded that there are significantly more collisions in darkness than in daylight when there is no restricted visibility reported and both ships are large.

As a theoretical sideline the reason why this test is equivalent to a test of a proportion using the normal distribution is that the χ^2 statistic on 1 degree of freedom is the square of a standard normal statistic. Thus, for instance, the critical value of $3{\cdot}84$ here $= (1{\cdot}96)^2$, the critical value for the two-tailed test we would otherwise have performed.

A third instance of the use of the goodness of fit test arises when we have to test if the observed distribution follows a given distribution but for which the parameters are unknown. Let us consider the opening example on the flow of ships past a lightvessel. To fit a Poisson distribution to these data we must first know the mean rate per hour, as this is the essential parameter for a Poisson distribution.

The mean of the data $\bar{x}$ is given by:

$$\bar{x} = \frac{\sum_i f_i x_i}{\sum_i f_i} = 4{\cdot}97$$

We take H_0: the data has a Poisson distribution and H_1: the data does not have a Poisson distribution, with $\alpha = 0{\cdot}05$. Then the probability function for a Poisson distribution is:

$$p(x) = \frac{e^{-\lambda} \lambda^x}{x!}$$

where λ is the mean rate of occurrence. If we let $\hat{\lambda} = \bar{x} = 5{\cdot}0$ then $p(0) = e^{-5{\cdot}0}$ and subsequent probabilities may be calculated by the recurrence relationship $p(x + 1) = p(x)\lambda/(x + 1)$. Thus in 500 hr the expected frequency for $x = 0$ is $500p(0)$ and hence subsequent expected frequencies may be calculated. The following results are obtained:

Marine Statistics

x	0	1	2	3	4	5	6	7	8	9	10	11 or more	Total
Observed frequency O_i	4	17	42	68	81	98	76	50	32	19	11	2	500
Expected frequency E_i	3	17	42	70	88	88	73	52	33	18	9	7	500

Before proceeding with the calculation of the χ^2 statistic it may be necessary to combine adjacent cells if there are expected frequencies of less than 5. This is because small expected frequencies can lead to misleadingly large contributions to the χ^2 statistic. In this case we have to combine cells at the lower extreme of the distribution, thus reducing our number of cells to 11.

Our degrees of freedom are given by $\nu = 11 - 1 - 1 = 9$, since we lose an additional one because both the observed and expected distributions are constrained to have the same mean.

Our value of χ^2 is given by:

$$\chi^2 = \sum_i \frac{(O_i - E_i)^2}{E_i} = 0.05 + 0.00 + 0.06 + 0.56 + 1.14 + 0.12 \\ + 0.08 + 0.03 + 0.06 + 0.44 + 3.57 \\ = 6.11$$

The critical region is given by $\chi^2 > \chi^2_{9, 5\%} = 16.92$; $6.11 < 16.92 \Rightarrow$ non-significant result $\Rightarrow$ accept H_0. Hence our conclusion is that it is reasonable to assume that the distribution of the number of ships per hour passing the lightvessel follows a Poisson distribution.

As a final example on this type of test let us consider a continuous distribution and for this we will take the example in Chapter 7 where it was thought that an observed distribution might be an exponential distribution. Readers who have no knowledge of calculus can skip this example and go onto the next section.

It was observed that the number of minutes that a ship was late in 30 trips had the following distribution:

	Frequency
Under 10 min late	9
10 but under 20 min late	7
20 but under 30 min late	4
30 but under 40 min late	3
40 but under 50 min late	3
50 but under 60 min late	2
60 min late and over	2
Total	30

χ^2 Distribution Tests

The mean of the distribution is 24 min and hence it was suggested that the exponential distribution with probability density function $f(x) = 1/24\ e^{-x/24}$ would best describe the distribution. We need to know the expected number occurring in each 10 min interval and this may be found by integrating the probability density function between the limits defined for each class and then multiplying by 30. In general,

$$\int_a^b \frac{1}{24} e^{-x/24}\ dx = [e^{-a/24} - e^{-b/24}]$$

Hence the expected frequencies are as follows:

Under 10 min late	$30(1 - e^{-10/24}) = 10$
10 but under 20 min late	$30(e^{-10/24} - e^{-20/24}) = 7$
20 but under 30 min late	$30(e^{-20/24} - e^{-30/24}) = 5$
30 but under 40 min late	$30(e^{-30/24} - e^{-40/24}) = 3$
40 but under 50 min late	$30(e^{-40/24} - e^{-50/24}) = 2$
50 but under 60 min late	$30(e^{-50/24} - e^{-60/24}) = 1$
60 min late and over	$30\ e^{-60/24} = 2$

Combining the last four classes to get an expected frequency of at least 5 gives the following comparison between the observed and expected frequencies:

Observed frequency: O_i	Expected frequency: E_i	$(O_i - E_i)^2 / E_i$
9	10	0·1
7	7	0·0
4	5	0·2
10	8	0·5
		0·8

Hence the calculated value of χ^2 is 0·8.

The number of degrees of freedom is $2(4 - 1(\text{total}) - 1(\text{mean}))$. Performing the test at the 5% level of significance we have a critical value of 5·99, which since $0·8 < 5·99$ gives a non-significant result. Thus we may conclude that the exponential distribution does provide a good model for this data set.

Steps to Performing a Goodness of Fit Test

1. Calculate the expected frequencies under the null hypothesis.
2. Combine adjacent cells to make all expected frequencies greater than five.
3. Calculate $\chi^2 = \Sigma_{\text{cells}} [(O_i - E_i)^2]/E_i$, where O_i is the observed frequency in the ith cell and E_i is the expected frequency.

Marine Statistics

4. Number of degrees of freedom, v = number of cells after combining minus 1 minus number of parameters in common.

5. The critical region is hence defined by $\chi^2 > \chi^2_{v, 100\alpha\%}$, where α is the predetermined level of significance.

10.5 Contingency Tables

Another use of the χ^2 statistic is in determining whether two classification variables may be considered as independent. Suppose we take a random sample of cadets from each of three shipping companies and classify them according to whether they pass or fail their first examination at the first attempt. The results are shown in Table 10.1, which is known as a contingency table.

Table 10.1. Analysis of results of cadets in their first attempt at an examination by shipping company

Company	A	B	C	Total
Pass	25	37	29	91
Fail	50	25	20	95
Total	75	62	49	186

The question of interest is whether the pass rate differs significantly from company to company or, put alternatively, is the success of a cadet in the examination at the first attempt independent of his company? To answer this question we set up the following pair of hypotheses:

H_0: The success of a cadet in the examination at the first attempt is independent of company

H_1: The success of a cadet in the examination at the first attempt is not independent of company

The null and alternative hypotheses for a test on a contingency table are always of this form with the null hypothesis stating that the new classification is independent of the column classification and the alternative hypotheses stating no independence. We are now in a position to calculate the frequencies we would expect in each cell of the table under the null hypothesis:

Let P be the event a cadet passes and F be the event he fails

Let A, B and C be the events that a cadet belongs to shipping company A, B and C respectively

χ^2 Distribution Tests

Then the probability that a cadet belongs to company A and passes is: $p(A \cap P) = p(A) \times p(P)$ under the null hypothesis using the multiplication law of probability for independent events. But

$$p(A) = \frac{\text{number of cadets in company } A}{\text{total number}} = \frac{75}{186}$$

Similarly

$$p(P) = \frac{\text{number passing}}{\text{total number}} = \frac{91}{186}$$

Thus under H_0:

$$p(A \cap P) = \frac{75}{186} \times \frac{91}{186}$$

Hence the expected number in the cell represented by $A \cap P$ if N is total number:

$$N \times p(A \cap P) = n(A \cap P) = 186 \times \frac{75}{186} \times \frac{91}{186} = \frac{75 \times 91}{186} = 37$$

Using a similar argument for the other cells, since $p(B) = 62/186$, $p(C) = 49/186$ and $p(F) = 95/186$:

$$n(A \cap F) = N \times p(A \cap F) = \frac{75 \times 95}{186} = 38$$

$$n(B \cap P) = N \times p(B \cap P) = \frac{62 \times 91}{186} = 30$$

$$n(B \cap F) = N \times p(B \cap F) = \frac{62 \times 95}{186} = 32$$

$$n(C \cap P) = N \times p(C \cap P) = \frac{49 \times 91}{186} = 24$$

$$n(C \cap F) = N \times p(C \cap F) = \frac{49 \times 95}{186} = 25$$

In each case we are multiplying the appropriate row total by the column total and then dividing by the overall total to get the expected number in a cell. Hence our original table of observed frequencies can now be rewritten showing the expected frequencies for each cell in brackets, which is given in Table 10.2.

Table 10.2. Observed and expected frequencies

Company	A	B	C	Total
Pass	25 (37)	37 (30)	29 (24)	91
Fail	50 (38)	25 (32)	20 (25)	95
Total	75	62	49	186

Marine Statistics

Obviously the row and column totals, known as the marginal totals, for each category are preserved with the expected frequencies. We now calculate a χ^2 statistic as for a goodness of fit test to measure the disparity between the observed and expected frequencies. Thus

$$\chi^2 = \sum_{\text{cells}} \frac{(O_i - E_i)^2}{E_i}$$

where O_i is the observed frequency in the ith cell, E_i is the expected frequency in the ith cell, and the summation for i is over all cells. Thus

$$\chi^2 = \frac{(25-37)^2}{37} + \frac{(50-38)^2}{38} + \frac{(37-30)^2}{30} + \frac{(25-32)^2}{32}$$

$$+ \frac{(29-24)^2}{24} + \frac{(20-25)^2}{25}$$

$$= 3 \cdot 89 + 3 \cdot 79 + 1 \cdot 63 + 1 \cdot 53 + 1 \cdot 04 + 1 \cdot 00$$

$$= 12 \cdot 88$$

We have to compare this value with a value from the χ^2 distribution on ν degrees of freedom where ν is the number of independent choices we had for filling the expected frequency cells. In this situation we had two independent choices, say for $n(A \cap P)$ and $n(B \cap P)$, since once these are fixed the other cells are fixed as the marginal totals remain constant.

As a general rule, if a contingency table has r rows and c columns then the number of degrees of freedom $\nu = (r-1) \times (c-1)$. In this case $r = 2$ and $c = 3$, hence $\nu = 1 \times 2 = 2$. We perform a one-tailed test for a large discrepancy between observed and expected frequencies and if we opt for a 5% level of significance, our critical region is defined in this case by $\chi^2 > \chi^2_{2,\,5\%}$. Thus numerically our critical region is $\chi^2 > 5 \cdot 99$; $12 \cdot 88 > 5 \cdot 99 \Rightarrow$ significant result $\Rightarrow$ reject H_0. Hence our statistical conclusion is that the success of a cadet in his first examination at the first attempt does not seem to be inde-

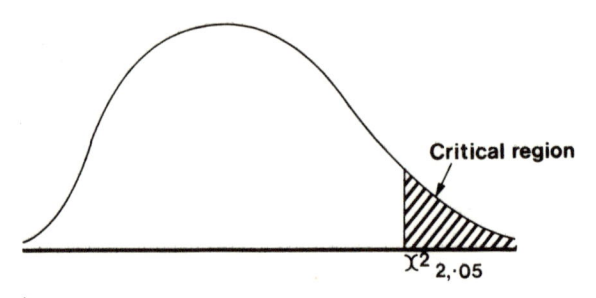

Fig. 10·5 Critical region for the cadet examination problem

pendent of the company to which he belongs. If we examine the data we can see that the main discrepancy lies in company *A* where the pass rate is much lower than in the other two companies. It is now up to company *A* or the person originally conducting the survey, if they want to, to investigate further to see if any explanation can be found. However, the main purpose of this type of analysis is to see whether further investigation might be warranted on the results of the test. It cannot show us any causal relationships but shows us where association might lie. It is also very useful as it can be used on the simplest form of qualitative classification data and does not rely on extensive measurements being taken.

Returning to this particular example, many readers will have realised that this one is really only an extension to three companies of one we considered in the previous chapter when we were looking at a test based on the normal distribution for the difference between two proportions. In fact contingency tables can be used to analyse the difference between the proportion of a sample having a given attribute from two or more populations.

Although the χ^2 distribution is a continuous distribution, we build up our test value by taking a series of discrete values, one from each cell. However the approximation works very well especially with a large number of degrees of freedom. If we only have one degree of freedom, which occurs in a contingency table with only 2 rows and 2 columns, and if in addition our expected frequencies are small, then it is quite usual to employ a correction known as Yates' correction for continuity. Our new χ^2 statistic is calculated as

$$\chi^2 = \sum_{\text{cells}} \frac{(|O_i - E_i| - \tfrac{1}{2})^2}{E_i}$$

In other words, once the difference between the observed and the expected frequencies has been calculated, the sign is ignored and 0·5 subtracted from the size of the difference before squaring it. The rest of the test proceeds as before. Since this procedure is only an approximation there is a lot of differing advice about how much it should be used. The recommendation here is that it need only be bothered with if the expected frequencies are between 5 and 10 and if there is only one degree of freedom. If the expected frequencies with one degree of freedom are less than 5 then a special test known as the Fisher–Irwin exact test should be used. This involves a certain amount of computation so details will not be given in this book. One of the main advantages of using contingency tables for testing the difference between two proportions is that, compared to the test based on the normal distribution, they may be used for very much smaller sample sizes, say a minimum of 10 from each of the two populations.

As an example of the use of the continuity correction consider the following situation. In a study to determine whether there is any relationship between temperament and speed of professional advancement for ship engineers, a random sample of 35 men is taken and classified as follows:

Marine Statistics

	Speed of advancement		
	Slow	Fast	Total
Good self control	5	14	19
Poor self control	10	6	16
Total	15	20	35

H_0: Temperament and speed of professional advancement are independent
H_1: Temperament and speed of professional advancement are dependent
$\alpha = 0.05$

Expected frequencies under H_0: Let G be those having good self-control, P be those having poor self-control, S be those having slow advancement, and F be those having fast advancement.

$$n(G \cap S) = \frac{19 \times 15}{35} = 8 \qquad n(G \cap F) = \frac{19 \times 20}{35} = 11$$

$$n(P \cap S) = \frac{16 \times 15}{35} = 7 \qquad n(P \cap F) = \frac{16 \times 20}{35} = 9$$

Hence

$$\chi^2 = \frac{(|5-8| - \tfrac{1}{2})^2}{8} + \frac{(|14-11| - \tfrac{1}{2})^2}{11}$$
$$+ \frac{(|10-7| - \tfrac{1}{2})^2}{7} + \frac{(|6-9| - \tfrac{1}{2})^2}{9}$$
$$= 0.78 + 0.57 + 0.89 + 0.69$$
$$= 2.93$$

(The Yates' continuity correction is used as there is only one expected frequency greater than 10 and that is only 11.) Since there are 2 rows and 2 columns, the degrees of freedom are 1. Critical region is $\chi^2 > \chi^2_{1,\,5\%} = 3.84$; $2.93 < 3.84 \Rightarrow$ non-significant result $\Rightarrow$ accept H_0. Hence we conclude that we have no evidence to suggest that temperament and speed of professional advancement are related for ships' engineers.

If we had not employed the continuity correction we would have had a value of 4.24 and we would have made the opposite conclusion. However with a small sample size it seems better to conclude that the evidence does not suggest an association if the result is marginal as in this case.

10.5.1 If a sample is classified into r categories on one variable and c categories on another then the resulting table with r rows and c columns is known as an $r \times c$ **contingency table**.

χ^2 Distribution Tests

Steps to Analysing an $r \times c$ Contingency Table

1. Set up H_0: row categories are independent of column categories and
   H_1: row categories are dependent on column categories.
2. Calculate expected frequencies using the relationship that for each cell the expected frequency is

$$\frac{\text{the row total} \times \text{the column total}}{\text{the overall total}}$$

3. Calculate $\chi^2 = \Sigma_{\text{cells}} [(O_i - E_i)^2]/E_i$, where O_i is the observed frequency in the ith cell and E_i is the expected frequency. Apply Yates' correction if necessary.
4. Number of degrees of freedom $\nu = (r-1)(c-1)$.
5. The critical region is hence defined by $\chi^2 > \chi^2_{\nu, 100\alpha\%}$, where α is the predetermined level of significance.

10.6 Tests for a Poisson Distribution

As a final section it is worth mentioning briefly that there is an alternative test which may be used instead of the goodness of fit test if one thinks a distribution might follow a Poisson distribution. It is a more powerful test than the goodness of fit test because it is based on a χ^2 statistic with a large number of degrees of freedom. The test makes use of the fact that in a Poisson distribution the mean and variance are equal. Hence we can estimate the population variance by the sample mean and the statistic $\chi^2 = [(n-1)s^2]/\bar{x}$ may be compared to a χ^2 distribution on $n-1$ degrees of freedom (s^2 is the sample variance, $\bar{x}$ the sample mean and n the sample size).

If we take as our example the initial set of data based on 500 hr of observations of ships passing a lightvessel then we proceed as follows:

H_0: the distribution is Poisson with mean λ
H_1: the distribution is not Poisson with mean λ
$\alpha = 0.05$

We calculate:

$$\bar{x} = \hat{\lambda} = 5.0 \qquad s^2 = 4.6 \qquad n = 500$$

Hence
$$\chi^2 = 466$$

Critical region is $\chi^2 > \chi^2_{499, 5\%}$.

As our degrees of freedom are so large we have to use the approximation to the χ^2 distribution discussed at the start of the chapter. Thus we calculate $z = (\sqrt{2\chi^2} - \sqrt{2\nu - 1})/1$ and compare this to a standard normal distribution:

$$z = 30.52 - 31.58 = -1.06$$

Marine Statistics

Our critical region is $-1.96 < z$ and $z > 1.96$. Hence as $-1.96 < -1.06 < 1.96$ we have a non-significant result and we conclude that our distribution fits a Poisson distribution.

10.7 Summary

In this chapter we have considered some of the most useful tests in statistical inference; those based on the χ^2 distribution for dealing with frequencies. In the next chapter we will consider another important group of tests, known as non-parametric tests.

Exercises

1. The mean speed of ships transitting a channel in good visibility was observed to be 12·98 knots, with a standard deviation of 2·45 knots. The sample taken was so large that we assume that this is a reasonable estimate of the population standard deviation. The mean speed of 85 ships transitting the same channel during a 48 hr traffic survey, when visibility was less than 1 km, was observed to be 12·48 knots with a standard deviation of 2·30 knots. Is it reasonable, at 0·05 level of significance, to assume that the new population standard deviation is the same as the old one? What are the confidence limits for an estimate of the standard deviation of the speed in restricted visibility?

2. Using the data given in Ex. 6 of Chapter 9, test whether it is reasonable to assume that each of the two sets of times are samples drawn from populations which each have a standard deviation equal to 1·84 min.

3. In 60 measurements of miles steamed, by a ship's log, it was found that the reading was too high on 45 occasions and too low on 15 occasions. Is this sufficient evidence to conclude that the log has a systematic bias at the 1% level?

4. A consignment of scrap tyres is loaded, and these are arbitrarily classed as small, medium and large. The cargo description indicates that $\frac{1}{4}$ of the consignment consists of small tyres, $\frac{1}{2}$ consists of medium-sized tyres and the remaining $\frac{1}{4}$ of large tyres. During the course of loading, 40 tyres were chosen at random and classed according to the following table:

	Number of tyres
Small	14
Medium	15
Large	11
Total	40

χ^2 Distribution Tests

Is this sufficient evidence to suggest, at the 5% level, that the proportions of the three classes, as loaded, are different from those specified?

5. A shipping company investigated past records to find that the number of defective cases found in 120 similar consignments of cargo were as follows:

Number of defective cases	Recorded frequency
0	24
1	38
2	28
3	16
4	10
5	2
6	1
7	0
8	1
> 8	0
Total	120

It is suggested that the Poisson distribution gives a reasonable approximation to the frequency with which different numbers of defective cases are found in consignments of this type. Use a goodness of fit test to assess whether it is valid to make this assumption at the 5% level.

6. In a ten-day traffic survey the number of ships crossing the survey line during each hour were observed with the following frequencies:

Number of ships per hr	Observed frequency
0	48
1	80
2	55
3	30
4	19
5	6
6	1
> 6	1

Use a goodness of fit test to decide whether the Poisson distribution gives a reasonable approximation to the distribution of the hourly numbers of ships crossing the survey line.

Marine Statistics

7. (Note: This question should only be tackled by readers who are familiar with calculus.)

    In order to investigate the performance of an electro-mechanical component, engine-room records for a number of ships were examined and the intervals between failures of the component were noted as follows:

Interval between failures in thousands of hours	Frequency of occurrence
0 but under 1	20
1 but under 2	14
2 but under 3	10
3 but under 4	6
4 but under 5	4
5 but under 6	3
6 but under 7	2
7 but under 8	1
8 and over	0

    It is required to test whether the exponential distribution provides a good fit for these observed frequencies.

8. Marine traffic surveys were conducted over a 48 hr period to establish the type of traffic flowing along two adjacent routes A and B. Three classes of ship were identified according to size in gross registered tons and the results were as follows:

	< 500 grt	500–9,999 grt	⩾ 10,000 grt
Route A	40	61	49
Route B	43	29	28

    In the face of this evidence, is it reasonable to believe that the traffic mix is similar for each route?

9. In collision avoidance tests in a radar simulator the actions of the subjects were categorised as: turn right (R), turn left (L), increase speed (I), or decrease speed (D). The performance of 50 navigators with 5 or more years experience and 60 navigators with less than 5 years experience was noted and the results are summarised in the table below:

Experience	R	L	I	D
⩾ 5 yr	20	14	8	8
< 5 yr	38	6	3	13

    Is this sufficient evidence to suggest, at the 5% level, that the action categories are distributed differently for the two levels of experience?

χ^2 Distribution Tests

10. To test the efficiency of two look-out stations, the look-out men were stationed on the forecastle for two nights and on the monkey island for two nights. Whilst on the forecastle, out of a total of 18 lights seen, 6 were reported by the look-out men before they were seen by the officer of the watch on the bridge. Whilst on the monkey island, out of 15 lights seen, 11 were reported by the look-out men before they were seen by the officer of the watch. Does this information indicate a significant difference (at the 0·05 level) between the efficiency of the two look-out stations?

11. Test whether the data given in Ex. 5 can reasonably be approximated by a Poisson distribution, making use of the fact that the mean and the variance of a Poisson distribution are equal.

Answers

1. We have:
$$s = 2·30 \qquad n = 85$$
We let σ = population standard deviation in restricted visibility. We set up:
$$H_0: \sigma = 2·45$$
$$H_1: \sigma \neq 2·45$$
If the population is normally distributed, we have:
$$\chi^2 = \frac{(n-1)s^2}{\sigma^2} \sim \chi^2 \text{ on 84 d.f.} = \frac{84 \times 2·3^2}{2·45^2}$$
$$= \underline{74·03}$$
Since a two-tailed test is appropriate, we find the limits between which 95% of the distribution lies as:
$$\chi^2_{84, 0·975} = 60·55 \quad \text{and} \quad \chi^2_{84, 0·025} = 111·20$$
The calculated value of 74·03 lies in the acceptance region between these limits. Hence we accept H_0 and conclude that it is reasonable to assume that the standard deviation of the ships' speeds is the same in restricted visibility as it is in good visibility.

To set up confidence limits for the estimated standard deviation in restricted visibility we use the same limits for χ^2 and set up:
$$60·55 \leqslant \frac{84s^2}{\sigma^2} \leqslant 111·2$$
Hence:
$$\sigma^2 \leqslant \frac{84 \times 2·3^2}{60·55} = 7·34 \quad \text{i.e. } \sigma \leqslant 2·71$$
or
$$\sigma^2 \geqslant \frac{84 \times 2·3^2}{111·2} = 4·00 \quad \text{i.e. } \sigma \geqslant 2·00$$
Thus we are 95% certain that the standard deviation in restricted visibility is between 2·00 knots and 2·71 knots.

Marine Statistics

2. For crew A:

$$n_A = 8 \qquad \sum x_A = 101 \cdot 6 \qquad \sum x_A^2 = 1{,}305 \cdot 5$$

$$s_A = \sqrt{\frac{1}{n-1}\left[\sum x^2 - \frac{(\sum x)^2}{n}\right]} = \sqrt{\frac{1}{7}\left[1{,}305 \cdot 5 - \frac{101 \cdot 6^2}{8}\right]} = \underline{1 \cdot 47}$$

For crew B:

$$n_B = 9 \qquad \sum x_B = 92 \cdot 4 \qquad \sum x_B^2 = 959 \cdot 06$$

$$s_B = \sqrt{\frac{1}{n-1}\left[\sum x^2 - \frac{(\sum x)^2}{n}\right]} = \sqrt{\frac{1}{8}\left[959 \cdot 06 - \frac{92 \cdot 4^2}{9}\right]} = \underline{1 \cdot 14}$$

To test whether sample A can be assumed to be drawn from a population with standard deviation 1·84, we set up:

$$H_0: \sigma = 1 \cdot 84$$
$$H_1: \sigma \neq 1 \cdot 84$$

We then calculate:

$$\chi^2 = \frac{(n-1)s^2}{\sigma^2} = \frac{7 \times 1 \cdot 47^2}{1 \cdot 84^2} = 4 \cdot 48$$

Since a two-tailed test is appropriate, we find the limits between which 95% of the distribution lies as:

$$\chi^2_{7, 0 \cdot 975} = 1 \cdot 69 \quad \text{and} \quad \chi^2_{7, 0 \cdot 025} = 16 \cdot 01$$

Since our calculated value of χ^2 lies between these limits, we accept H_0. We conclude that the set of times for crew A could reasonably have been drawn from a population with standard deviation 1·84 min.

A similar conclusion may be found for crew B since the calculated value of $\chi^2 = 3 \cdot 08$ which lies between $\chi^2_{8, 0 \cdot 975} = 2 \cdot 18$ and $\chi^2_{8, 0 \cdot 025} = 17 \cdot 53$.

3. We set up our null hypothesis that there is no systematic bias in the log readings and that it is therefore equally likely that the log will under-read as that it will over-read. Our alternative hypothesis is that under-reading and over-reading are not equally likely.

This is basically a binomial-type problem, but the large sample size makes binomial calculations unwieldy and a χ^2 test is a practical alternative.

Under H_0, we would expect 30 under-readings and 30 over-readings in 60 measurements. Hence we may calculate χ^2 as:

$$\chi^2 = \sum_{i=1}^{2} \frac{[O_i - E_i]^2}{E_i} = \frac{[45 - 30]^2}{30} + \frac{[15 - 30]^2}{30}$$

$$= 15 \sim \chi^2 \text{ on 1 d.o.f. under } H_0$$

Using a one-tailed test, we find that:

$$\chi^2_{1, 0 \cdot 01} = 6 \cdot 63$$

χ^2 Distribution Tests

Hence our calculated value of χ^2 is in the rejection region since $15 > 6 \cdot 63$. We thus reject H_0 and conclude that there is a systematic bias in the log causing it generally to read too high.

4. We set up the null hypothesis that the population proportions are as stated:

$H_0: \pi_S : \pi_M : \pi_L = \frac{1}{4} : \frac{1}{2} : \frac{1}{4}$
$H_1: \pi_S : \pi_M : \pi_L \neq \frac{1}{4} : \frac{1}{2} : \frac{1}{4}$

Under H_0, the expected frequencies for the 40 tyres of the sample are thus:

$E_S = 10 \quad E_M = 20 \quad E_L = 10$

We calculate:

$$\chi^2 = \sum_{i=1}^{i=3} \frac{[O_i - E_i]^2}{E_i} = \frac{[14 - 10]^2}{10} + \frac{[15 - 20]^2}{20} + \frac{[11 - 10]^2}{10}$$
$$= 1 \cdot 6 + 1 \cdot 25 + 0 \cdot 1 = 2 \cdot 95$$

Using a one-tailed test, and noting that we have 2 degrees of freedom, we find that:

$\chi^2_{2, 0 \cdot 05} = 5 \cdot 99$

Our calculated value of χ^2 is less than 5·99 and thus within the acceptance region. We therefore accept H_0 and conclude that we do not have sufficient evidence to suggest that the mix of tyre sizes is different from that indicated in the cargo description.

5. We calculate the mean of the data as:

$$\bar{x} = \frac{\sum fx}{\sum f} = \frac{206}{120} = 1 \cdot 72$$

We set up:

H_0: The data have a Poisson distribution
H_1: The data do not have a Poisson distribution
$\alpha = 0 \cdot 05$

We calculate the probabilities for frequencies of x defectives under H_0 as:

$p(x) = \frac{e^{-\lambda} \lambda^x}{x!}$ [letting $\lambda = \bar{x}$] $\quad p(0) = \frac{e^{-1 \cdot 72} 1 \cdot 72^0}{0!} = 0 \cdot 179$

$p(1) = \frac{p(0) \cdot 1 \cdot 72}{1} = 0 \cdot 308 \quad\quad p(2) = \frac{p(1) \cdot 1 \cdot 72}{2} = 0 \cdot 265$

$p(3) = \frac{p(2) \cdot 1 \cdot 72}{3} = 0 \cdot 152 \quad\quad p(4) = \frac{p(3) \cdot 1 \cdot 72}{4} = 0 \cdot 065$

$p(5) = \frac{p(4) \cdot 1 \cdot 72}{5} = 0 \cdot 022 \quad\quad p(6) = \frac{p(5) \cdot 1 \cdot 72}{6} = 0 \cdot 006$

$p(7) = \frac{p(6) \cdot 1 \cdot 72}{7} = 0 \cdot 002 \quad\quad p(8) = 1 - \sum_{x=0}^{7} p(x) = 0 \cdot 001$

The expected frequencies are then calculated as $120p(x)$, to give the following table:

219

Marine Statistics

Number of defective cases	Expected frequency
0	120 × 0·179 = 21·48 ≃ 21
1	120 × 0·308 = 36·96 ≃ 37
2	120 × 0·265 = 31·80 ≃ 32
3	120 × 0·152 = 18·24 ≃ 18
4	120 × 0·065 = 7·80 ≃ 8
5	120 × 0·022 = 2·64 ≃ 3
6	120 × 0·006 = 0·72 ≃ 1
7	120 × 0·002 = 0·24 ≃ 0
8	120 × 0·001 = 0·12 ≃ 0

For this example, we combine the last five expected frequencies since we should not use χ^2 where any one frequency is less than 5. We then calculate χ^2 as:

$$\chi^2 = \sum_{i=1}^{5} \frac{[O_i - E_i]^2}{E_i} = \frac{(24 - 21)^2}{21} + \frac{(38 - 37)^2}{37} + \frac{(28 - 32)^2}{32}$$

$$+ \frac{(16 - 18)^2}{18} + \frac{(14 - 12)^2}{12}$$

$$= 1 \cdot 51$$

The number of degrees of freedom is equal to the number of cells minus two, since the observed and expected frequencies have two parameters (the total and the mean) in common. We thus find:

$$\chi^2_{3, 0 \cdot 05} = 7 \cdot 81$$

Clearly, our calculated value of 1·51 is well within the acceptance region. Hence we accept H_0 and conclude that a Poisson distribution is a good fit for our given data.

6. $\bar{x} = \dfrac{399}{240} = 1 \cdot 66 \quad [n = 240] \quad p(x) = \dfrac{e^{-\lambda}\lambda^x}{x!}$

$p(0) = \dfrac{e^{-1 \cdot 66} \cdot 1 \cdot 66^0}{1} = 0 \cdot 190$

$p(1) = 0 \cdot 190 \times \dfrac{1 \cdot 66}{1} = 0 \cdot 315$

		np	E_i	O_i	$= (\ )$
$p(0)$	0·190	45·60	46	48	−2
$p(1)$	0·315	75·60	76	80	−4
$p(2)$	0·262	62·88	63	55	8
$p(3)$	0·145	34·80	35	30	5
$p(4)$	0·060	14·40	14	19	5
$p(5)$	0·020	4·80	⎧5	⎧6	⎧0
$p(6)$	0·006	1·44	6⎨1	8⎨1	2⎨1
$p(7)$	0·002	0·48	⎩0	⎩1	⎩1
Total	1·000		240		

χ^2 Distribution Tests

$$\chi^2 = 4\cdot48 \qquad \chi^2_{4, 0\cdot05} = 9\cdot49$$

Hence accept H_0.

The method of working this exercise is identical to the method for Ex. 5. The calculated value of χ^2 is 4·48 which is less than the critical value of $\chi^2_{4, 0\cdot05} = 9\cdot49$. Hence we accept H_0 and conclude that the Poisson distribution provides a good fit for the given data.

7. We calculate:

$$\bar{x} = \frac{\Sigma fx}{\Sigma x} = \frac{132}{60} = 2\cdot2$$

We set up:

H_0: An exponential distribution with mean $\hat{\mu} = \bar{x}$ is a good fit to the above data

H_1: The exponential distribution is not a good fit to the data

Under H_0, the probability density function of the exponential distribution is given by:

$$f(x) = \frac{1}{\mu}e^{-x/\mu} = \frac{1}{2\cdot2}e^{-x/2\cdot2}$$

To find the expected frequency occurring in each interval of 1,000 hr, we integrate $f(x)$ between the time limits a and b thus:

$$\int_a^b \frac{1}{2\cdot2}e^{-x/2\cdot2}\,dx = e^{-a/2\cdot2} - e^{-b/2\cdot2}$$

Taking appropriate values for a and b, we calculate the expected frequencies and tabulate them with the observed frequencies below:

Interval	Expected frequency	Observed frequency
0 but under 1	$60(e^{-0/2\cdot2} - e^{-1/2\cdot2}) = 22$	20
1 but under 2	$60(e^{-1/2\cdot2} - e^{-2/2\cdot2}) = 14$	17
2 but under 3	$60(e^{-2/2\cdot2} - e^{-3/2\cdot2}) = 9$	10
3 but under 4	$60(e^{-3/2\cdot2} - e^{-4/2\cdot2}) = 6$	6
4 but under 5	$60(e^{-4/2\cdot2} - e^{-5/2\cdot2}) = 4$	3
5 but under 6	$60(e^{-5/2\cdot2} - e^{-6/2\cdot2}) = 2$	3
6 but under 7	$60(e^{-6/2\cdot2} - e^{-7/2\cdot2}) = 1$	2
7 but under 8	$60(e^{-7/2\cdot2} - e^{-8/2\cdot2}) = 1$	1
8 but under 9	$60(e^{-8/2\cdot2} - e^{-9/2\cdot2}) = 1$	0
9 but under 10	$60(e^{-9/2\cdot2} - e^{-10/2\cdot2}) = 0$	0

Marine Statistics

We calculate:

$$\chi^2 = \sum \frac{(O_i - E_i)^2}{E_i} \quad \text{(pooling the results from 4,000 to 9,000 hr to avoid having less than 5 in a cell)}$$

$$= \frac{(20-22)^2}{22} + \frac{(17-14)^2}{14} + \frac{(10-9)^2}{9} + \frac{(6-6)^2}{6} + \frac{(9-9)^2}{9}$$

$$= \underline{0.99}$$

We have both the total and the mean in common for the two distributions so there are $5 - 2 = 3$ degrees of freedom. Hence we find:

$$\chi^2_{3, 0.05} = \underline{7.81}$$

Our calculated value of χ^2 is quite low and well within the acceptance region. Hence we accept H_0 and conclude that the exponential distribution appears to be a very good fit for the given data.

8. We re-tabulate the given data to include marginal totals and expected values under the null hypothesis in brackets. The null hypothesis, H_0, being that the proportion of ships in each class is independent of the route and the alternative hypothesis, H_1, being that these proportions are different for the two routes. We take $\alpha = 0.05$ as a reasonable level.

	< 500	500–9,999	≥ 10,000	
Route A	40 (50)	61 (54)	49 (46)	150
Route B	43 (33)	29 (36)	28 (31)	100
	83	90	77	

We calculate:

$$\chi^2 = \sum_{\text{cells}} \frac{(O-E)^2}{E} = \frac{(40-50)^2}{50} + \frac{(61-54)^2}{54} + \frac{(49-46)^2}{46}$$

$$+ \frac{(43-33)^2}{33} + \frac{(29-36)^2}{36} + \frac{(28-31)^2}{31}$$

$$= \underline{7.78}$$

We have two degrees of freedom and we use χ^2 tables to find:

$$\chi^2_{2, 0.05} = \underline{5.99}$$

Since our calculated value exceeds the critical value, we reject H_0 and accept H_1. We conclude that there appears to be a significant difference between the proportions of ships of our three classes using routes A and B.

χ^2 Distribution Tests

9. We set up:

    H_0: That the proportions of subjects taking the four categories of action are the same for each level of experience
    H_1: That the proportions are different for each level of experience
    $\alpha = 0{\cdot}05$

    We re-tabulate the given data to include marginal totals and the expected values under the null hypothesis in brackets. Thus:

Experience	R	L	I	D	
$\geqslant 5$ yr	20 (26)	14 (9)	8 (5)	8 (10)	50
< 5 yr	38 (32)	6 (11)	3 (6)	13 (11)	60
	58	20	11	21	110

    We calculate:
    $$\chi^2 = \sum \frac{(O-E)^2}{E} = \frac{(20-26)^2}{26} + \frac{(14-9)^2}{9} + \frac{(8-5)^2}{5}$$
    $$+ \frac{(8-10)^2}{10} + \frac{(38-32)^2}{32} + \frac{(6-11)^2}{11}$$
    $$+ \frac{(3-6)^2}{6} + \frac{(13-11)^2}{11}$$
    $$= \underline{11{\cdot}63}$$

    We have three degrees of freedom and we use χ^2 tables to find:
    $$\chi^2_{3,\,0{\cdot}05} = 7{\cdot}81$$

    Our calculated value exceeds the critical value and hence we reject H_0 and accept H_1. We conclude that navigators of less than 5 years experience take different patterns of collision avoidance manoeuvres than navigators of more than 5 years experience.

10. We set up:

    H_0: That there is no difference between the proportions of first sightings between the two positions
    H_1: That there is such a difference
    $\alpha = 0{\cdot}05$

    We tabulate the data to include marginal totals and with the expected values under the null hypothesis in brackets:

Lookout position	Seen first	Seen second	
Forecastle	6 (9)	12 (9)	18
Monkey Island	11 (8)	4 (7)	15
	17	16	33

Marine Statistics

Since we have small expected frequencies, we use Yates' correction to calculate χ^2 as:

$$\chi^2 = \sum \frac{(|O - E| - \frac{1}{2})^2}{E} = \frac{(|6 - 9| - \frac{1}{2})^2}{9} + \frac{(|12 - 9| - \frac{1}{2})^2}{9}$$

$$+ \frac{(|11 - 8| - \frac{1}{2})^2}{8} + \frac{(|4 - 7| - \frac{1}{2})^2}{7}$$

$$= \frac{2 \cdot 5^2}{9} + \frac{2 \cdot 5^2}{9} + \frac{2 \cdot 5^2}{9} + \frac{2 \cdot 5^2}{9}$$

$$= 2 \cdot 78$$

In this example, we have one degree of freedom and we use tables to find:

$$\chi^2_{1, 0 \cdot 05} = 3 \cdot 84$$

Since our calculated value of χ^2 is less than the critical value, we are in the acceptance region and we therefore accept H_0. We conclude that we have insufficient evidence to state that there is a significant difference between the look-out men's performance at the two stations.

Note: Without Yates' correction an apparently significant value of 4·00 would have been calculated for χ^2.

11. As before, we calculate the mean of the data as:

$$\bar{x} = \frac{\sum fx}{\sum f} = \frac{206}{120} = 1 \cdot 72$$

We also calculate the standard deviation as:

$$s^2 = \frac{1}{n-1}\left[\sum fx^2 - \frac{\sum (fx)^2}{n}\right] = \frac{1}{119}\left[604 - \frac{206^2}{120}\right]$$

$$= 2 \cdot 10$$

We calculate:

$$\chi^2 = \frac{(n-1)s^2}{\bar{x}} = \frac{119 \times 2 \cdot 1}{1 \cdot 72}$$

$$= 145 \cdot 29$$

The critical region is $\chi^2 > \chi^2_{119, 0 \cdot 05}$ but, because our degrees of freedom are so large, we use the normal approximation to the χ^2 distribution as a basis for the test. Thus:

$$z = \sqrt{2\chi^2} - \sqrt{2\nu - 1}$$
$$= \sqrt{2 \times 145 \cdot 29} - \sqrt{2 \times 119 - 1}$$
$$= 1 \cdot 65$$

Our critical region is $-1 \cdot 96 < z < +1 \cdot 96$. The calculated value of z is clearly within this region and hence we accept H_0 and conclude that the Poisson distribution provides an acceptable fit to the given data.

Chapter 11
Non-Parametric Tests

11.1 Introduction

A new davit for launching liferafts was required to be tested in comparison with an existing equipment. It was fitted to a passenger ferry in addition to the existing gear and over a period of two months comparisons were made between the time taken for launching and embarkation using both systems simultaneously, on each occasion. Fifteen tests were made, using different crew teams, and undertaken in differing weather conditions and at different times of the day and night. The results, measured in minutes, were as follows:

Trial Number	Existing davit	New davit
1	10	8
2	15	12
3	8	9
4	12	9
5	7	7
6	12	11
7	18	15
8	9	12
9	15	13
10	14	10
11	16	12
12	9	7
13	11	10
14	16	14
15	8	10

Is this sufficient evidence to support the hypothesis that the new system gives significantly faster embarkation times than the old system?

During the two months that the system was under trial it was only possible to test the system on fifteen different occasions which means that we are faced with a fairly small sample of results on which to make our conclusions. Rather than go on making the tests, it is necessary to make a decision as soon as we can, so we want to analyse the evidence from this small sample. Since each trial was carried out under differing weather and light conditions, but each piece of equipment was tested simultaneously, we should look at the differences in time taken on each occasion. It must be assumed that the crews were assigned at random to the pieces of equipment and that no systematic bias was introduced that way. However, it is not very reasonable to assume that the difference in times has a normal distribution and in fact this sort of distribution is usually a skewed distribution. As the sample size is small it is not appropriate to use a test based on the normal distribution, nor can we use a 't' statistic as this assumes that the underlying distribution for the population is normal. Thus we have to use a non-parametric test for this situation.

Strictly speaking there are two different terms required here. A distribution-free test makes no assumptions about the underlying distribution, while a non-parametric test makes no assumptions about the parameters of the distribution, but we term both these types 'non-parametric' tests. Although there are situations when we can only use non-parametric methods, in fact the calculations involved are usually easier and hence we sometimes use these methods to give a quick answer whatever the circumstances. It can be shown theoretically that in a situation for which a 't' test is appropriate, a non-parametric test will be less powerful, i.e. one is less likely to reject H_0 when in fact it should be rejected, but often the difference is minimal. There are many non-parametric tests which have been developed now, and extensive literature has been written about them. We will consider in this chapter three types of test to illustrate this approach rather than attempting to give an exhaustive comprehensive account of the variety of non-parametric tests available.

11.2 Sign Test

The sign test is used when we have paired observations from two samples and want to test if the means of the two populations, from which the samples were drawn, are equal. Typical of this is the example on the testing of the liferaft davit. Since both the new and existing davits were tested on each occasion we can treat the two simultaneous observations as a pair. The level of both values in the pair will reflect the overall conditions prevailing at the time but the interesting statistic is in each case the difference between the two members of the pair. The easiest way of looking at the difference is simply to take the sign, and this produces the following results, subtracting the time for the new davit from the time for the existing davit.

Hence out of 15 differences we have 11 positive values, 3 negative values and one value which implies no difference. The usual practice is to ignore the values of 0, and to reduce the sample size accordingly. If there are several values of 0, such that the sample size changes considerably, then the implication is that the recording or measurement device was not giving enough

Non-parametric Tests

Trial number	Sign of difference	Trial number	Sign of difference
1	+	8	−
2	+	9	+
3	−	10	+
4	+	11	+
5	0	12	+
6	+	13	+
7	+	14	+
		15	−

accuracy and hence it is difficult to make a decision. It must only be assumed that this problem is sorted out at the stage when the experiment was designed and the order of accuracy for the measurement chosen to reflect worthwhile differences. Thus in this example it was probably considered that the new davit would only be worthwhile if it made a difference of minutes to the time taken. In these results we are ignoring one value in fifteen which should cause no problems.

The hypotheses and significance level for this test could be as follows:

H_0: The mean launch time with the new davit is the same as the mean launch time with the old davit
H_1: The mean launch times are different
$\alpha = 0.05$

This is equivalent to saying that in the population of differences between the launch times, the proportion of positive differences should be $\frac{1}{2}$ under H_0 and different from $\frac{1}{2}$ under H_1. These new hypotheses may be tested either by using a test for proportions based on the binomial distribution, as described in Chapter 9, or a χ^2 goodness of fit test as described in Chapter 10. Using the second method we can write down the following table:

	Number of positive differences	Number of negative differences	Total
Observed frequencies: O	11	3	14
Expected frequencies: E	7	7	14

$$\chi^2 = \sum \frac{(|O - E| - \frac{1}{2})^2}{E} \quad \text{using Yates' correction}$$

$$= \frac{(|11 - 7| - \frac{1}{2})^2}{7} + \frac{(|3 - 7| - \frac{1}{2})^2}{7} = 3.50$$

Marine Statistics

There is 1 degree of freedom and hence our critical region is $\chi^2 > \chi^2_{1, 5\%} = 3.84$, $3.50 < 3.84$, which implies that we have a non-significant result and hence we accept our null hypothesis that there is no difference between the launch times using the two types of davit.

As an extra note on this particular example it should be mentioned that in fact the original question asked was whether the new davit was faster than the old davit. If we do a one-tailed test, using the binomial probabilities, then the probability of getting only 3 or less negative signs in 14 trials is given by:

$$\binom{14}{3}\left(\frac{1}{2}\right)^3 \left(\frac{1}{2}\right)^{11} + \binom{14}{2}\left(\frac{1}{2}\right)^2 \left(\frac{1}{2}\right)^{12} + \binom{14}{1}\left(\frac{1}{2}\right)^1 \left(\frac{1}{2}\right)^{13} + \left(\frac{1}{2}\right)^{14}$$

$$= 364\left(\frac{1}{2}\right)^{14} + 91 \cdot \left(\frac{1}{2}\right)^{14} + 14 \cdot \left(\frac{1}{2}\right)^{14} + \left(\frac{1}{2}\right)^{14} = 470 \cdot \left(\frac{1}{2}\right)^{14} = 0.029$$

Hence, using a 5% level of significance, we would conclude that we had a significant result on the one-tailed test, and so would conclude further that the launch time for the new davit was significantly faster than the launch time for the old davit. This illustrates again how important it is to be really clear, before starting an experiment, exactly which alternative hypothesis and which level of significance is required.

Steps to Performing a Sign Test

1. This test is used for paired observations and usually when we only have a small sample and cannot assume a normal distribution for the population of differences.

2. Take the difference between each pair and, ignoring values of 0, count the number of positive signs and the number of negative signs.

3. Perform a test to determine whether the proportion of positive signs in the population of differences is $\frac{1}{2}$ against a suitable alternative. This can be done either using a χ^2 goodness of fit test or a binomial test of proportion.

11.3 *Wilcoxon Signed-ranks Test*

The sign test is obviously a very easy one to apply but it is a very crude test as it makes no allowance for the size of the difference between the readings, only for the direction of the difference. In many situations it is possible to rank the differences, in ascending order say, and if this can be done it is appropriate to use the *Wilcoxon signed-ranks* test also known as the Wilcoxon matched pairs test. In the previous example the size of time difference was not completely independent of the conditions under which the test was made. If the two davits had been compared under the same weather conditions on each trial it would have meant more to rank the size of difference, but it is felt that a 3-minute time difference when the shorter time is 9 min (trial 4) is not the

Non-parametric Tests

same as a 3-minute time difference when the shorter time is 15 min (trial 7).
In the following situation it is, however, more realistic to rank the size of the differences. It is desired to see what effect there is on examination results by doing a set series of sample problems rather than by studying independently. A group of 12 students studying for their second mate's examination were divided into 6 pairs, matched on their previous examination performances. One of each pair was given a set series of sample problems and the other was left to study alone after the lectures. In the examination the following grades were obtained:

	\multicolumn{6}{c}{Pair}
	1	2	3	4	5	6
With sample problems:	54	79	51	36	67	55
No extra problems:	48	76	53	41	62	55

We set up the following pair of hypotheses:

H_0: There is no difference in the mean level of grades depending on whether the students have sample problems or not
H_1: There is a difference
$\alpha = 0.05$

The first step is to calculate the difference between each pair giving the following values:

	\multicolumn{6}{c}{Pair}
	1	2	3	4	5	6
Difference	6	3	−2	−5	5	0

Ignoring any differences of 0 and not worrying about the signs the differences are ranked, calling the smallest 1, the next 2 and so on. If two or more values are the same they are given the mean of the ranks they would have shared:

	\multicolumn{6}{c}{Pair}
	1	2	3	4	5	6
Difference	6	3	−2	−5	5	0
Rankings	5	2	1	$3\frac{1}{2}$	$3\frac{1}{2}$	—

The sum of the ranks for positive differences is calculated and similarly for negative differences and the test statistic W is the smaller of these two.

Positive sum $= 5 + 2 + 3\frac{1}{2} = 10\frac{1}{2}$
Negative sum $= 1 + 3\frac{1}{2} = 4\frac{1}{2}$
Hence $W = 4\frac{1}{2}$

Marine Statistics

If we assume that the population means are equal, each difference between the pairs is equally likely to be positive or negative. Hence there are two equally likely ways for a given rank to receive a sign, and for n ranks there are 2^n equally likely ways for the n ranks to receive signs. We then count up the number of ways, N, in which we could assign signs to the n ranks such that the total of the negative ranks, say, is less than or equal to our value of W. We then calculate the probability of getting a value as low as W or less, given H_0 is true $= N/2^n$.

In this case $n = 5$, since there are 5 ranks.

If $W = 0$, there is 1 way this can arise if no differences are negative
If $W = 1$, there is 1 way this can arise if the difference marked 1 is negative
If $W = 2$, there is 1 way this can arise if the difference ranked 2 is negative
If $W = 3$, there are 2 ways this can arise: if the difference ranked 3 is negative or if the differences ranked 1 and 2 are negative
If $W = 4$, there are 2 ways again: if the negative differences are ranked 1 and 3 or 4
If $W = 5$, there are 3 ways: if the negative differences are ranked 1 and 4 or 2 and 3 or 5

For $W = 5 \quad N = 10$ and $\dfrac{N}{2^n} = \dfrac{10}{32} = 0.31$

$W = 4 \quad N = 7$ and $\dfrac{N}{2^n} = \dfrac{7}{32} = 0.22$

With a two-tailed test and a significance level of 5% we would reject H_0 if $N/2^n < 0.025$ for the appropriate W value. However in this case, for $W = 4$, $N/2^n$ is much larger than 0.025, so we conclude that there is no significant difference between the means. Hence our conclusion is that we do not have sufficient evidence to distinguish between any effects on examination results of giving sample problems or relying on independent study.

Obviously it can get rather tedious working out the number of ways in which the rank sums can arise, so tables giving critical values of W are published (Table 5).

Steps to Performing a Wilcoxon Signed Ranks Test

1. Calculate the difference between each pair.
2. Rank the differences in ascending order ignoring signs and omitting zero differences.
3. Calculate the sum of the ranks for the positive differences and similarly for the negative differences and let the test statistic, W, be the smaller of these values.

Non-parametric Tests

4. Either refer the value of W to tables of critical values, or calculate from first principles the probability of getting a value as extreme as W given that H_0 is true. This probability is given by the number of ways in which each value as extreme as W can arise, divided by 2^n, where n is the effective number of pairs.

11.4 Mann–Whitney Test

The Mann–Whitney test is used to test whether two means are equal when the samples are not matched and again when a parametric method is not appropriate or required. The following situation is one for which it would be used. Fourteen new cargo runners were supplied to a ship, six being of type A and eight of type B. The chief officer was requested to see that each runner was subjected to approximately the same level of service. A runner was to be withdrawn from service when three or more wires were found to be broken in any 2 m length. The resulting lengths of time in weeks for which each runner lasted were as follows:

A1	80	B1	58
A2	122	B2	49
A3	75	B3	108
A4	110	B4	67
A5	84	B5	53
A6	97	B6	45
		B7	52
		B8	76

Is this sufficient evidence to conclude that the runners of type A are on the whole more durable than the runners of type B?

We cannot assume that the underlying distributions for the duration of a runner were normal for either type A or type B and hence with such small samples a parametric test is not appropriate at all. To see why the underlying distributions cannot be taken as normal, it must be realised that although longer periods can always be associated with greater wear, corrosion, etc., similar intervals of time do not necessarily indicate similar increments of wear. For example, during the five-week period from week 53 to week 58 the ship may have called at only two ports, in one of which the cargo may have been handled by shore cranes, whilst during the five-week period from week 75 to week 80 the ship may have called at six ports, in all of which the cargo was handled by the ship's equipment.

The first step is to calculate the means of the two samples, which gives $\bar{x}_1(\text{type } A) = 94.7$ and $\bar{x}_2(\text{type } B) = 63.5$. It is not essential that these be worked out exactly as we only need to know which is the larger value. It is

231

Marine Statistics

customary, though, to call sample 1 the one with the smaller number (n_1) of members, and sample 2 the one with the larger number (n_2) of members. In this case $n_1 = 6$ and $n_2 = 8$.

The next step in performing a Mann–Whitney test is to put the two samples together and then rank the values as if it were one large sample. If $\bar{x}_1 > \bar{x}_2$ then the ranking is done in descending order but if $\bar{x}_1 < \bar{x}_2$ then the ranking is done in ascending order. Our initial test statistic, w_1, is the sum of the ranks of the members of the smaller sample and it makes life simpler if we work with the smaller values. In this example $\bar{x}_1 > \bar{x}_2$ so we rank the combined sample in descending order as follows:

Rank	Value	Reading
1	122	A2
2	110	A4
3	108	B3
4	97	A6
5	84	A5
6	80	A1
7	76	B8
8	75	A3
9	67	B4
10	58	B1
11	53	B5
12	52	B7
13	49	B2
14	45	B6

W_1 is the sum of the ranks of the members of the smaller sample, the A type, so

$$W_1 = 1 + 2 + 4 + 5 + 6 + 8 = 26$$

If we had had two or more values the same we would again have given them each the mean of the shared ranks.

Let us take as hypotheses:

$\left. \begin{array}{l} H_0: \mu_1 = \mu_2 \\ H_1: \mu_1 > \mu_2 \end{array} \right\}$ where μ_1 is the mean for runners of type A and μ_2 is the mean for runners of type B

$\alpha = 0\cdot05$

We will first perform the test from first principles to illustrate the reasoning behind the test. The number of possible ways of getting a value of W_1 if the two populations have equal means is the number of ways in which $n_1 + n_2$ ranks could be assigned to n_1 observations $= n_1\binom{n_1+n_2}{n_2}$ in this case:

$$\binom{14}{6} = \frac{14!}{8!\,6!} = \frac{14.13.12.11.10.9}{6.5.4.3.2.1} = 3{,}003$$

Non-parametric Tests

We want to know the number of ways, N, in which a value as extreme as ours (i.e. 26 or less) can arise given H_0 is true. The probability of getting a value of $W_1 \leq 26$ if H_0 is true will then be $N/\binom{14}{6}$. The smallest value that W_1 can take is if the n_1 observations from the small sample were the first n_1 observations in the combined sample, i.e.

$$\min W_1 = 1 + 2 + \cdots + n_1$$
$$= 1 + 2 + 3 + \cdots + 6 = 21 \text{ in this case}$$

We can then draw up the following table:

W_1	Ways it arises	Number of ways
21	$1 + 2 + 3 + 4 + 5 + 6$	1
22	$1 + 2 + 3 + 4 + 5 + 7$	1
23	$1 + 2 + 3 + 4 + 6 + 7; 1 + 2 + 3 + 4 + 5 + 8$	2
24	$1 + 2 + 3 + 4 + 6 + 8; 1 + 2 + 3 + 4 + 5 + 9;$ $1 + 2 + 3 + 5 + 6 + 7$	3
25	$1 + 2 + 3 + 4 + 6 + 9; 1 + 2 + 3 + 4 + 5 + 10;$ $1 + 2 + 3 + 5 + 6 + 8; 1 + 2 + 3 + 4 + 7 + 8;$ $1 + 2 + 4 + 5 + 6 + 7$	5
26	$1 + 2 + 3 + 4 + 6 + 10; 1 + 2 + 3 + 4 + 5 + 11;$ $1 + 2 + 3 + 5 + 6 + 9; 1 + 2 + 3 + 4 + 7 + 9;$ $1 + 3 + 4 + 5 + 6 + 7; 1 + 2 + 3 + 5 + 7 + 8;$ $1 + 2 + 4 + 5 + 6 + 8$	7
	Total	$19 = N$

Hence $p(W_1 < 26 \mid H_0 \text{ is true}) = 19/3{,}003 = 0.006$.

This probability is obviously very much smaller than 0·05, our set α level, and so we can conclude that it is very likely that runners of type A are more durable than runners of type B.

We have performed this test from first principles to illustrate the basis of the test but it would obviously be a very long-winded affair if the probability of such a value of W_1 had to be evaluated each time. Tables are therefore readily available to speed up the test. These are usually in the form of Tables 6 and 7 at the end of this book and are based on a test statistic U. U is defined by the following formula: $U = W_1 - \min W_1$. But

$$\min W_1 = 1 + 2 + \cdots + n_1 = \frac{n_1(n_1 + 1)}{2}$$

Hence

$$U = \text{sum of ranks of smaller sized sample} - \frac{n_1(n_1 + 1)}{2}$$

233

or

$$U = W_1 - \frac{n_1(n_1 + 1)}{2}$$

In our sample $U = 26 - 21 = 5$.

If we look in the table with $n_1 = 6$, $n_2 = 8$ and $U = 5$ we see that the probability of getting a value of $U \leqslant 5$ is 0·006 (as before). We therefore, in general, reject H_0 if this probability is less than α for a one-tailed test or less than $\alpha/2$ for a two-tailed test.

In practice, therefore, the tables of values of U will always be used. For situations in which $n_2 > 8$ but $\leqslant 20$ instead of giving the exact probability associated with each value of U, the tables are given in an alternative form and show the critical value for U for a given probability. If n_2 lies between 20 and 30 then we may use a normal approximation for our test statistic. It has been shown that for large samples the mean of U, $\mu_u = n_1 n_2 / 2$, and the standard deviation of U, $\sigma_u = [n_1 . n_2(n_1 + n_2 + 1)]/12$, where U has a normal distribution.

Hence we can take as test statistic

$$z = \frac{U - \mu_u}{\sigma_u} = \frac{U - \frac{n_1 n_2}{2}}{\sqrt{\frac{n_1 n_2 (n_1 + n_2 + 1)}{12}}}$$

and refer this to a standard normal distribution. Once the sample sizes get larger than 30 then we can use a parametric test anyway.

Steps to Performing a Mann–Whitney Test

1. Call sample 1 the one with the smaller number of members n_1 and the other has n_2 members.
2. Calculate $\bar{x}_1$, the mean of sample 1, and $\bar{x}_2$, the mean of sample 2.
3. Put the observations in one combined sample and rank in ascending order if $\bar{x}_1 < \bar{x}_2$ and in descending order if $\bar{x}_1 > \bar{x}_2$.
4. Calculate W_1, the sum of the ranks of the members of the first sample.
5. Calculate $U = W_1 - \min W_1 = W_1 - [n_1(n_1 + 1)]/2$.
6. Refer the value of U to the appropriate table for the given values of n_1 and n_2.

11.5 Summary

In this chapter brief descriptions of three of the more common non-parametric tests have been given. They are intended as an illustration of how this approach to statistical significance testing can be used and if the reader is in a situation where he believes this type of test would be appropriate it may be necessary

Non-parametric Tests

to consult a more comprehensive book on the subject. In the following chapter we are going to return to the more conventional parametric methods of testing and will consider how to test for differences between the means of several populations simultaneously.

Exercises

1. In order to make a comparison between cargo runners supplied by manufacturer A and those supplied by manufacturer B, a shipowner supplies one of each type to each of ten ships in his fleet. In each case, the chief officer is asked to see that both runners are given the same amount of usage and to take a runner out of service when three or more wires are found to be broken in any 2 m length. The resulting lengths of time (in weeks) for which the runners lasted are summarised below:

Ship	Type A	Type B
1	65	50
2	115	95
3	76	86
4	204	170
5	25	16
6	69	72
7	146	129
8	113	93
9	97	83
10	162	120

   Would you conclude that there is any difference between the two types of runners?

2. Two identical buoys were moored close together and, for test purposes, were painted with different specifications of green paint. To compare the visibility of the two paints, a ship made a number of approaches to the buoys under different conditions of lighting, sea-state, etc., and in 25 observations it was found that the buoy coated with paint A was identified first on 19 occasions and the buoy coated with paint B was identified first on 6 occasions. Is this sufficient evidence to conclude, at the 5% level, that paint A has generally better visibility than paint B?

3. Two models of radio direction finding equipment, model A and model B, were tested for the ease with which bearings could be obtained. Twelve experienced navigators were asked to acquire the signals and measure the

235

Marine Statistics

bearings of three radio beacons in succession using each of the two models and their times were recorded as follows:

Navigator	Time in minutes using model A	Time in minutes using model B
1	4·5	2·7
2	3·8	2·9
3	2·9	2·7
4	3·4	2·9
5	3·8	4·1
6	4·2	2·8
7	3·3	3·1
8	3·0	3·0
9	3·8	2·6
10	3·4	3·5
11	4·4	3·3
12	3·2	3·6

Use the Wilcoxon signed-ranks test to establish whether there is a difference between the mean times for using the two models at the 0·05 level of significance.

4. In trading between two ports, ships have the option of going through passage *A* or passage *B*. It is suggested by shipmasters that the visibility in passage *A* is generally better than the visibility in passage *B*. In order to ascertain whether this is in fact the case at the 5% level of significance records of visibility were studied over the course of a year. Days on which the visibility in both channels was more than 5 km were not considered, which left 20 days on which the visibility in at least one of the channels was 5 km or less according to the following table:

Day	Channel A visibility (km)	Channel B visibility (km)	Day	Channel A visibility (km)	Channel B visibility (km)
1	2·3	1·8	11	3·6	4·2
2	4·5	2·1	12	0·8	0·4
3	1·7	1·9	13	2·9	2·1
4	5·5	3·7	14	5·5	3·5
5	3·4	2·5	15	2·9	1·8
6	1·2	1·0	16	3·4	2·9
7	4·3	2·8	17	0·2	0·5
8	3·2	3·9	18	2·7	1·3
9	0·5	0·5	19	4·4	5·5
10	2·5	1·2	20	3·0	2·5

Non-parametric Tests

Use a Wilcoxon signed-rank test at the 0·01 significance level to decide whether, on the above evidence, the visibility in passage A can be considered generally superior to that in passage B.

5. Twelve lifejackets of type A and eight lifejackets of type B were moored in an exposed harbour area with test weights of iron suspended from each lifejacket. The lengths of times before the lifejackets sank were recorded as follows:

Type A	Length survival, hr	Type B	Length survival, hr
1	101	1	104
2	98	2	86
3	122	3	110
4	150	4	74
5	108	5	96
6	158	6	105
7	116	7	114
8	129	8	90
9	110		
10	139		
11	84		
12	132		

Use a Mann–Whitney test to decide whether there is a significant difference, at the 0·05 level, between the mean survival times for the two types of lifejacket.

6. In order to compare the quality of two brands of paint, A and B, 16 test pieces of steel were prepared, of which 6 were coated with brand A and 10 with brand B. The test pieces were then placed in an exposed position on the foredeck of a ship while it followed its normal trading pattern for a year. At the end of that period, the test pieces were ranked in descending order according to the severity of corrosion found, and the paint brand was noted against each rank as shown in the following table:

Rank	Paint brand	Rank	Paint brand
1	A	9	B
2	A	10	A
3	B	11	B
4	A	12	B
5	A	13	B
6	B	14	B
7	A	15	B
8	B	16	B

Is this sufficient evidence to conclude, using a 5% level of significance, that, under the test conditions, the mean performance of paint B (the more expensive brand) is superior to that of A?

Answers

1. The argument for using a non-parametric test is that similar intervals of time on different ships do not necessarily correspond to similar increments of wear. Thus the 9-week difference in the longevity of the runners in ship 5 may actually represent a greater increment of wear than the 15-week difference for ship 1, depending upon the intensity of usage in the two periods.

   For the two runners on each ship, however, it is clear which is the more durable and so a sign test is appropriate. We tabulate the differences below:

Ship	Interval, $A - B$
1	+15
2	+20
3	−10
4	+34
5	+9
6	−3
7	+17
8	+20
9	+14
10	+42

   Since the values of the differences are not comparable we consider only the signs. If the runner types were equally durable we would expect that such a test would yield approximately 50% each of plus and minus signs and that they would be distributed binomially with $p = 0.5$. We set up:

   H_0: The intervals of time, $A - B$, are equally likely to be positive or negative

   H_1: The intervals of time, $A - B$, are not equally likely to be positive or negative

   Working from first principles, we find that, when $n = 10$, the probability of a split as extreme as the one observed of 2:8 is $0.055 \times 2 = 0.11$, using a two-tailed test since H_1 does not predict which sense the differences should have.

   On this basis, we accept H_0 and conclude that the collected data gives insufficient evidence to accept that the two types of runner differ in durability.

Non-parametric Tests

2. In this case, we have only sufficient evidence to perform a sign test, allocating a positive sign to the case where buoy A is seen first and a negative sign to the case where buoy B is seen first. This is essentially a binomial problem, but it may be dealt with conveniently by means of a χ^2 test. We set up:

   H_0: Positive and negative results are equally likely
   H_1: Positive and negative results are not equally likely
   $\alpha = 0.05$

   We form a table:

	Positive results	Negative results
Observed frequencies	19	6
Expected frequencies under H_0	12.5	12.5

   $$\chi^2 = \sum \frac{(O-E)^2}{E} = \frac{(19-12\cdot5)^2}{12\cdot5} + \frac{(6-12\cdot5)^2}{12\cdot5}$$
   $$= \underline{6\cdot76}$$

   The critical value for $\chi^2_{1,\,0\cdot05} = 3\cdot84$.

   Hence our calculated value of χ^2 lies in the rejection region. We therefore reject H_0 and accept H_1 and conclude that there is evidence to suggest that buoy A is more easily seen than buoy B.

3. We set up:

   H_0: There is no difference in the mean times for using models A and B
   H_1: There is a difference in the mean times for using models A and B
   $\alpha = 0.05$

   We re-tabulate the data and add columns for the differences between the times for each subject and the ranks for those differences. Thus:

Navigator	Time A	Time B	$A - B$	Rank
1	4.5	2.7	+1.8	11
2	3.8	2.9	+0.9	7
3	2.9	2.7	+0.2	2½
4	3.4	2.9	+0.5	6
5	3.8	4.1	−0.3	4
6	4.2	2.8	+1.4	10
7	3.3	3.1	+0.2	2½
8	3.0	3.0	0.0	—
9	3.8	2.6	+1.2	9
10	3.4	3.5	−0.1	1
11	4.4	3.3	1.1	8
12	3.2	3.6	−0.4	5

Marine Statistics

The sum of the ranks for positive differences is 56. The sum of the ranks for negative differences is 10. We take the smaller of these, i.e. 10, as our test statistic, W.

We find the two-tailed critical value of W from tables against $N = 11$ and for 0·05 level of significance as $W = 11$.

Since our calculated value is less than 11, it lies in the rejection region. Hence we accept H_1, and conclude that the mean time for using model B is shorter than the mean time for using model A.

4. We set up:

   H_0: There is no difference in the mean visibility between the two passages
   H_1: The mean visibility in passage B is lower than that in passage A
   $\alpha = 0.01$

We re-tabulate the data, adding columns for the differences between the observed visibilities and the ranks of those differences:

Ship	Vis. A	Vis. B	A − B	Rank
1	2·3	1·8	+0·5	6
2	4·5	2·1	+2·4	19
3	1·7	1·9	−0·2	1½
4	5·5	3·7	+1·8	17
5	3·4	2·5	+0·9	11
6	1·2	1·0	+0·2	1½
7	4·3	2·8	+1·5	16
8	3·2	3·9	−0·7	9
9	0·5	0·5	0·0	—
10	2·5	1·2	+1·3	14
11	3·6	4·2	−0·6	8
12	0·8	0·4	+0·4	4
13	2·9	2·1	+0·8	10
14	5·5	3·5	+2·0	18
15	2·9	1·8	+1·1	12½
16	3·4	2·9	+0·5	6
17	0·2	0·5	−0·3	3
18	2·7	1·3	+1·4	15
19	4·4	5·5	−1·1	12½
20	3·0	2·5	+0·5	6

We calculate the sum of the positive difference ranks = 156 and the sum of the negative difference ranks = 34. We take the smaller of these, 34, as our test statistic, W.

We find the one-tailed critical value of W from tables against $N = 19$ and for 0·01 level of significance as $W = 38$.

Non-parametric Tests

The observed value of $W = 34$ is thus in the rejection region. Hence we reject H_0 and accept H_1, concluding that the evidence does suggest that the mean visibility in passage A is better than that in passage B.

5. We set up:

$H_0: \mu_A = \mu_B$ (where μ_A and μ_B are the mean survival times for
$H_1: \mu_A \neq \mu_B$ lifejackets of types A and B respectively)
$\alpha = 0.05$

We rank the observations in ascending order thus:

Rank	Time	Number	Rank	Time	Number
1	74	B4	$11\frac{1}{2}$	110	B3
2	84	A11	$11\frac{1}{2}$	110	A9
3	86	B2	13	114	B7
4	90	B8	14	116	A7
5	96	B5	15	122	A3
6	98	A2	16	129	A8
7	101	A1	17	132	A12
8	104	B1	18	139	A10
9	105	B6	19	150	A4
10	108	A5	20	158	A6

W_1 is the sum of the ranks of the smaller sample, i.e. the B-type jackets. Thus:

$$W_1 = 54.5$$

We calculate:

$$U = W_1 - \frac{n_B(n_B + 1)}{2} = 54.5 - \frac{8 \times 9}{2}$$
$$= \underline{18.5}$$

From tables, we find that the critical value of U for the given $n_A = 12(n_2)$, $n_B = 8(n_1)$ at 0.05 level and for a two-tailed test is 22. Hence our calculated value is within the critical region and we conclude that we have evidence to say that there is a significant difference between the mean survival time of the two types of lifejacket.

6. We set up:

$H_0: \mu_A = \mu_B$ (where μ_A and μ_B are the mean corrosion levels for
$H_1: \mu_A < \mu_B$ paints A and B respectively)
$\alpha = 0.05$

Marine Statistics

W_1 is the sum of the ranks of the smaller sample, i.e. the brand A paint treated test pieces. Thus:

$$W_1 = 29$$

$$U = W_1 - \frac{n_A(n_A + 1)}{2} = 30 - \frac{6 \times 7}{2}$$

$$= \underline{9}$$

The critical value of U for a one-tailed test at $\alpha = 0.05$ is found to be 14, using $n_1 = 6$ and $n_2 = 10$.

The calculated value is thus within the rejection region of $U \leqslant 14$. We thus reject H_0 and accept H_1, concluding that the tests suggest that paint B has better corrosion-resistant properties than paint A.

Chapter 12
Analysis of Variance

12.1 Introduction

In order to compare the performance of four models of a radio navigation aid of varying levels of sophistication, a sample of each was fitted to a ship which was then anchored at a surveyed position. Eight readings were taken at random intervals during a 24-hour period from each of the four models, but only one model was read at each occasion. The displacements of the observed position lines from the surveyed position were as follows (with northerly displacements being named positive and southerly displacements being named negative):

	Equipment Model			
	A	B	C	D
Distance in metres	−19	+227	+148	−24
	−75	+85	−37	+12
	−105	+263	+224	+186
	+3	+287	+258	+243
	−296	−24	−143	+220
	−127	+158	+84	+84
	−24	+183	−108	+141
	−253	−12	−174	+23
Mean	$\bar{x}_A = -112$	$\bar{x}_B = +146$	$\bar{x}_C = +32$	$\bar{x}_D = +111$

We want to know if there is any significant difference between the performance of the four equipment models.

For each model we may calculate the mean displacement and we can use this figure to represent the performance of the model. As we are comparing four means we would have to perform 6 different tests, 6 't' tests say, if we assume normal distributions for each of the populations. In fact if we compare n means we would have to perform $n(n-1)/2$ tests which could result in a considerable amount of work. Worse than this is the fact that as each test has a type one error attached to it, say a 1 in 20 chance of rejecting a true null hypothesis, then the chances of making at least one type one error when a

Marine Statistics

whole series of tests are performed becomes considerably larger than this. It would therefore be expedient if we could have one test which enabled us to compare all four sample means simultaneously to see if they do suggest significantly a difference between the respective four population means. The technique which has been devised to do this is termed the analysis of variance, often referred to as Anova for short. There are various different adaptations of this technique depending on how the raw data have been collected and in this book we will only have space to cover some of the more simple forms, but the fundamental principles remain the same so a reader requiring a more complicated form should have no trouble in using it from a more specialised book. However before proceeding with more details on even the simplest form we must first introduce a new statistic known as an 'F' statistic which forms the basis of the test.

12.2 'F' Distribution

Theoretically an 'F' distribution is defined as the ratio of two independent chi-squared distributions each divided by their degrees of freedom. Thus if a value of the first chi-squared distribution is χ_1^2 on ν_1 degrees of freedom and of the second is χ_2^2 on ν_2 degrees of freedom then we define a value of an F distribution as

$$F = \frac{\chi_1^2/\nu_1}{\chi_2^2/\nu_2}$$

Practically the distribution arises from considering the variances of two normally distributed populations estimated from a random sample from each of them. It will be recalled from Chapter 10 that if we take a random sample of size n_1 from a normally distributed population with variance σ_1^2 and calculate an estimate of σ_1^2 from the sample s_1^2 then

$$\chi_1^2 = \frac{(n_1 - 1)s_1^2}{\sigma_1^2} \sim \chi^2 \text{ on } n_1 - 1 \text{ d.o.f.}$$

Hence, since $\nu_1 = n_1 - 1$, $\chi_1^2/\nu_1 = s_1^2/\sigma_1^2$.

Similarly if we take a random sample of size n_2 from a second normally distributed population with σ_2^2 and s_2^2 correspondingly defined, then:

$$\chi_2^2 = \frac{(n_2 - 1)s_2^2}{\sigma_2^2} \quad \text{and} \quad \frac{\chi_2^2}{\nu_2} = \frac{s_2^2}{\sigma_2^2}$$

Thus we can define F as

$$\frac{s_1^2/\sigma_1^2}{s_2^2/\sigma_2^2} = \frac{s_1^2/s_2^2}{\sigma_1^2/\sigma_2^2}$$

Analysis of Variance

Since the actual value of each of the χ^2 distributions depends on their respective number of degrees of freedom, then the value of the particular F distribution depends on both these degrees of freedom. Hence we say that

$$F = \frac{s_1^2/\sigma_1^2}{s_2^2/\sigma_2^2} \sim F \text{ on } \nu_1, \nu_2 \text{ degrees of freedom}$$

where in this case $\nu_1 = n_1 - 1$ and $\nu_2 = n_2 - 1$. It is important when using an F distribution to remember that the values depend not only on both degrees of freedom but on their order too. The first value, ν_1, refers to the numerator and the second value, ν_2, to the denominator. Figure 12.1 shows a typical F curve.

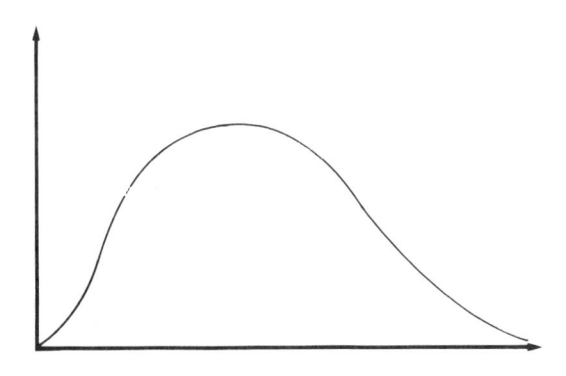

Fig. 12·1 A typical F distribution curve

As the F distribution is derived from the χ^2 distribution there are again no negative values of F, and the F distribution 'hump' is not symmetrical in any way.

Tables are published of percentage points of the F distribution, of which Table 4 at the back of the book is an example. However the situation is even worse space-wise than it was for the 't' and χ^2 distributions as the two degrees of freedom mean that each percentage point itself requires a whole page. Thus it is usual only to find the top 5% and 1% and possibly $2\frac{1}{2}$% and 0·1% points of the curves published, i.e. the points such that 5% or 1% or $2\frac{1}{2}$% or 0·1% of the curves lie above them. The most usual format is to have the degrees of freedom associated with the numerator, ν_1, across the columns of the page and the degrees of freedom associated with the denominator, ν_2, down the rows of the page.

Thus for example $F_{9,4,\,5\%} = 6·00$, i.e. the value of the F distribution on 9 and 4 degrees of freedom such that 5% of the curve lies above it is 6·00. In

Marine Statistics

contrast $F_{4, 9, 5\%} = 3.63$, so it is very important to get the correct ordering of the degrees of freedom.

The F distribution can be used to provide confidence limits for the ratio of two population variances and to provide a test statistic for hypotheses tests on two population variances, assuming that the populations are both normally distributed. If we are calculating confidence limits then we need to know the bottom percentage points of a distribution as well as the top ones. However there is a useful relationship which enables us to find lower percentage points and that is:

$$F_{v_1, v_2, 100\alpha\%} = \frac{1}{F_{v_2, v_1, (100-\alpha)\%}}$$

Hence if we required to know $F_{6, 7, 97.5\%}$, then

$$F_{6, 7, 97.5\%} = \frac{1}{F_{7, 6, 2.5\%}} = \frac{1}{5.70} = 0.18$$

As an example of a hypothesis test on two population variances consider the example in Chapter 9, Section 9.7, where we were discussing the 't' test for determining whether the means of two normally distributed populations could be considered equal or not. Two observers each took a series of 20 astro observations and we had details of their errors from a known position. To perform the test on the means, we had to assume the population variances were equal but we are now in a position to verify this. Let σ_1^2 be the variance of the population and s_1^2 be the variance of the sample of size n_1 for the first observer. Similarly, let σ_2^2, s_2^2 and n_2 be the corresponding quantities for the second observer.

The readings for the first observer were as follows:

+0.1 −0.1 +0.1 +1.1 +1.8 +0.2 +2.8 +0.3 −1.9 +0.3
+1.2 −3.2 −1.3 +0.8 −0.6 +0.9 +0.2 +2.7 +0.7 +0.0

Hence using the formula

$$s_1^2 = \frac{1}{n_1 - 1}\left[\sum x_i^2 - \frac{(\sum x_i)^2}{n_1}\right]$$

where the sumation is over the values of the first sample:

$$s_1^2 = \frac{1}{19}[39.15 - 1.86] = 1.96$$

The readings for the second observer were as follows:

−1.2 −0.2 −0.1 −2.7 +0.6 +1.6 +1.1 +0.6 −0.4 −0.8
−1.9 +1.7 +0.1 −1.0 −0.4 −1.4 −0.7 +2.3 −3.1 +0.1

Analysis of Variance

and using the formula again but summing over the values of the second sample:

$$s_2^2 = \frac{1}{19}[39\cdot10 - 1\cdot68] = 1\cdot97$$

$H_0: \sigma_1^2 = \sigma_2^2$
$H_1: \sigma_1^2 \neq \sigma_2^2$
$\alpha = 0\cdot05$

Test statistic F.

When we are performing a two-tail test like this with $H_0: \sigma_1^2 = \sigma_2^2$, or even a one-tail test, it is a good idea if we define our test statistic:

$$F = \frac{\text{larger value of sample variance}}{\text{smaller value of sample variance}}$$

By doing this we ensure that our test value is larger than one and hence the only critical value we need worry about is the upper one. Hence,

$$F = \frac{s_2^2/\sigma_2^2}{s_1^2/\sigma_1^2} = \frac{s_2^2}{s_1^2} \quad \text{under } H_0 \quad \sim F \text{ on } n_2 - 1, n_1 - 1 \text{ d.o.f.}$$

Our value of $F = 1\cdot97/1\cdot96 = 1\cdot01$.

As we are performing a two-tail test at a 5% level of significance, we only want 2·5% of the distribution in the upper critical region: $F_{19,19,2\cdot5\%} = 2\cdot61$. It should be noted that if we are not given values for all degrees of freedom then we simply use linear interpolation. Hence the upper critical region is $F \geqslant 2\cdot61$, but $1\cdot01 < 2\cdot61$, giving a non-significant result. Hence we can conclude that there is no evidence to suggest that the variances of the two populations are different.

Despite the initially complicated format of the F statistic when we define it theoretically, it is in fact fairly straight-forward to calculate, particularly when used, as it most frequently is, to test if two population variances are equal since the population variances cancel under the null hypothesis of equality. In this situation the test statistic to be calculated is simply the ratio of the two sample variances.

12.3 *One Way Analysis of Variance: Equal Sample Sizes*

The most straight-forward situation for an analysis of variance is when we wish to compare the means of several different samples and where the samples are of equal size. In general, suppose we have k samples each consisting of n different members. The introductory example to this chapter provides such a situation with 4 samples each consisting of 8 values, as we are comparing 4 models of the equipment on the basis of 8 readings on each. The 8 readings for each model differ from each other as we would expect in any set of statistical data and we can think of this as simply random variation. However,

Marine Statistics

what we are interested in is whether the 4 sets of readings differ from each other by very much more than could be expected from random processes. Our interest is in whether there is a significant variation between the means of the 4 samples—we are less directly concerned with any variation within the samples. Hence we use the term one-way analysis of variance. Before proceeding with further details it will be helpful to establish some notation to make the description of the process easier. In general, we will have a situation with k samples each consisting of n readings which we can write in a matrix form thus:

		Sample				
		1	2	3	$\cdots$	k
Reading	1	x_{11}	x_{12}	x_{13}	$\cdots$	x_{1k}
	2	x_{21}	x_{22}	x_{23}	$\cdots$	x_{2k}
	3	x_{31}	x_{32}	x_{33}	$\cdots$	x_{3k}
	$\vdots$	$\vdots$	$\vdots$	$\vdots$		$\vdots$
	n	x_{n1}	x_{n2}	x_{n3}	$\cdots$	x_{nk}
	Total	$T_{.1}$	$T_{.2}$	$T_{.3}$	$\cdots$	$T_{.k}$
	Mean	$\bar{x}_{.1}$	$\bar{x}_{.2}$	$\bar{x}_{.3}$	$\cdots$	$\bar{x}_{.k}$

Thus x_{32} is the third reading in the second sample, for example, and in general x_{ij} will be the ith reading in the jth sample where $i = 1, 2, \ldots, n$ and $j = 1, 2, \ldots, k$.

We let $T_{.1}$ be the total of readings in the first sample where $T_{.1} = \sum_{i=1}^{n} x_{i1}$. Similarly the totals for all k samples may be defined where

$$T_{.j} = \sum_{i=1}^{n} x_{ij} \quad j = 1, \ldots, k$$

As a general rule with this type of notation, a . signifies that a summation has taken place over the suffix which it replaces. The means of each of the samples may then be indicated in a similar way. Thus

$$\bar{x}_{.j} = \frac{T_{.j}}{n} = \frac{\sum_{i=1}^{n} x_{ij}}{n} \quad j = 1, \ldots, k \text{ is the mean of the } i\text{th sample}$$

If we sum all the observations over all the samples we get the grand total $T_{..}$, where

$$T_{..} = \sum_{j=1}^{k} T_{.j} = \sum_{j=1}^{k} \sum_{i=1}^{n} x_{ij}$$

The overall mean of the whole data set is written as

$$\bar{x}_{..} = \frac{T_{..}}{nk} = \frac{\sum_{j=1}^{k} \sum_{i=1}^{n} x_{ij}}{nk}$$

Analysis of Variance

As an example of this notation, if we refer back to the initial example, then by adding the first column we get that $T_{.1} = -896$ and $\bar{x}_{.1} = \bar{x}_A = -112$. Similarly, $T_{.2} = 1,167$ and $\bar{x}_{.2} = \bar{x}_B = 146$, $T_{.3} = 252$ and $\bar{x}_{.3} = \bar{x}_C = 32$ and $T_{.4} = 885$ with $\bar{x}_{.4} = \bar{x}_D = 111$. The grand total of all the readings $T_{..} = 1,408$ and the overall mean $\bar{x}_{..} = 44 = 1,408/32$.

Having established some of the basic notation there are some assumptions which must be made. Firstly, we assume that each of the populations from which the samples are drawn have a normal distribution and, secondly, we assume that they have an equal variance, say σ^2. These assumptions are very similar to the small sample two-population situation whereby to perform the parametric 't' test we had to assume normal distributions and equal variances. If we call μ_j the mean of the jth population with $j = 1, 2, \ldots, k$, then the algebraic model which we can assume for x_{ij}, the ith reading in the jth sample, is $x_{ij} = \mu_j + \varepsilon_{ij}$, where ε_{ij} (ε = epsilon) is the difference between the reading and the appropriate population mean. ε_{ij} is often called the error term because it represents the natural error that is inherent in data sets. Algebraically we may write that $x_{ij} \sim N(\mu_j, \sigma^2)$ or its equivalent is that $\varepsilon_{ij} \sim N(0, \sigma^2)$. Many authors prefer to write $\mu_j = \mu + \alpha_j$, $j = 1, 2, \ldots, k$, where μ is the overall mean treating all the populations together and α_j is termed the jth population effect. With this alternative notation $x_{ij} = \mu + \alpha_j + \varepsilon_{ij}$. It does not matter which notation is used but it is important to note that we are assuming that the error term adds on to the term, or terms, representing the mean level and hence does not depend on what the mean level is. If we use the alternative notation of $x_{ij} = \mu + \alpha_j + \varepsilon_{ij}$, then we also have to assume that the individual population effects, the α_j's, sum to be 0, i.e. $\sum_{j=1}^{k} \alpha_j = 0$. The null hypothesis which we are interested in testing is that the means of the k populations are all equal. The alternative hypothesis to this is that there is at least one that is different from the others. It is important to note that this logically is the alternative hypothesis if the null hypothesis is disproved and not that the means are all different. Symbolically we may write this as:

$H_0: \mu_1 = \mu_2 = \cdots = \mu_k$
H_1: At least one $\mu_j, j = 1, \ldots, k$, is different

If we use the alternative notation the alternative hypothesis is slightly simpler to write:

$H_0: \alpha_1 = \alpha_2 = \cdots = \alpha_k = 0$
$H_1: \alpha_j \neq 0$ for at least one $j = 1, \ldots, k$

The test is based on a comparison of two independent estimates of the common population variance σ^2. These estimates are obtained by splitting the total variability of our data into two components. If we imagine our data pooled into one large sample then the total variability is the sum of the

Marine Statistics

squared deviations of each individual item from the overall mean, i.e.

$$\sum_{i=1}^{n} \sum_{j=1}^{k} (x_{ij} - \bar{x}_{..})^2$$

It may be shown algebraically that this total sum of squares is equal to two separate components, i.e.

$$\sum_{i=1}^{n} \sum_{j=1}^{k} (x_{ij} - \bar{x}_{..})^2 = n \sum_{j=1}^{k} (\bar{x}_{.j} - \bar{x}_{..})^2 + \sum_{i=1}^{n} \sum_{j=1}^{k} (x_{ij} - \bar{x}_{.j})^2$$

The proof of this is not difficult but is not essential to actually performing an analysis of variance.

The first component, $n \sum_{j=1}^{k} (\bar{x}_{.j} - \bar{x}_{..})^2$, measures how each of the sample means differs from the overall mean, i.e. the variability of the sample means, also termed the between samples variability. We often think of the different populations arising because of different treatments being compared, hence this term is referred to as the between treatments variability or between treatments sum of squares.

The second component is $\sum_{i=1}^{n} \sum_{j=1}^{k} (x_{ij} - \bar{x}_{.j})^2$. If we take $j = 1$ then $\sum_{i=1}^{n} (x_{i1} - \bar{x}_{.1})^2$ is the variability within the first sample. Similarly, if $j = 2$ then $\sum_{i=1}^{n} (x_{i2} - \bar{x}_{.2})^2$ is the variability within the second sample. Thus, as we sum over all the values of j we are measuring the variability within the samples or the within treatments variability. This is also known as the error variability or error sum of squares. If we let SST be the total sum of squares, SSC be the sum of squares between treatments and SSE be the error sum of squares, then we can rewrite this relationship as

$$SST = SSC + SSE$$

If we wanted to estimate the common variance σ^2 treating the situation as one large population then we would calculate:

$$\hat{\sigma}^2 = \frac{\sum_{i=1}^{n} \sum_{j=1}^{k} (x_{ij} - \bar{x}_{..})^2}{nk - 1} \equiv \hat{\sigma}^2 = \frac{SST}{nk - 1}$$

The divisor is $nk - 1$ since there are nk observations but one unknown parameter, so there are only $nk - 1$ degrees of freedom. If we wanted to estimate the value of σ^2 from the error variability within each sample then we would have

$$\hat{\sigma}^2 = \frac{\sum_{i=1}^{n} \sum_{j=1}^{k} (x_{ij} - \bar{x}_{.j})^2}{nk - k} = \frac{SSE}{nk - k}$$

The divisor is $nk - k$ this time as we again have nk observations but k unknown parameters estimated by the sample means $\bar{x}_{.1}, \ldots, \bar{x}_{.k}$. Under H_0 if the population (or treatment) means do not differ from each other, then the

Analysis of Variance

variability which we are observing between the sample means can be no different than the ordinary error variance. Hence under H_0 another estimate of σ^2 would be given by

$$\hat{\sigma}^2 = \frac{\sum_{j=1}^{k}(\bar{x}_{.j} - \bar{x}_{..})^2}{k-1} = \frac{SSC}{k-1}$$

It is useful to note that the partitioning of the total sum of squares has led to a similar partitioning of the total number of degrees of freedom, i.e. $nk - 1 = nk - k + k - 1$. If we let $s_1^2 = \hat{\sigma}_1^2 = SSC/(k-1)$, the estimate from our between treatments variation, and we let $s_2^2 = \hat{\sigma}_2^2 = SSE/(nk - k)$, the estimate from our error variation, then our test becomes a test of the hypotheses

$$H_0: \sigma_1^2 = \sigma_2^2 \quad \text{and} \quad H_1: \sigma_1^2 > \sigma_2^2$$

H_1 is $\sigma_1^2 > \sigma_2^2$ since, in the original alternative hypothesis, if one of the means at least is significantly different from the others then the estimate of σ^2 from the between treatments variation must be an overestimate. This pair of hypotheses is tested using an 'F' statistic with

$$F = \frac{s_1^2/\sigma_1^2}{s_2^2/\sigma_2^2} = \frac{s_1^2}{s_2^2} \quad F \text{ on } (k-1, nk-k) \text{ d.o.f. under } H_0$$

We can set out the above procedure in what is termed an analysis of variance table or ANOVA table for short:

Anova

Source of variation	S.S	D.o.f.	M.S.	F
Between treatments	SSC	$k-1$	$SSC/k - 1 = s_1^2$	s_1^2/s_2^2
Within treatments	SSE	$nk - k$	$SSE/nk - k = s_2^2$	
Total	SST	$nk - 1$		

The abbreviations in the table are S.S.—sum of squares, D.o.f.—degrees of freedom, and M.S.—mean squares, i.e. sum of squares divided by degrees of freedom, and F—'F' statistic. Before we actually calculate the values for an example it is useful to note that the sum of squares terms may be written in an easier form for calculation. Firstly, the total sum of squares,

$$SST = \sum_{i=1}^{n}\sum_{j=1}^{k}(x_{ij} - \bar{x}_{..})^2 = \sum_{i=1}^{n}\sum_{j=1}^{k}x_{ij}^2 - \frac{T_{..}^2}{nk}$$

Secondly, the treatment sum of squares,

$$SSC = \sum_{j=1}^{k}(\bar{x}_{.j} - \bar{x}_{..})^2 = \frac{\sum_{j=1}^{k}T_{.j}^2}{n} - \frac{T_{..}^2}{nk}$$

Marine Statistics

where the terms $T_{.j}$ and $T_{..}$ are the total terms defined at the start of this section. Both these forms are analogous to the computational form for the variance which we have used in earlier chapters.

Finally, since SST = SSC + SSE, the easiest way to calculate the error sum of squares is to subtract the treatment sum of squares from the total sum of squares, i.e. SSE = SST − SSC.

Let us now consider the example on the four models of the radio navigation aid described at the start of the chapter. We will assume that the readings from each of the models form a normal distribution with a mean error of 0 and a common variance of σ^2. We will further assume that x_{ij}, the ith reading from the jth model, can be expressed as follows:

$$x_{ij} = \mu + \alpha_j + \varepsilon_{ij} \quad \begin{array}{l} i = 1, \ldots, 8 \text{ (8 readings)} \\ j = 1, \ldots, 4 \text{ (4 models)} \end{array}$$

where μ is the overall mean reading, α_j is the jth model effect and ε_{ij} is the error term.

We want to test

$$H_0: \alpha_1 = \alpha_2 = \alpha_3 = \alpha_4 = 0$$
$$H_1: \alpha_j \neq 0 \text{ for at least one } j = 1, \ldots, 4$$
$$\alpha = 0 \cdot 05$$

Total sum of squares:

$$\text{SST} = \sum_{i=1}^{8} \sum_{j=1}^{4} x_{ij}^2 - \frac{T_{..}^2}{8.4}$$

where

$$T_{..} = \sum_{j=1}^{4} \sum_{i=1}^{8} x_{ij}$$

$$= [-19 - 75 - 105 \cdots + 23]$$
$$= 1{,}408$$

$$\text{SST} = [(-19)^2 + (-75)^2 + (-105)^2 + \cdots + (+23)^2] - \frac{1{,}408^2}{32}$$

$$= 834{,}504 - 61{,}952$$
$$= 772{,}552$$

$$\text{SSC} = \frac{\sum_{j=1}^{4} T_{.j}^2}{8} - \frac{T_{..}^2}{8.4}$$

$$T_{.1} = [-19 - 75 + \cdots - 253] = -896$$
$$T_{.2} = [227 + 85 + \cdots - 12] = 1{,}167$$
$$T_{.3} = [148 - 37 + \cdots - 174] = 252$$
$$T_{.4} = [-24 + 12 + \cdots + 23] = 885$$

252

Analysis of Variance

Hence

$$SSC = \frac{(-896)^2 + (1{,}167)^2 + (252)^2 + (885)^2}{8} - \frac{1{,}408^2}{32}$$

$$= 376{,}429 - 61{,}952$$
$$= 314{,}477$$

$$SSE = SST - SSC$$
$$= 772{,}552 - 314{,}477$$
$$= 458{,}075$$

Hence the Anova table is:

Source	S.S.	D.o.f.	M.S.	F
Between treatments	314,477	3	104,826	6·41
Within treatments	458,075	28	16,360	
Total	772,552	31		

Critical region $F > F_{3,28,5\%}$, i.e. $F > 2{\cdot}95$; $6{\cdot}41 > 2{\cdot}95$ hence we have a significant result.

We may thus conclude that it is likely that at least one of the treatment effects is significantly different from zero. In other words it is likely that at least one model produces readings whose mean is significantly different from the others.

At this stage this is the only conclusion we can make. If we had had a non-significant result for our analysis of variance then we would stop any further analysis here and conclude that there was no difference between the means of the various populations. However, as we have got a significant result we can go on and try to ascertain what differences there are.

12.4 Least Significant Difference

The least significant difference method enables us to distinguish between pairs of means to find significant differences. If we consider the previous example then we have that the means of the readings for the various equipment models were as follows:

$$\bar{x}_A = -112 \quad \bar{x}_B = +146 \quad \bar{x}_C = +32 \quad \bar{x}_D = +111$$

Rearranging them in order of magnitude we have:

$$\bar{x}_A = -112 \quad \bar{x}_C = +32 \quad \bar{x}_D = +111 \quad \bar{x}_B = +146$$

We want to perform a series of tests to test $H_0: \mu_i = \mu_j$ against $H_1: \mu_i \neq \mu_j$, where μ_i and μ_j are the population means corresponding to the sample means, $\bar{x}_i$ and $\bar{x}_j$, and i and j range over all the different combinations of the four

Marine Statistics

letters A, B, C and D. Since, to perform the analysis of variance, we assumed that all the populations were normally distributed with equal variances, σ^2, then we can use a 't' test to compare two means. Our test statistic would be:

$$t = \frac{\bar{x}_i - \bar{x}_j}{\hat{\sigma}\sqrt{\frac{1}{n_i} + \frac{1}{n_j}}} \sim t \text{ on } \nu \text{ d.o.f. under } H_0$$

n_i and n_j would be the sample sizes corresponding to $\bar{x}_i$ and $\bar{x}_j$ but in this case all the samples are equal to n. Hence the statistic may be rewritten as

$$t = \frac{\bar{x}_i - \bar{x}_j}{\hat{\sigma}\sqrt{\frac{2}{n}}}$$

The best estimate we can get for σ is the square root of the error mean square from the analysis of variance table, and the degrees of freedom for the 't' statistic will be the degrees of freedom associated with this. In this example,

$$\hat{\sigma} = \sqrt{16{,}360} = 127 \cdot 91$$

with 28 degrees of freedom. Hence

$$t = \frac{\bar{x}_i - \bar{x}_j}{127 \cdot 91\sqrt{\frac{2}{8}}}$$

since $n = 8$. Thus

$$t = \frac{\bar{x}_i - \bar{x}_j}{63 \cdot 95}$$

If we perform the test at the 5% level of significance, then the critical region will be $|t| > t_{28, 0 \cdot 025}$, i.e. $|t| > 2 \cdot 048$. Thus for the difference $\bar{x}_i - \bar{x}_j$ to be significant we want $|\bar{x}_i - \bar{x}_j|/63 \cdot 95 > 2 \cdot 048$, i.e. $|\bar{x}_i - \bar{x}_j| > 131$. Hence we may conclude in this example that there is likely to be a significant difference in the population means considered in pairs of the sample means are more than 131 m different.

Since $\bar{x}_A = -112$, $\bar{x}_C = +32$, $\bar{x}_D = +111$, $\bar{x}_B = +146$, we can see that $\bar{x}_A$ is significantly lower than all of the other three but there is no significant difference between the others since even the largest difference of 114 between $\bar{x}_C$ and $\bar{x}_B$ is less than 131.

Hence our conclusion would be that equipment model A produces a mean reading that is significantly different from the other three. In fact by inspection it can be seen that equipment A has a southerly bias, whereas equipments B, C and D all have northerly biases. Although when we first saw the results it looked as though equipment A differed from the others, without the analysis it would have been impossible to tell if it was worth considering as important

Analysis of Variance

the difference between the other three or even the difference between A and the others.

Returning to the least significant difference method, the reader has probably noticed, as the example progressed, that the difference between this test and a 't' test for comparing the means taking the samples in pairs is that we have been able to use an estimate for the common variance, σ, based on all 32 observations rather than having to recalculate it for each pair based on only 16 observations. The more observations we have the more precise our answer is likely to be and hence we are less likely statistically to make the wrong conclusions. However, as an extra safeguard it is recommended that the least significant difference method should only be used when the analysis of variance produces a significant result. Otherwise one still runs an increased risk of finding a significant difference when in fact it does not exist.

12.5 One Way Analysis of Variance: Unequal Sample Sizes

In the previous section we dealt with the situation where we wanted to compare the means of k samples, where each sample consisted of an equal number of readings, say n. However, in practice we often end up with having unequal sample sizes. The actual philosophy of the analysis of variance remains unaltered from the first case, but the calculations become slightly more complicated.

We will suppose that the first sample has n_1 observations in it, the second n_2 and so on, with the kth sample having n_k observations. Using the same notation as before, the calculation formulae for the sums of squares may be shown to be as follows:

Total sum of squares,

$$\text{SST} = \sum_{j=1}^{k} \sum_{i=1}^{n_j} x_{ij}^2 - \frac{T_{..}^2}{N}$$

where n_j is the number in the jth sample. $N = \sum_{j=1}^{k} n_j$ is the total number of observations, and $T_{..} = \sum_{j=1}^{k} \sum_{i=1}^{n_j} x_{ij}$ is the sum of all the observations.

Column sum of squares (between treatments),

$$\text{SSC} = \sum_{j=1}^{k} \frac{T_{.j}^2}{n_j} - \frac{T_{..}^2}{N}$$

where $T_{.j} = \sum_{i=1}^{n_j} x_{ij}$ is the sum of observations in the jth sample.

Error sum of squares (within treatments),

$$\text{SSE} = \text{SST} - \text{SSC}$$

Although the formulae look more complicated, they are very similar to the previous case when it comes to using them. To calculate the total sum of

Marine Statistics

squares, it is again a matter of finding the sum of the squares of each observation in the total data set and subtracting the correction factor of the squared sum of all the observations divided by the total number of observations. The between treatments sum of squares is found by totalling each sample, squaring the result and then dividing by the number of observations in the sample, before finding the sum of these figures for all the samples and then subtracting the correction factor as for the total sum of squares. The error sum of squares is calculated, as before, by taking the difference between the total sum of squares and the between treatments sum of squares. The analysis of variance table is constructed as follows:

Anova

Source of variation	S.S.	D.o.f.	M.S.	F
Between treatments	SSC	$k - 1$	$SSC/k - 1 = s_1^2$	s_1^2/s_2^2
Within treatments	SSE	$N - k$	$SSE/N - k = s_2^2$	
Total	SST	$N - 1$		

An example should help to illustrate these points.

For the purpose of a survey of marine traffic in a particular area, the ships rounding a buoy marking a shoal patch were classed as being ferries, freighters, tankers or fisherman. During a survey period of 4 hr around high tide the distances at which ships rounded the buoy were measured from the shore by range-finder with the following results given in cables (tenths of a nautical mile):

Ferries	Freighters	Tankers	Fishermen
8	11	13	5
7	8	9	2
10	12	16	4
7	15	12	1
	10	14	
	13		

Is this sufficient evidence that there is a systematic difference in the range at which different classes of ship pass the buoy? Assuming that, for each type of ship, the distribution of distances from the buoy is normal and that all the variances are equal, we can use the analysis of variance to test if there is any significant difference between the mean distances. We will assume that a general reading x_{ij} can be written as:

$$x_{ij} = \mu + \alpha_j + \varepsilon_{ij} \quad i = 1, \ldots, n_j; j = 1, \ldots, 4$$

and
$$n_1 = 4 \quad n_2 = 6 \quad n_3 = 5 \quad n_4 = 4$$
μ is the overall mean, α_j is the effect of the jth ship type and ε_{ij} is the random error where $\varepsilon_{ij} \sim N(0, \sigma^2)$.

We proceed as follows:
$$H_0: \alpha_1 = \alpha_2 = \alpha_3 = \alpha_4 = 0$$
$$H_1: \alpha_j \neq 0 \text{ for at least one } j = 1, \ldots, 4$$
$$\alpha = 0.05$$

Total sum of squares,
$$\text{SST} = \sum_{j=1}^{4} \sum_{i=1}^{n_j} x_{ij}^2 - \frac{T_{..}^2}{N}$$

where $N = \sum_{j=1}^{4} n_j = 4 + 6 + 5 + 4 = 19$.

$$T_{..} = \sum_{j=1}^{4} \sum_{i=1}^{n_j} x_{ij} = 8 + 7 + \cdots + 1 = 177$$

$$\text{SST} = [8^2 + 7^2 + \cdots + 1^2] - \frac{177^2}{19}$$

$$= 1{,}977 - 1{,}648 \cdot 89 = 328 \cdot 11$$

$$T_{.1} = \sum_{i=1}^{4} x_{i1} = 32 \qquad T_{.2} = \sum_{i=1}^{6} x_{i2} = 69$$

$$T_{.3} = \sum_{i=1}^{5} x_{i3} = 64 \qquad T_{.4} = \sum_{i=1}^{4} x_{i4} = 12$$

Between treatments sum of squares,
$$\text{SSC} = \sum_{j=1}^{4} \frac{T_{.j}^2}{n_j} - \frac{T_{..}^2}{N} = \frac{32^2}{4} + \frac{69^2}{6} + \frac{64^2}{5} + \frac{12^2}{4} - \frac{177^2}{19}$$

$$= 256 \cdot 00 + 793 \cdot 50 + 819 \cdot 20 + 36 \cdot 00 - 1{,}648 \cdot 89$$
$$= 255 \cdot 81$$

SSE = SST − SSC = 328·11 − 255·81 = 72·30.

Anova

Source of variation	S.S.	D.o.f.	M.S.	F
Between treatments	255·81	3	85·27	17·69
Within treatments	72·30	15	4·82	
Total	328·11	18		

$F_{(3,15), 5\%} = 3 \cdot 29 \qquad 17 \cdot 69 > 3 \cdot 29 \Rightarrow$ significant result

Hence we may conclude that there seems to be a significant difference between the distances at which different types of ships pass the buoy. To establish how the differences arise we will look at a modified form of the least significant

Marine Statistics

difference method. We cannot apply it exactly as before as we have different sample sizes but we may still take as an estimate of the common variance, σ^2, the within treatments mean square, i.e. 4·82. Our test statistic is:

$$t = \frac{|\bar{x}_i - \bar{x}_j|}{\sqrt{\left[\frac{1}{n_i} + \frac{1}{n_j}\right]}} \sim t \text{ on 15 d.o.f. under } H_0$$

At the 5% level of significance, then, the critical region will be given by:

$$|t| > t_{15, 0.025} \text{ i.e. } |t| > 2.13$$

This is equivalent to saying

$$\frac{|\bar{x}_i - \bar{x}_j|}{\sqrt{\frac{1}{n_i} + \frac{1}{n_j}}} > 2.13 \times \sqrt{4.82} = 4.68$$

Now

$\bar{x}_1(\text{ferries}) = 8.0$ $\bar{x}_2(\text{freighters}) = 11.5$
$\bar{x}_3(\text{tankers}) = 12.8$ $\bar{x}_4(\text{fishermen}) = 3.0$

Thus, comparing the types of ships in pairs we get the following results:

Ferries and freighters,

$$\frac{|\bar{x}_1 - \bar{x}_2|}{\sqrt{\frac{1}{4} + \frac{1}{6}}} = \frac{3.5}{0.65} = 5.38 > 4.68 \Rightarrow \text{significant difference}$$

Ferries and tankers,

$$\frac{|\bar{x}_1 - \bar{x}_3|}{\sqrt{\frac{1}{4} + \frac{1}{5}}} = \frac{4.8}{0.67} = 7.16 > 4.68 \Rightarrow \text{significant difference}$$

Ferries and fishermen,

$$\frac{|\bar{x}_1 - \bar{x}_4|}{\sqrt{\frac{1}{4} + \frac{1}{4}}} = \frac{5}{0.71} = 7.04 > 4.68 \Rightarrow \text{significant difference}$$

Freighters and tankers,

$$\frac{|\bar{x}_2 - \bar{x}_3|}{\sqrt{\frac{1}{6} + \frac{1}{5}}} = \frac{1.3}{0.61} = 2.13 < 4.68 \Rightarrow \text{no significant difference}$$

Freighters and fishermen,

$$\frac{|\bar{x}_2 - \bar{x}_4|}{\sqrt{\frac{1}{6} + \frac{1}{4}}} = \frac{8.5}{0.65} = 13.08 > 4.68 \Rightarrow \text{significant difference}$$

Tankers and fishermen,

$$\frac{|\bar{x}_3 - \bar{x}_4|}{\sqrt{\frac{1}{5} + \frac{1}{4}}} = \frac{9.8}{0.67} = 14.62 > 4.68 \Rightarrow \text{significant difference}$$

Analysis of Variance

Summarising these results we can see firstly that there appears to be no difference between the behaviour of freighters and tankers on the distance at which they pass the buoy. Fishermen take the smallest passing distance of all and this is significantly smaller than all other types of ship. Ferries have a significantly larger passing distance than fishermen but one that is significantly smaller than freighters and tankers.

12.6 Two Way Analysis of Variance: One Observation per Cell

Suppose that the examination marks of five students, chosen at random from a first year group on a maritime studies degree, are as follows for four different papers in mathematics, physics, navigation and naval architecture respectively:

Student	Mathematics	Physics	Navigation	Naval architecture
1	68	57	73	61
2	83	94	91	86
3	72	81	63	59
4	55	73	77	66
5	92	68	75	87

We want to test the following hypotheses:

(i) the courses are of equal difficulty as measured by the mean mark;
(ii) the students have equal ability.

This example differs from the previous situations in that this time there is a connection between the observations in each column and in each row. In the first example, comparing equipment models, if we had taken the observations on each model simultaneously we would have been able to compare between occasions as well as between equipment. In this new example, as we have the marks for each examination measured on the same students, we may compare between students as well as between examinations. This situation is suitable for analysis by a two-way analysis of variance. We need to introduce some extra notation for this as follows: Let us assume that we have r rows and c columns. This time we will only consider the most straight-forward case where we have no missing observations. Let $T_{i.} = \sum_{j=1}^{c} x_{ij}$ be the sum of the ith row, $i = 1, \ldots, r$, and $\bar{x}_{i.} = T_{i.}/c$ be the mean of the ith row. $T_{.j}$ and $\bar{x}_{.j}$ will be the sum and mean respectively of the jth column, $j = 1, \ldots, c$, defined as before, and $T_{..}$ will be the grand total with $\bar{x}_{..}$ the overall mean defined as before.

Marine Statistics

Our total variation is given by $\sum_{i=1}^{r}\sum_{j=1}^{c}(x_{ij}-\bar{x}_{..})^2$ as before and this may be split as follows:

$$\sum_{i=1}^{r}\sum_{j=1}^{c}(x_{ij}-\bar{x}_{..})^2 = c\sum_{i=1}^{r}(\bar{x}_{i.}-\bar{x}_{..})^2 + r\sum_{j=1}^{c}(\bar{x}_{.j}-\bar{x}_{..})^2$$
$$+ \sum_{i=1}^{r}\sum_{j=1}^{c}(x_{ij}-\bar{x}_{i.}-\bar{x}_{.j}+\bar{x}_{..})^2$$

The first term on the right-hand side shows us how the row means vary from each other, the second term how the column means vary and the final term is the residual variation, or what we have termed the error variation. Hence if SST, SSC and SSE are defined as before and we let SSR denote the between rows sum of squares we have the identity that:

$$\text{SST} = \text{SSR} + \text{SSC} + \text{SSE}$$

We will assume that each observation x_{ij} may be written in the form

$$x_{ij} = \mu + \alpha_i + \beta_j + \varepsilon_{ij} \quad i=1,\ldots,r;\, j=1,\ldots,c$$

where μ is the overall mean, α_i is the effect of being in the ith row, β_j is the effect of being in the jth row, ε_{ij} is the error term, and $\sum_{i=1}^{r}\alpha_i = 0$, $\sum_{j=1}^{c}\beta_j = 0$, and $\varepsilon_{ij} \sim N(0,\sigma^2)$.

We have rc observations in all; hence if we wished to estimate σ^2 from all of them we would have:

$$\sigma^2 = \frac{\text{SST}}{rc - 1}$$

We can split up the degrees of freedom as before, giving:

$$rc - 1 = (r-1) + (c-1) + (rc - r - c + 1)$$

If we want to test for differences between row means we would take as hypotheses:

$H_0: \alpha_1 = \alpha_2 = \cdots = \alpha_r = 0$
$H_1: \alpha_i \neq 0$ for at least one $i = 1,\ldots,r$

Using similar arguments as in the one-way case, we can test this by considering whether an estimate of σ^2 based on the row variation, i.e. $\hat{\sigma}^2 = \text{SSR}/(r-1)$, is significantly different from an estimate of σ^2 based on the error term only, i.e. $\hat{\sigma}^2 = \text{SSE}/(rc - r - c + 1)$. This is tested using an F statistic $= [\text{SSR}/(r-1)]/[\text{SSE}/(rc - r - c + 1)]$ which, under H_0, is distributed as F on $r - 1$ and $rc - r - c + 1$ degrees of freedom. If we want to test for differences between column means we would take as hypotheses:

$H_0': \beta_1 = \beta_2 = \cdots = \beta_c = 0$
$H_1': \beta_j \neq 0$ for at least one β_j, $j = 1,\ldots,c$

Analysis of Variance

Our test statistic here would be $F = [\text{SSC}/(c-1)]/[\text{SSE}/(rc-r-c+1)]$ which, under H_0', is distributed as F on $c-1$ and $rc-r-c+1$ degrees of freedom.

The computational sums of squares are as follows:

$$\text{SST} = \sum_{i=1}^{r}\sum_{j=1}^{c} x_{ij}^2 - \frac{T_{..}^2}{rc} \quad \text{(total sum of squares)}$$

$$\text{SSR} = \frac{\sum_{i=1}^{r} T_{i.}^2}{c} - \frac{T_{..}^2}{rc} \quad \text{(row sum of squares)}$$

$$\text{SSC} = \frac{\sum_{j=1}^{c} T_{.j}^2}{r} - \frac{T_{..}^2}{rc} \quad \text{(column sum of squares)}$$

and $\text{SSE} = \text{SST} - \text{SSR} - \text{SSC}$.

The analysis of variance table is as shown:

Source of variation	S.S.	D.o.f.	M.S.	F
Between rows	SSR	$r-1$	$\frac{\text{SSR}}{r-1} = s_1^2$	s_1^2/s_3^2
Between columns	SSC	$c-1$	$\frac{\text{SSC}}{c-1} = s_2^2$	s_2^2/s_3^2
Error	SSE	$rc-r-c+1$	$\frac{\text{SSE}}{rc-r-c+1} = s_3^2$	
Total	SST	$rc-1$		

If we calculate these values for the student examination problem we get the following results ($r = 5$, $c = 4$):

$$T_{..} = \sum_{i=1}^{5}\sum_{j=1}^{4} x_{ij} = 68 + 57 + \cdots + 87 = 1{,}481$$

It does not matter if the rows are summed first and then the columns, or vice versa:

$$\text{SST} = \sum_{i=1}^{5}\sum_{j=1}^{4} x_{ij} - \frac{T_{..}^2}{20} = 68^2 + 57^2 + \cdots + 87^2 - 1{,}481^2$$
$$= 112{,}441 - 109{,}668 \cdot 5 = 2{,}772 \cdot 95$$

$$\text{SSR} = \sum_{i=1}^{5} \frac{T_{i.}^2}{4} - \frac{T_{..}^2}{20}$$

$T_{1.} = 68 + 57 + 73 + 61 = 259$
$T_{2.} = 83 + 94 + 91 + 86 = 354$
$T_{3.} = 72 + 81 + 63 + 59 = 275$
$T_{4.} = 55 + 73 + 77 + 66 = 271$
$T_{5.} = 92 + 68 + 75 + 87 = 322$

Marine Statistics

Thus

$$\text{SSR} = \frac{259^2 + 354^2 + 275^2 + 271^2 + 322^2}{4} - \frac{1{,}481^2}{20}$$

$$= 111{,}286 \cdot 75 - 109{,}668 \cdot 05 = 1{,}618 \cdot 70$$

$$\text{SSC} = \sum_{i=1}^{4} \frac{T_{.j}^2}{5} - \frac{T_{..}^2}{20}$$

$T_{.1} = 68 + 83 + 72 + 55 + 92 = 370$
$T_{.2} = 57 + 94 + 81 + 73 + 68 = 373$
$T_{.3} = 73 + 91 + 63 + 77 + 75 = 379$
$T_{.4} = 61 + 86 + 59 + 66 + 87 = 359$

Hence

$$\text{SSC} = \frac{370^2 + 373^2 + 379^2 + 359^2}{5} - \frac{1{,}481^2}{20}$$

$$= 109{,}710 \cdot 20 - 109{,}668 \cdot 05 = 42 \cdot 15$$

$$\text{SSE} = \text{SST} - \text{SSR} - \text{SSC}$$

$$= 2{,}772 \cdot 95 - 1{,}618 \cdot 70 - 42 \cdot 15 = 1{,}112 \cdot 10$$

This gives the following analysis of variance table:

Source of variation	S.S.	D.o.f.	M.S.	F
Between rows (students)	1,618·70	4	404·68	4·37
Between columns (courses)	42·15	3	14·05	0·15
Error	1,112·10	12	92.68	
Total	2,772·95	19		

$F_{(4,12),\,5\%} = 3 \cdot 26 \quad \quad 4 \cdot 37 > 3 \cdot 26 \Rightarrow$ significant result
$F_{(3,12),\,5\%} = 3 \cdot 49 \quad \quad 0 \cdot 15 < 3 \cdot 49 \Rightarrow$ non-significant result

The first question we considered was whether the courses are of equal difficulty, which is answered by considering the between columns test which gave a non-significant result. The second question concerned the students, which is answered by considering the between rows test which gave a significant result. Hence our conclusions would be that on the basis of the mean mark per course, there appears to be no significant difference in the courses and on the basis of the mean mark per student, there does appear to be a significant difference in the students' attainments (which might reflect the amount of work they had done rather than their ability!).

With this two-way analysis of variance not only have we been able to test two factors simultaneously but we might have been able to get a more precise result on either one of them. If both the row and column effects are significant then the resultant error mean squares is smaller than it would be if only a one-way analysis of variance is performed.

12.7 Two Way Analysis of Variance: Several Observations per Cell

The final form which we will consider in this book is the situation whereby we have variation in two directions and several observations per cell, i.e. on the particular combination of the two factors. The advantage of this type of situation is that we are able to detect interactions which might arise between the factors. In the previous examples we have assumed that the terms all add on to each other which is the concept known as *additivity*. However it may well be that the row and column effects interact.

Suppose we conduct an experiment to compare the efficiency of seamen in steering by day and by night, by gyro and by magnetic compass. If the measure of efficiency shows that the difference between day and night steering is the same whether gyro or magnetic compass is used, then the assumption of additivity is correct. However, if there is a marked change in the difference between day and night steering when the gyro compass is used rather than the magnetic compass, then the assumption of additivity is false and we say we have *interaction*. To illustrate this further, let us take some numerical data for this situation.

A measure of efficiency of steering was taken by noting, from the course recorder, the length of time that was spent more than two degrees off course during a one-hour period with no change of helmsman. In all, 24 separate hours were assessed, spread equally between daytime and nighttime watches and between the use of magnetic compass only and gyro compass only. The results were as follows:

Time off course in minutes

	Type of compass	
	Gyro	Magnetic
Daytime	12	14
	14	17
	5	12
	10	7
	16	10
	9	18
Nighttime	8	19
	10	15
	7	18
	11	21
	6	18
	4	16

We say that the experiment has been replicated six times as each combination of factors is repeated for six different headings.

Marine Statistics

Our notation becomes slightly more involved as we have the addition of the number of the replicate to consider as well as the row and column. We will assume that in general we have r rows and c columns giving rc cells but in each cell we will have p replicates.

Let x_{ijk} denote a general observation, where $i = 1, 2, \ldots, r, j = 1, 2, \ldots, c, k = 1, 2, \ldots, p$. Hence in our example $x_{125} = 10$ since it is the fifth observation in the first row, second column. We will assume that within each cell the observations are taken from a normal distribution with a common variance σ^2. The model we assume is that:

$$x_{ijk} = \mu + \alpha_i + \beta_j + (\alpha\beta)_{ij} + \varepsilon_{ijk}$$

where μ is the overall mean, α_i is the ith row effect, β_j is the jth column effect, $(\alpha\beta)_{ij}$ is the ith row–jth column interaction effect, and ε_{ijk} is the error term, with $\sum_{i=1}^{r} \alpha_i = 0$, $\sum_{j=1}^{c} \beta_j = 0$, $\sum_{i=1}^{r} (\alpha\beta)_{ij} = 0$, $\sum_{j=1}^{c} (\alpha\beta)_{ij} = 0$, and $\varepsilon_{ijk} \sim N(0, \sigma^2)$.

The three sets of hypotheses to be tested are as follows:

(i) $H_0: \alpha_1 = \alpha_2 = \cdots = \alpha_r = 0$
$H_1:$ At least one of the $\alpha_i \neq 0$ } row effects

(ii) $H_0': \beta_1 = \beta_2 = \cdots = \beta_c = 0$
$H_1':$ At least one of the $\beta_j \neq 0$ } column effects

(iii) $H_0'': (\alpha\beta)_{11} = (\alpha\beta)_{12} = \cdots = (\alpha\beta)_{rc}$
$H_1'':$ At least one of the $(\alpha\beta)_{ij} \neq 0$ } interaction effects

This time we split the total variation into four parts which we can write symbolically as:

$$\text{SST} = \text{SSR} + \text{SSC} + \text{SS(RC)} + \text{SSE}$$

where SS(RC) is the row–column interaction sum of squares and the other symbols are as defined for the previous examples.

The computational sums of squares are as follows:

$$\text{SST} = \sum_{i=1}^{r} \sum_{j=1}^{c} \sum_{k=1}^{p} x_{ijk}^2 - \frac{T_{\ldots}^2}{rcp} \quad \text{where } T_{\ldots} = \sum_{i=1}^{r} \sum_{j=1}^{c} \sum_{k=1}^{p} x_{ijk}$$

This is the same form as before, with each observation squared and then added less the correction factor of the total sum of the observations divided by the number of observations.

$$\text{SSR} = \frac{\sum_{i=1}^{r} T_{i\ldots}^2}{cp} - \frac{T_{\ldots}^2}{rcp} \quad \text{where } T_{i\ldots} = \sum_{j=1}^{c} \sum_{k=1}^{p} x_{ijk}$$

Thus $T_{i\ldots}$ is the sum of all the observations in the ith row.

$$\text{SSC} = \frac{\sum_{j=1}^{c} T_{\cdot j \cdot}^2}{rp} - \frac{T_{\ldots}^2}{rcp} \quad \text{where } T_{\cdot j \cdot} = \sum_{i=1}^{r} \sum_{k=1}^{p} x_{ijk}$$

$T_{.j.}$ is the sum of all the observations in the jth column.

$$SS(RC) = \frac{\sum_{i=1}^{r}\sum_{j=1}^{c} T_{ij.}^2}{p} - \frac{\sum_{i=1}^{r} T_{i..}^2}{cp} - \frac{\sum_{j=1}^{c} T_{.j.}^2}{rp} + \frac{T_{...}^2}{rcp}$$

$T_{ij.}$ is the sum of all the observations in the ijth cell.

$$SSE = SST - SSR - SSC - SS(RC)$$

The Anova table showing the appropriate degrees of freedom and the corresponding 'F' ratios is given below:

Source of variation	S.S.	D.o.f.	M.S.	F
Between rows	SSR	$r - 1$	$SSR/r - 1 = s_1^2$	s_1^2/s_4^2
Between columns	SSC	$c - 1$	$SSC/c - 1 = s_2^2$	s_2^2/s_4^2
Interaction	SS(RC)	$rc - r - c + 1$	$\frac{SS(RC)}{rc - r - c + 1} = s_3^2$	s_3^2/s_4^2
Error	SSE	$rcp - rc$	$\frac{SSE}{rcp - rc} = s_4^2$	
Total	SST	$rcp - 1$		

For the steering example,
$r = 2 \quad c = 2 \quad$ and $\quad p = 6$

$$SST = \sum_{i=1}^{2}\sum_{j=1}^{2}\sum_{k=1}^{6} x_{ijk}^2 - \frac{T_{...}^2}{24} \quad \text{where } T_{...} = \sum_{i=1}^{2}\sum_{j=1}^{2}\sum_{k=1}^{6} x_{ijk}$$

$$= [12^2 + 14^2 + \cdots + 18^2 + 16^2]$$
$$- \frac{[12 + 14 + \cdots + 18 + 16]^2}{24}$$

$$= 4{,}221 - \frac{297^2}{24} = 4{,}221 - 3{,}675 \cdot 38 = 545 \cdot 62$$

$$SSR = \frac{\sum_{i=1}^{2} T_{i..}^2}{12} - \frac{T_{...}^2}{24}$$

$T_{1..} = [12 + 14 + \cdots + 10 + 18] = 144$
$T_{2..} = [8 + 10 + \cdots + 18 + 16] = 153$

$$SSR = \frac{144^2 + 153^2}{12} - \frac{297^2}{24} = 3{,}678 \cdot 75 - 3{,}675 \cdot 38 = 3 \cdot 37$$

$$SSC = \frac{\sum_{j=1}^{2} T_{.j.}^2}{12} - \frac{T_{...}^2}{24}$$

Marine Statistics

$$T_{.1.} = [12 + 14 + \cdots + 6 + 4] = 112$$
$$T_{.2.} = [14 + 17 + \cdots + 18 + 16] = 185$$

$$SSC = \frac{112^2 + 185^2}{12} - \frac{297^2}{24} = 3{,}897{\cdot}42 - 3{,}675{\cdot}38 = 222{\cdot}04$$

$$SS(RC) = \frac{\sum_{i=1}^{2}\sum_{j=1}^{2} T_{ij.}^2}{6} - \frac{\sum_{i=1}^{2} T_{i..}^2}{12} - \frac{\sum_{j=1}^{2} T_{.j.}^2}{12} + \frac{T_{...}^2}{24}$$

$$T_{11.} = [12 + 14 + \cdots + 9] = 66$$
$$T_{12.} = [14 + 17 + \cdots + 18] = 78$$
$$T_{21.} = [8 + 10 + \cdots + 4] = 46$$
$$T_{22.} = [19 + 15 + \cdots + 16] = 107$$

$$SS(RC) = \frac{66^2 + 78^2 + 46^2 + 107^2}{6}$$
$$- 3{,}678{\cdot}75 - 3{,}897{\cdot}42 + 3{,}675{\cdot}38$$
$$= 4{,}000{\cdot}83 - 3{,}678{\cdot}75 - 3{,}897{\cdot}42 + 3{,}675{\cdot}38 = 100{\cdot}04$$

$$SSE = SST - SSR - SSC - SS(RC)$$
$$= 545{\cdot}62 - 3{\cdot}37 - 22{\cdot}04 - 100{\cdot}04 = 220{\cdot}17$$

Anova

Source	S.S.	D.o.f.	M.S.	F
Between rows (time of day)	3·37	1	3·37	0·31
Between columns (type of compass)	222·04	1	222·04	20·17
Interaction	100·04	1	100·04	9·09
Error	220·17	20	11·01	
Total	545·62	23		

$F_{(1,20),\,5\%} = 4{\cdot}35$
$0{\cdot}31 < 4{\cdot}35 \Rightarrow$ non-significant result (time of day)
$20{\cdot}17 > 4{\cdot}35 \Rightarrow$ significant result (type of compass)
$9{\cdot}09 > 4{\cdot}35 \Rightarrow$ significant result (interaction)

The conclusions here are that there appears to be no difference between night and day time performance considered alone, but there is a significant difference between performance using gyro or magnetic compass considered alone. However, probably the most interesting result is the significant interaction result. If we look at the readings we can see that there are high values associated with the magnetic compass at night and low values associated with the gyro compass at night. In practical terms the increased performance using gyro compass at night may be ascribed to the greater effectiveness of the

Analysis of Variance

illuminated scale and the absence of distractions so that the audible cue of the clicking of the gyro with change of heading becomes more effective. The decrement of performance of the magnetic compass at night may be ascribed to its poor illumination and the generally lower level of arousal of the helmsman at night.

As a final point on this section it should be mentioned that the interaction effect should be tested first. If this proves to be non-significant then before testing the row and column effects, the error term should be recalculated by adding the interaction sum of squares into the existing error sum of squares and similarly for the degrees of freedom, to improve the error estimate for σ^2, the unknown variance.

12.8 Summary

In this chapter we have considered some of the simpler forms of the analysis of variance. For more complicated results it will be necessary to consult more specialised books but the principles for application and computation should be easy to understand as they are simply extensions of the results considered here.

Exercises

1. Using the data given in Ex. 6 of Chapter 9, use the F test to decide whether it is reasonable, at the 0.05 level, that each of the two sets of times are samples drawn from populations with a common standard deviation.

2. Using the data given in Ex. 5 of Chapter 11, use the F test to decide whether it is reasonable, at the 0.05 level, to take each of the sets of times for lifejackets A and B as samples drawn from populations with the same variance.

3. Six samples each of 1-cm diameter ropes from manufacturers A, B and C were tested to destruction. The breaking loads were measured in kilograms as follows:

Make A	Make B	Make C
592	621	576
647	634	550
628	607	594
580	642	602
612	618	580
606	622	592

Marine Statistics

(a) Is this sufficient evidence to conclude, at the 5% level, that there is a difference between the breaking loads of the three brands of ropes?

(b) Find the least significant difference between the means of these three sets of breaking loads.

4. Four tanker sister ships arrive frequently to discharge crude oil at a particular terminal. During the course of a year, ship A arrives eight times, ship B ten times, ship C six times and ship D six times. The times taken in hours to discharge the full cargo on each occasion are listed below:

Ship A	Ship B	Ship C	Ship D
24·5	23·8	22·6	20·2
21·6	25·4	19·5	23·4
20·4	22·2	19·9	19·6
25·2	25·7	21·2	21·2
22·4	24·5	22·7	23·5
23·7	26·3	20·4	22·0
19·8	27·4		
21·5	24·2		
	26·8		
	25·1		

(a) Is this sufficient evidence to suggest that there is a difference, at the 0·05 significance level, between the mean discharge times for any of these ships?

(b) Between which ships does a significant difference apply?

5. Six ships are engaged in operating a regular transatlantic container service. As part of an investigation prior to rescheduling the service, the mean voyage time for each ship was calculated for each of the three-month periods corresponding to spring, summer, autumn and winter. These are listed in days as follows:

Ship	Spring	Summer	Autumn	Winter
1	6.31	6·14	6·28	6·35
2	6·05	6·11	6·12	6·18
3	6·24	6·02	6·26	6·38
4	6·36	6·13	6·31	6·42
5	6·21	5·98	6·13	6·29
6	6·25	6·08	6·22	6·40

It is required to ascertain, at the 5% level, whether there is a difference between the voyage times for different ships and for different seasons.

Analysis of Variance

6. A ship is fitted with two models of radar equipment, A and B, which are quite independent of one another. During the course of six voyages, the master posted each of six officers in turn to each of the two radars as the ship approached her home port, and noted the distances at which they first reported picking up the echo of a particular landfall buoy as follows:

Officer	Radar A distance, cables	Radar B, distance, cables
1	3·5	4·8
2	3·9	3·7
3	3·3	4·5
4	2·8	3·6
5	3·7	4·2
6	3·0	4·0

Is this sufficient evidence for the master to conclude, at the 5% level, that there is:
(a) A difference in the mean performance of the radars?
(b) A difference in the mean performance of the officers?

7. It has been found that there was excessive damage to boxes of tomatoes carried in a particular trade. In order to investigate this problem, two new types of boxes were designed and, as an experiment, four standard consignments were shipped in each of the three box designs: old (O), new design one (N1) and new design two (N2). For each consignment, half was shipped in the ship's lower hold and half in the 'tween deck. The numbers of damaged boxes in each category were as shown in the following table:

Consignment	Lower hold O	$N1$	$N2$	'Tween deck O	$N1$	$N2$
1	12	8	30	5	3	10
2	14	10	22	10	0	15
3	22	4	27	8	6	18
4	18	9	34	12	4	8

Does this information suggest, at the 1% level, that there is a significant difference in the robustness of the box designs? Or that there is a useful advantage to be gained by stowing the tomatoes in the 'tween deck rather than the lower hold?

Marine Statistics

Answers

1. For crew A:

$$n_A = 8 \qquad \sum x_A = 101 \cdot 6 \qquad \sum x_A^2 = 1{,}305 \cdot 5$$

$$s_A^2 = \frac{1}{n_A - 1}\left[\sum x_A^2 - \frac{(\sum x_A)^2}{n}\right] = \frac{1}{7}\left[1{,}305 \cdot 5 - \frac{101 \cdot 6^2}{8}\right] = \underline{2 \cdot 17}$$

For crew B:

$$n_B = 9 \qquad \sum x_B = 92 \cdot 4 \qquad \sum x_B^2 = 959 \cdot 06$$

$$s_B^2 = \frac{1}{n_B - 1}\left[\sum x_B^2 - \frac{(\sum x_B)^2}{n}\right] = \frac{1}{8}\left[959 \cdot 06 - \frac{92 \cdot 4^2}{9}\right] = \underline{1 \cdot 30}$$

We set up:

$$H_0: \sigma_A^2 = \sigma_B^2$$
$$H_1: \sigma_A^2 \neq \sigma_B^2$$
$$\alpha = 0 \cdot 05$$

We calculate:

$$F = \frac{\text{larger value of sample variance}}{\text{smaller value of sample variance}} = \frac{2 \cdot 17}{1 \cdot 3} = \underline{1 \cdot 67}$$

We find from tables:

$$F_{7, 8, 0 \cdot 025} = \underline{4 \cdot 53}$$

Our calculated value of $1 \cdot 67$ is less than the critical value and hence in the acceptance region. We accept H_0 and conclude that it is reasonable to assume that the two samples were drawn from populations with the same variance and therefore also the same standard deviations.

2. For type A lifejackets:

$$n_A = 12 \qquad \sum x_A = 1{,}447 \qquad \sum x_A^2 = 179{,}815$$

$$s_A^2 = \frac{1}{n - 1}\left[\sum x^2 - \frac{(\sum x)^2}{n}\right] = \frac{1}{11}\left[179{,}815 - \frac{1{,}447^2}{12}\right] = \underline{484 \cdot 6}$$

For type B lifejackets:

$$n_B = 8 \qquad \sum x_B = 779 \qquad \sum x_B^2 = 77{,}125$$

$$s_B^2 = \frac{1}{n - 1}\left[\sum x^2 - \frac{(\sum x)^2}{n}\right] = \frac{1}{7}\left[77{,}125 - \frac{779^2}{8}\right] = \underline{181 \cdot 4}$$

We set up:

$$H_0: \sigma_A^2 = \sigma_B^2$$
$$H_1: \sigma_A^2 \neq \sigma_B^2$$
$$\alpha = 0 \cdot 05$$

We calculate:

$$F = \frac{\text{larger value of sample variance}}{\text{smaller value of sample variance}} = \frac{484 \cdot 6}{181 \cdot 4} = \underline{2 \cdot 67}$$

Analysis of Variance

We find from tables:

$$F_{11,7,0\cdot 025} = 4\cdot 72$$

Our calculated value of 2·67 is less than the critical value and hence in the acceptance region. We therefore accept H_0 and conclude that it is reasonable to assume that the two samples were drawn from populations with the same variance.

3. (a) We set up:

$H_0: \alpha_1 = \alpha_2 = \alpha_3 = 0$ (using the notation of Chapter 12)
$H_1: \alpha_j \neq 0$ for at least one $j = 1, \ldots, 3$
$\alpha = 0\cdot 05$

$$\text{SST} = \sum_{i=1}^{6}\sum_{j=1}^{3} x_{ij}^2 - \frac{T_{..}^2}{6.3} = 6{,}615{,}035 - \frac{10{,}903^2}{18}$$
$$= 10{,}846$$

$$\text{SSC} = \frac{\sum_{j=1}^{3} T_{.j}^2}{6} - \frac{T_{..}^2}{6.3} = \frac{3{,}665^2 + 3{,}744^2 + 3{,}494^2}{6} - \frac{10{,}903^2}{18}$$
$$= 5{,}443$$

$$\text{SSE} = \text{SST} - \text{SSC} = 10{,}846 - 5{,}443$$
$$= 5{,}403$$

Hence our Anova table is:

Source	S.S.	D.o.f.	M.S.	F
Between treatments	5,443	2	2,721	7·56
Within treatments	5,403	15	360	
Total	10,846	17		

Critical region: $F > F_{2,15,0\cdot 05}$, i.e. $F > 3\cdot 68$

We thus reject H_0 and conclude that at least one of the three makes of rope has a mean breaking load significantly different from the others.

(b) To find the least significant difference for this example, we use:

$$t = \frac{\bar{x}_i - \bar{x}_j}{\hat{\sigma}\sqrt{\dfrac{1}{n_1} + \dfrac{1}{n_2}}}$$

where:

$\hat{\sigma} = \sqrt{360} = 18\cdot 97$
$n_1 = n_2 = 6$
$|t| > t_{15,0\cdot 025} = 2\cdot 13$

Marine Statistics

thus:

$$\frac{|\bar{x}_i - \bar{x}_j|}{18.97\sqrt{\frac{2}{6}}} > 2.13 \quad \text{i.e.} \quad |\bar{x}_i - \bar{x}_j| > 23.33$$

We thus conclude that there is a significant difference between the mean breaking loads for the three manufacturers if the difference exceeds 23·33 kg.

We have:

$$\bar{x}_A = \frac{3,665}{6} = 610.8 \text{ kg}$$

$$\bar{x}_B = \frac{3,744}{6} = 624.0 \text{ kg}$$

$$\bar{x}_C = \frac{3,494}{3} = 582.3 \text{ kg}$$

We thus conclude that ropes from manufacturer *C* have a significantly smaller mean breaking load than ropes from manufacturers *A* and *B*, but that there is no significant difference between the mean breaking load for ropes from manufacturers *A* and *B*.

4. (a) We set up:

$$H_0: \alpha_1 = \alpha_2 = \alpha_3 = \alpha_4 = 0$$
$$H_1: \alpha_j \neq 0 \text{ for at least one of } j = 1, \ldots, 4$$
$$\alpha = 0.05$$

$$\text{SST} = \sum_{j=1}^{k} \sum_{i=1}^{n_j} x_{ij}^2 - \frac{T_{..}^2}{N} = 15,870.63 - \frac{687.7^2}{30}$$
$$= 152.06$$

$$\text{SSC} = \sum_{j=1}^{k} \frac{T_{.j}^2}{n_j} - \frac{T_{..}^2}{N}$$
$$= \left(\frac{179.1^2}{8} + \frac{251.4^2}{10} + \frac{126.3^2}{6} + \frac{129.9^2}{6}\right) - \frac{686.7^2}{30}$$
$$= 82.18$$

$$\text{SSE} = \text{SST} - \text{SSC} = 152.06 - 82.18$$
$$= 69.88$$

Hence our Anova table is:

Source	S.S.	D.o.f.	M.S.	F
Between ships	82·18	3	27·39	10·18
Within ships	69·88	26	2·69	
Total	152·06	29		

The critical region is $F > F_{3, 26, 0.05}$, i.e. $F > 2.98$.

Analysis of Variance

Since our calculated value of 10·18 is greater than the critical value, we reject H_0 and conclude that the mean discharge time for at least one of the four ships is significantly different from the others.

(b) For a consideration of least significant differences, our test statistic is:

$$t = \frac{|\bar{x}_i - \bar{x}_j|}{\hat{\sigma}\sqrt{\frac{1}{n_i} + \frac{1}{n_j}}} \sim t \text{ on 26 d.f.}$$

where $\hat{\sigma} = \sqrt{2\cdot 69} = 1\cdot 64$ and $|t| > t_{26, 0\cdot 025}$, i.e. $|t| > 2\cdot 05$. Hence

$$\frac{|\bar{x}_i - \bar{x}_j|}{\sqrt{\frac{1}{n_1} + \frac{1}{n_2}}} > 2\cdot 05 \times 1\cdot 64 = 3\cdot 36$$

but

$$\bar{x}_A = \frac{179\cdot 1}{8} = 22\cdot 39$$

$$\bar{x}_B = \frac{251\cdot 4}{10} = 25\cdot 14$$

$$\bar{x}_C = \frac{126\cdot 3}{6} = 21\cdot 05$$

$$\bar{x}_D = \frac{129\cdot 9}{6} = 21\cdot 65$$

Thus, comparing these mean times in pairs, we find:

A v. B: $\quad \dfrac{|\bar{x}_A - \bar{x}_B|}{\sqrt{\frac{1}{8} + \frac{1}{10}}} = \dfrac{2\cdot 75}{0\cdot 474} = 5\cdot 80$

This is a significant difference, since $5\cdot 80 > 3\cdot 36$.

A v. C: $\quad \dfrac{|\bar{x}_A - \bar{x}_C|}{\sqrt{\frac{1}{8} + \frac{1}{6}}} = \dfrac{1\cdot 34}{0\cdot 54} = 2\cdot 48$

This is not a significant difference, since $2\cdot 48 < 3\cdot 36$.

A v. D: $\quad \dfrac{|\bar{x}_A - \bar{x}_D|}{\sqrt{\frac{1}{8} + \frac{1}{6}}} = \dfrac{0\cdot 74}{0\cdot 54} = 1\cdot 37$

This is not a significant difference, since $1\cdot 37 < 3\cdot 36$.

B v. C: $\quad \dfrac{|\bar{x}_B - \bar{x}_C|}{\sqrt{\frac{1}{10} + \frac{1}{6}}} = \dfrac{4\cdot 09}{0\cdot 516} = 7\cdot 93$

Marine Statistics

This is a significant difference, since 7·93 > 3·36.

$$B \text{ v. } D: \quad \frac{|\bar{x}_B - \bar{x}_D|}{\sqrt{\frac{1}{10} + \frac{1}{6}}} = \frac{3·49}{0·516} = 6·76$$

There is a significant difference, since 6·76 > 3·36.

$$C \text{ v. } D: \quad \frac{|\bar{x}_C - \bar{x}_D|}{\sqrt{\frac{1}{6} + \frac{1}{6}}} = \frac{0·60}{0·577} = 1·04$$

This is not a significant difference, since 1·04 < 3·36.

Summarising these results, we find that ship B has a mean discharge time which is significantly longer than any of the others, but there is no significant difference between the mean discharge times for the other three. Clearly, the relatively poor performance of ship B requires further investigation.

5. We set up:

H_0': $\alpha_1 = \alpha_2 = \alpha_3 = \alpha_4 = \alpha_5 = \alpha_6 = 0$
H_0'': $\beta_1 = \beta_2 = \beta_3 = \beta_4$

where α_1, etc., are the row effects and β_1, etc., are the column effects.

H_1': At least one of $\alpha_1, \ldots, \alpha_6 \neq 0$
H_1'': At least one of $\beta_1, \ldots, \beta_4 \neq 0$
$\alpha = 0·05$

$$\text{SST} = \sum_{i=1}^{r} \sum_{j=1}^{c} x_{ij}^2 - \frac{T_{..}^2}{rc} \quad (\text{where } r = 6 \text{ and } c = 4)$$

$$= 928·13 - \frac{149·22^2}{6 \times 4} = 0·355$$

$$\text{SSR} = \frac{\sum_{i=1}^{r} T_{i.}^2}{c} - \frac{T_{..}^2}{rc}$$

$$= \frac{(25·08^2 + 24·46^2 + 24·90^2 + 25·22^2 + 24·61^2 + 24·95^2)}{4}$$

$$- \frac{149·22^2}{24} = 927·878 - 927·775 = 0·103$$

$$\text{SSC} = \frac{\sum_{j=1}^{c} T_{.j}^2}{r} - \frac{T_{..}^2}{rc}$$

$$= \frac{(37·42^2 + 36·46^2 + 37·32^2 + 38·02^2)}{6} - \frac{149·22^2}{24}$$

$$= 927·982 - 927·775 = 0·207$$

$\text{SSE} = \text{SST} - \text{SSR} - \text{SSC}$
$= 0·355 - 0·103 - 0·207 = 0·045$

Analysis of Variance

We tabulate these results as:

Source	S.S.	D.o.f.	M.S.	F
Between rows (ships)	0·103	5	0·0206	6·87
Between columns (seasons)	0·207	3	0·0690	23·00
Error	0·045	15	0·0030	
Total	0·355	23		

We find: $F_{5,15} = 2·90$.

The between rows (ships) value of F is $6·87 > 2·90$. We therefore reject H_0' and conclude that at least one of the ships has a mean voyage time significantly different from the others. We find:

$$F_{3,15} = 3·29$$

The between columns (seasons) value of F is $23·00 > 3·29$. We therefore reject H_0'' and conclude that at least one of the seasons gives voyage times significantly different from the others.

6. The working of this exercise is almost identical with the working of Ex. 5. This should lead to an Anova table as below:

Source	S.S.	D.o.f.	M.S.	F
Between rows	1·18	5	0·236	1·53
Between columns	1·76	1	1·760	11·43
Error	0·77	5	0·154	
Total	3·71	11		

The critical value for $F_{5,5,\,0·05} = 5·05$

Hence the between rows (Officers) effect is not significant and the master may conclude that there is no apparent difference between the performance of the six officers.

The critical value for $F_{1,5,\,0·05} = 6·61$

Hence the between columns (radars) effect is significant and the master may conclude that there is an apparent difference between the performance of the two radars.

Marine Statistics

7. We recast the data in row and column form:

	O	N1	N2
Lower hold	12	8	30
	14	10	22
	22	4	27
	18	9	34
'Tween deck	5	3	10
	10	0	15
	8	6	18
	12	4	8

Using the notation of Chapter 12, we set up:

$H_0: \alpha_1 = \alpha_2 = 0$
$H_1:$ At least one of α_1, etc. $\neq 0$ } row effects

$H_0': \beta_1 = \beta_2 = \beta_3 = 0$
$H_1':$ At least one of β_1, etc. $\neq 0$ } column effects

$H_0'': (\alpha\beta)_{11} = (\alpha\beta)_{12} = \cdots = (\alpha\beta)_{23} = 0$
$H_1'':$ At least one of $(\alpha\beta)_{11}$, etc. $\neq 0$
$\alpha = 0.01$
$r = 2 \quad c = 3 \quad p = 4$

$$\text{SST} = \sum_{i=1}^{r}\sum_{j=1}^{c}\sum_{k=1}^{p} x_{ijk}^2 - \frac{T_{...}^2}{rcp} = 5{,}785 - \frac{309^2}{2 \times 3 \times 4}$$
$$= 5{,}785 - 3{,}978.4 = 1{,}86.60$$

$$\text{SSR} = \frac{\sum_{i=1}^{r} T_{i..}^2}{cp} - \frac{T_{...}^2}{rcp} = \frac{210^2 + 99^2}{12} - 3{,}978.4$$
$$= 4{,}491 - 3{,}978.4 = \underline{513.3}$$

$$\text{SSC} = \frac{\sum_{j=1}^{c} T_{.j.}^2}{rp} - \frac{T_{...}^2}{rcp} = \frac{101^2 + 44^2 + 164^2}{8} - 3{,}978.4$$
$$= 4{,}879.1 - 3{,}978.4 = \underline{900.7}$$

$$\text{SS(RC)} = \frac{\sum_{i=1}^{r}\sum_{j=1}^{c} T_{ij.}^2}{p} - \frac{\sum_{i=1}^{r} T_{i..}^2}{cp} - \frac{\sum_{j=1}^{c} T_{.j.}^2}{rp} + \frac{T_{...}^2}{rcp}$$
$$= \frac{66^2 + 31^2 + 113^2 + 35^2 + 13^2 + 51^2}{4}$$
$$- 4{,}491.7 - 4{,}879.1 + 3{,}978.4 = \underline{127.8}$$

Analysis of Variance

We construct our Anova table as follows:

Source	S.S.	D.o.f.	M.S.	F
Between rows	513·3	1	513·3	34·9
Between columns	900·7	2	450·3	30·6
Interaction	127·8	2	63·9	4·3
Error	264·8	18	14·7	
Total	1,806·6	23		

For the interaction effect, we find $F_{2,18,0\cdot01} = 6\cdot01$. Our calculated value of 4·3 is thus in the acceptance region. We accept H_0'' and conclude that there is no significant interaction between the type of boxes and the place of stowage.

For the between boxes (between columns) effect, we find $F_{2,18,0\cdot01} = 6\cdot01$. Our calculated value of 30·6 is therefore in the rejection region. Hence we reject H_0' and conclude that the mean number of damaged boxes is significantly different for at least one of the box designs.

For the place of stowage (between rows) effect, we find $F_{1,18,0\cdot01} = 8\cdot29$. Our calculated value of 34·9 is therefore in the rejection region. Hence we reject H_0 and conclude that there is a significant difference in the number of damaged boxes depending on whether a consignment is stowed in the 'tween deck or the lower hold. Clearly, the probability of damage is less for 'tween deck stowage.

Strictly speaking, as we have a non-significant interaction effect, we should have combined the interaction and error terms giving a sum of squares 392·6 on 20 d.o.f. and hence a mean square of 19·63 for the new error term. However in this case the results are not affected at all for the row and column effects.

Chapter 13
Regression and Correlation

13.1 *Introduction*

In order to investigate the safety of navigation in a particular port, the ships using the port over a period of two years were assigned to twelve classes (lettered *A–L*) which were based on gross tonnage, nature of cargo and power of main engines. The investigator then assigned ranks to these classes in what he considered to be the order of increasing potential danger resulting from vessel casualties. The relative frequency of casualties defined as the number of casualties per number of ships at risk over the two-year period for each class were as follows:

Class of ship	Rank for potential danger of casualties	Relative frequency of casualties
B	1 Low	0·025
C	2	0·019
A	3	0·030
E	4	0·028
G	5	0·014
D	6	0·032
F	7	0·016
I	8	0·021
J	9	0·017
H	10	0·024
K	11	0·011
L	12 High	0·015

Is this evidence sufficient to confirm the suggestion that the risk of casualty decreases as the potential danger of the consequences increases?

In this example we are concerned with a rather different situation than any we have met previously as we want to look at the relationship between two variables rather than concentrating on one variable only as we have done up until now. Here, one variable is qualitative, the potential danger of any casualty, and the other, the relative frequency of casualties, is quantitative.

Regression and Correlation

However we could have situations where both variables are quantitative and again situations where both are qualitative. If our data set consists of readings on two variables for each member of the sample then we say that we have *bivariate data*. It is possible to have more than two variables measured for each member of the sample; for instance, in this example we might have taken additionally the modal speed for each class. Thus, in general, if there are measurements on several variables we say we have *multivariate data*. For simplicity's sake we will restrict the discussion almost entirely to two variable situations since the fundamental concepts can be established for these cases without becoming involved in the more complex calculations.

There are essentially two types of questions that we look at when considering bivariate quantitative data. The first technique, known as *regression*, attempts to establish whether, given a value of one of the variables, we can predict by means of an algebraic equation the corresponding value of the other variable. The second technique, known as *correlation*, establishes the strength of relationship between the variables and considers to what extent they vary together. The two techniques are very closely linked as we shall see in discussing them. Inevitably there is also the fact that we will on most occasions be dealing with only a sample of data from a population, so the question of the statistical significance of the calculated results also has to be examined.

If at least one of our variables is qualitative then no form of prediction is possible; hence we cannot perform a regression analysis. However, it is possible to consider the strength of the relationship and for this we use a non-parametric form of correlation.

13.2 Scatter Diagrams

The most elementary form of analysis which we may perform on any set of data is to represent it diagrammatically and if we have bivariate data then we can make a two-dimensional representation of it, known as a *scatter diagram*. We start with a pair of axes and a scale is chosen along the horizontal axis to represent one of the variables, x, and similarly a scale is chosen along the vertical axis to represent the other variable, y. Each of the data pairs is then plotted as a point in the coordinate system. To illustrate this let us consider the following example.

It is suggested, for reasons of economy, that there is no need to have two weather reporting stations, A and B, covering a particular area. It is felt that the manning of station B should be discontinued on the grounds that, if visibility is considered, the visibility in the area of station B can reasonably be inferred from a knowledge of the visibility at station A. The visibility reported from the two stations A and B over a period was noted on 12 occasions as follows:

Marine Statistics

	Visibility (kilometres)	
Occasion	Station *A*	Station *B*
1	1·5	1·0
2	3·0	1·5
3	20·0	10·0
4	2·0	7·5
5	5·0	3·0
6	3·0	3·0
7	2·5	2·0
8	10·0	6·0
9	1·0	0·5
10	8·0	5·0
11	4·0	3·5
12	0·5	2·0

If we let variable x, plotted horizontally, be the visibility at station *A* and variable y, plotted vertically, be the visibility at station *B*, then taking the two visibility readings on each occasion as a pair of Cartesian coordinates we obtain the scatter diagram shown in Fig. 13.1.

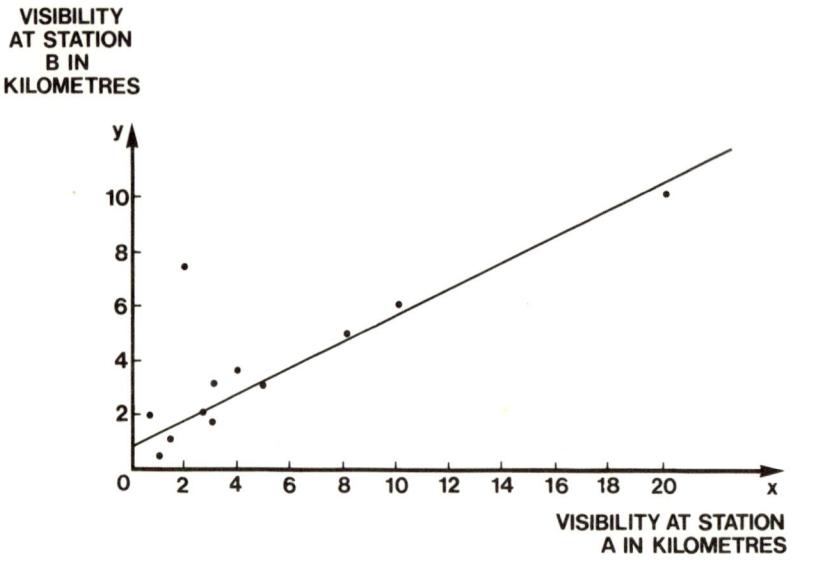

Fig 13·1 Scatter diagram to show the relationship between the visibility at two stations A and B.

The diagram gets its name as it shows the scatter of the points with respect to each other. Our main interest is to see if we can visually detect any recognisable shape in this scatter since we will then have some indication as to the possible form of mathematical relationship between the two variables. In the

Regression and Correlation

visibility example shown in Fig. 13.1 we can see that the points appear to lie roughly around a straight line which we call the line of best fit. In other cases the points may appear to lie around some curve such as is shown in Fig. 13.2.

In other situations the points may appear to be randomly scattered over the

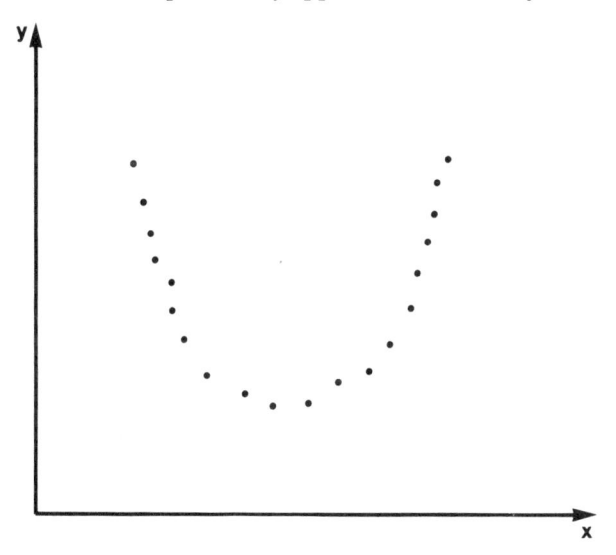

Fig 13·2 A non-linear relationship between x and y.

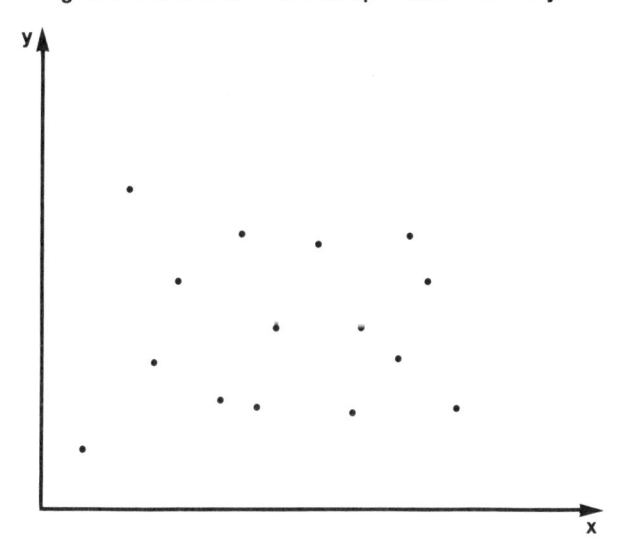

Fig 13·3 No relationship between x and y.

Marine Statistics

diagram, an example of this being shown in Fig. 13.3. If this appears to be the case there is probably little point in proceeding with any complicated analysis as the indications are that there is no relationship between the variables. However, the decision is often bound to be very subjective when based on graphical evidence only, so if one is specifically asked to examine the relationship between two variables one will need to produce a more objective result. If, on the other hand, one is at a very exploratory stage of some work, it is possible to save a considerable amount of time by plotting a scatter diagram, however roughly, and seeing if it appears to give any useful ideas for further analysis. For situations where we have three or more variables to be examined simultaneously the graphical approach obviously fails so, in such cases, we have to proceed with calculations straight away.

13.2.1 The *line of best fit* is the straight line drawn through a collection of points on a scatter diagram, which follows most suitably the general pattern in the points.

13.2.2 The *curve of best fit* is the curve drawn through a collection of points on a scatter diagram, which follows most suitably the general pattern in the points.

13.3 Regression Lines

The simplest form of relationship between two variables is a linear one so we will concentrate mainly on this and discuss first the technique known as *linear regression*. In the introduction we referred to regression as a technique to be used when we were hoping to predict values of one variable given values of another variable. Formally we say that of the two variables, one is considered to be a controlled variable in that we can choose which values of it we want to consider, and the other is considered to be a random variable in that its values arise by chance. In linear regression we are interested in establishing whether there is a linear relationship between the controlled variable and the random variable so that, given a value of the controlled variable, we can predict a corresponding value for the random variable.

We will assume that our controlled variable is x and our random variable is y, although in many practical situations it could well be vice versa. Given one particular value of x, say $x = X_1$, we will probably have in our sample more than one pair of values with $x = X_1$ but different y values appearing in each pair with $x = X_1$. Thus in our visibility example we have on occasions 2 and 6 that the visibility at A is equal to 3·0 km, but on occasion 2 the visibility at B is 1·5 km and on occasion 6 the visibility at B is 3·0 km. Thus for any one value of x there will be a whole distribution of values of y, because y is a random variable. If we had the whole population of values then we would see the distributions for y given each individual value of x. We write the population mean of each of these distributions as $\mu_{y|x}$ and read it as 'mu of y given x' or 'the mean of the y's given a value of x'. The symbol | is the same symbol as we used in conditional probability to describe the event A given B which we wrote as A|B. If there is a linear relationship between the variables y and x then we would expect the means of all these distributions to lie on a straight line. This can be represented algebraically by the equation

$$\mu_{y|x} = \alpha + \beta x$$

where α and β are constants. It is termed the regression line of y on x. Figure 13.4 attempts to represent this diagrammatically although it is difficult to do so as one is now dealing in three dimensions denoted by:

$$x, y \text{ and } f(y|x)$$

where $f(y|x)$ is the probability density function for $y|x$.

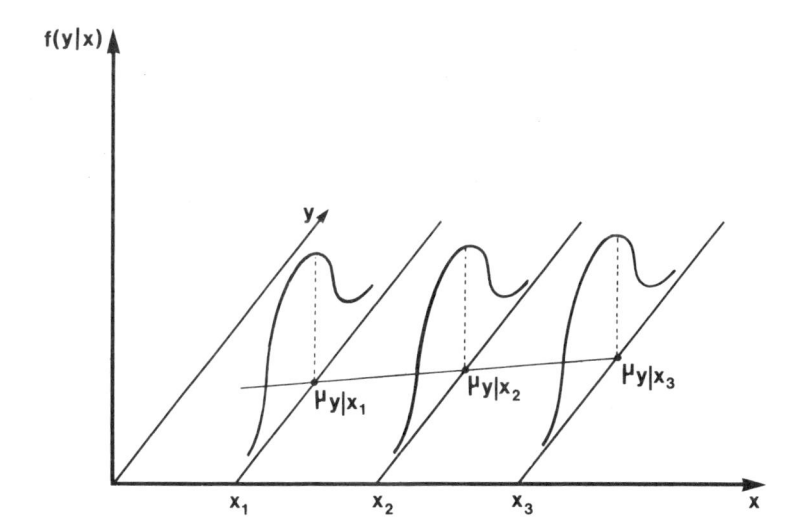

Fig 13·4 The distributions for y given x.

In Fig. 13.4 we have depicted each of the distributions of $y|x$ as normal distributions as this is a further assumption we will have to make when we come on to consider associated significance tests.

If we are only dealing with a sample of results then we estimate $\mu_{y|x}$ by $\bar{y}_x$ from the sample regression line $\bar{y}_x = a + bx$. In practice this is often written more simply as $y = a + bx$.

The only warning about doing this is that one must remember that this line is only appropriate for predicting y given a value of x. It must not be used to

Marine Statistics

predict a value of x given a value of y and therefore should NOT be written in the form of:

$$x = \frac{1}{b}y - \frac{a}{b}$$

If we want to treat x as a random variable and y as a controlled variable then we are dealing with a series of distributions for x given y with means $\mu_{x|y}$. Hence in the population if these means all lie on a straight line its equation will be of the form $\mu_{x|y} = \gamma + \delta y$ with γ and δ constants [where γ and δ are Greek letters pronounced 'gamma' and 'delta' respectively]. It is termed the regression line of x on y. Thus the two population lines $\mu_{y|x} = \alpha + \beta x$ and $\mu_{x|y} = \gamma + \delta y$ are two distinct lines.

Using sample results the line we calculate will be of the form $\bar{x}_y = c + dy$ which again in practice we may write as $x = c + dy$. However, the important result is that for any linear regression situation, except one, there are two distinct regression lines, one of the form $y = a + bx$, which is used for prediction of y, and one of the form $x = c + dy$, which is used for prediction of x. Both of these are also distinct from the line we draw on a scatter diagram, the line of best fit, as we shall see in the next section when we consider how the coefficients in the equations are calculated. The only exception is when the two variables are connected by an exact linear relationship, in which case the two regression lines are coincident together with the line of best fit.

An example of this would be if a series of temperature readings were taken in Centigrade (x) and in Fahrenheit (y). Assuming no experimental error for any one value of x, say x_0, the same value of y should appear for each reading of x_0 and the single regression line will have the equation

$$y = 1{\cdot}8x + 32$$

However in most practical experimental situations it is very unlikely that this case will arise.

13.3.1 **The regression line of y on x is the line connecting the means of the distributions of y for each value of x, $\mu_{y|x}$, with the population equation**

$$\mu_{y|x} = \alpha + \beta x \quad \alpha, \beta \text{ constants}$$

It is used to predict the value of y given a value of x.

13.3.2 **The regression line of x on y is the line connecting the means of the distributions of x for each value of y, $\mu_{x|y}$, with population equation**

$$\mu_{x|y} = \gamma + \delta y \quad \gamma, \delta \text{ constants}$$

It is used to predict the value of x given a value of y.

13.4 Method of Least Squares

The equations for calculating both the regression lines are derived using the method of least squares. We will suppose first of all that we are interested in estimating a value of y for a given value of x; hence we require the regression line of y on x, whose equation we can write in abbreviated form as $y = a + bx$.

For any one value of x, we will probably have in our sample several different values of y. Thus in Fig. 13.5 when $x = x_p$ there are three alternative

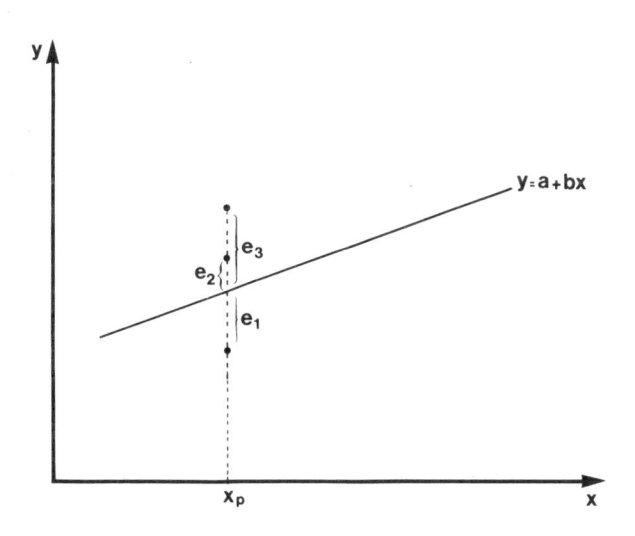

Fig 13·5 The vertical minimisation of errors.

values of y, shown by the three points all in the vertical line defined by $x = x_p$. However, once we have calculated the regression line of y on x, we will estimate the value of y when $x = x_p$ to be y_p, the point on the regression line for $x = x_p$. Thus $y_p = a + bx_p$. The difference between this value y_p and each of the three values of y we have observed when $x = x_p$ may be thought of as errors. Thus, if the lowest value of y is y_{p_1} when $x = x_p$ we can call the observed error or residual $e_1 = y_{p_1} - y_p$. Similarly, if the other two readings of y are y_{p_2} and y_{p_3} when x is x_p, then the observed errors here are $e_2 = y_{p_2} - y_p$ and $e_3 = y_{p_3} - y_p$ respectively. If we now consider a general point in the scatter diagram with coordinates (x, y) and if we write $\hat{y}$ to be the estimated value of y from the regression line, then the error $e = y - \hat{y}$. But $\hat{y} = a + bx$, thus $e = y - (a + bx)$. Some of these errors will have positive signs, like e_2 and e_3 in the diagram, and some will have negative signs. Hence to get a measure of the total error in a situation, one cannot sum the errors as the signs would have a cancelling effect. The easiest method mathematically

Marine Statistics

is again to sum the squares of the errors, as we did with variance to get a measure of error, which we may call E_y. Hence

$$E_y = \sum_x (y - \hat{y})^2 = \sum_x (y - (a + bx))^2$$

where the summation is over all the data points (x, y). Let us suppose we have n points of general form (x_i, y_i); then $E = \sum_{i=1}^{n} (y_i - a - bx_i)^2$. We now have a mathematical expression with two unknown quantities, a and b, and we want to find the particular values of a and b which will minimise this error term. There is a standard technique in differential calculus for finding the minimum of an algebraic expression. Applying it we get the following equations:

$$\sum_{i=1}^{n} y_i = na + b \sum_{i=1}^{n} x_i \tag{1}$$

and

$$\sum_{i=1}^{n} (x_i y_i) = a \sum_{i=1}^{n} x_i + b \left(\sum_{i=1}^{n} x_i^2 \right) \tag{2}$$

These are known as the normal equations. Solving them as a pair of simultaneous equations for a and b we get:

$$b = \frac{n \left(\sum_{i=1}^{n} x_i y_i \right) - \left(\sum_{i=1}^{n} x_i \right) \left(\sum_{i=1}^{n} y_i \right)}{n \left(\sum_{i=1}^{n} x_i^2 \right) - \left(\sum_{i=1}^{n} x_i \right)^2}$$

$$a = \frac{1}{n} \left(\sum_{i=1}^{n} y_i \right) - b \cdot \frac{1}{n} \left(\sum_{i=1}^{n} x_i \right)$$

Since $1/n \sum_{i=1}^{n} x_i = \bar{x}$, the mean of the x readings, and $1/n \sum_{i=1}^{n} y_i = \bar{y}$, the mean of the y readings, we may write the expression far more simply as:

$$a = \bar{y} - b\bar{x}$$

13.4.1 In the regression line of y on x with equation given by $y = a + bx$, the coefficient b, is the slope of the regression line known as the regression coefficient and has equation

$$b = \frac{n \sum_{i=1}^{n} (x_i y_i) - \left(\sum_{i=1}^{n} x_i \right) \left(\sum_{i=1}^{n} y_i \right)}{n \sum_{i=1}^{n} x_i^2 - \left(\sum_{i=1}^{n} x_i \right)^2}$$

and the coefficient a is the intercept of the regression line on the y axis and has equation

$$a = \bar{y} - b\bar{x}$$

where n is the number of pairs of observations.

Regression and Correlation

The equations for *a* and *b* are determined by minimising the sum of the errors squared; hence the method is known as the *method of least squares*.

Let us calculate the regression line of *y* on *x* for the visibility data given earlier in the chapter. We had 12 pairs of readings showing the visibility at station A, x, and the visibility at station B, y. We need to know $\sum_{i=1}^{12} x_i$—the sum of the x's, $\sum_{i=1}^{12} y_i$—the sum of the y's, $\sum_{i=1}^{12} x_i y_i$—the sum of the product of each x_i and y_i, and $\sum_{i=1}^{12} x_i^2$—the sum of the squares of the x_i's.

Occasion	x	y	xy	x^2
1	1·5	1·0	1·5	2·25
2	3·0	1·5	4·5	9·00
3	20·0	10·0	200·0	400·0
4	2·0	7·5	15·0	4·00
5	5·0	3·0	15·0	25·00
6	3·0	3·0	9·0	9·00
7	2·5	2·0	5·0	6·25
8	10·0	6·0	60·0	100·00
9	1·0	0·5	0·5	1·00
10	8·0	5·0	40·0	64·00
11	4·0	3·5	14·0	16·00
12	0·5	2·0	1·0	0·25
	$\sum_{i=1}^{12} x_i = 60\cdot 5$	$\sum_{i=1}^{12} y_i = 45\cdot 0$	$\sum_{i=1}^{12} x_i y_i = 365\cdot 5$	$\sum_{i=1}^{12} x_i^2 = 636\cdot 75$

If one is using an electronic calculator it is usually not necessary to show the individual values of xy and x^2 as the sums can be accumulated. The regression coefficient:

$$b = \frac{n \sum_{i=1}^{12} x_i y_i - \left(\sum_{i=1}^{12} x_i\right)\left(\sum_{i=1}^{12} y_i\right)}{n \sum_{i=1}^{12} x_i^2 - \left(\sum_{i=1}^{12} x_i\right)^2} \quad \text{where } n = 12$$

Hence

$$b = \frac{12 \times 365\cdot 5 - 60\cdot 5 \times 45}{12 \times 636\cdot 75 - 60\cdot 5 \times 60\cdot 5} = \frac{4{,}386 - 2{,}722\cdot 5}{7{,}641 - 3{,}660\cdot 25} = \frac{1{,}663\cdot 5}{3{,}980\cdot 75}$$
$$= 0\cdot 42$$

$$\bar{x} = \frac{\sum_{i=1}^{12} x_i}{12} = \frac{60\cdot 5}{12} = 5\cdot 04$$

$$\bar{y} = \frac{\sum_{i=1}^{12} y_i}{12} = \frac{45\cdot 0}{12} = 3\cdot 75$$

Thus the intercept:

$$a = \bar{y} - b\bar{x} = 3\cdot 75 - 0\cdot 42 \times 5\cdot 04$$
$$= 3\cdot 75 - 2\cdot 12 = 1\cdot 63$$

Marine Statistics

Hence the regression line of y on x has equation $y = 1 \cdot 63 + 0 \cdot 42x$. If we wanted to estimate the value of y, the visibility at B, when the value of x, the visibility at A, is 12 km our answer would be $1 \cdot 63 + 12 \times 0 \cdot 42 = 6 \cdot 67$, or to the accuracy given, say, 6·5 km.

The other regression line, for estimating x given y, has an equation of the form $x = c + dy$. The argument is similar to the previous case but we start with considering the discrepancies in values of x for any one value of y. In Fig. 13.6 if (x_q, y_q) is a point on the regression line then e_1 and e_2, errors

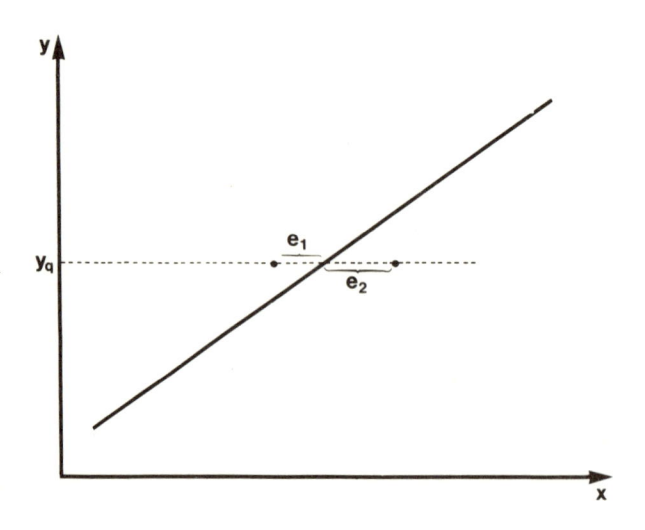

Fig 13·6 The horizontal minimisation of errors.

measured horizontally, are the errors in the observed values of x compared to the estimated value of x, when $y = y_q$. Thus this time we wish to minimise $E_x = \sum_{i=1}^{n} (x_i - \hat{x}_i)^2 = \sum_{i=1}^{n} (x_i - c - dy_i)^2$. This gives the normal equations as:

$$\sum_{i=1}^{n} x_i = nc + d \sum_{i=1}^{n} y_i \quad \text{and} \quad \sum_{i=1}^{n} (x_i y_i) = c \sum_{i=1}^{n} y_i + d \left(\sum_{i=1}^{n} y_i^2 \right)$$

Solving these as simultaneous equations for c and d we get:

$$d = \frac{n \sum_{i=1}^{n} (x_i y_i) - \left(\sum_{i=1}^{n} x_i \right) \left(\sum_{i=1}^{n} y_i \right)}{n \sum_{i=1}^{n} y_i^2 - \left(\sum_{i=1}^{n} y_i \right)^2} \quad \text{and} \quad c = \bar{x} - d\bar{y}$$

Comparing these formulae with the results for a and b, it is clear that they are of the same form but with x and y interchanged.

13.4.2 In the regression line of x on y with equation given by $x = c + dy$, the coefficient, d, is the slope of the regression line known as the regression coefficient and has equation

$$d = \frac{n \sum_{i=1}^{n}(x_i y_i) - \left(\sum_{i=1}^{n} x_i\right)\left(\sum_{i=1}^{n} y_i\right)}{n \sum_{i=1}^{n} y_i^2 - \left(\sum_{i=1}^{n} y_i\right)^2}$$

and the coefficient c is the intercept of the regression line on the x axis and has equation

$$c = \bar{x} - d\bar{y}$$

where n is the number of pairs of observations.

Returning to the visibility data, the only new piece of information which we need to calculate is the sum of the squares of the y_i's—$\sum_{i=1}^{12} y_i^2$. Thus $\sum_{i=1}^{12} y_i^2 = 1 \cdot 0^2 + 1 \cdot 5^2 + \cdots + 2 \cdot 0^2 = 259 \cdot 0$. Hence since

$$d = \frac{n \sum_{i=1}^{12} x_i y_i - \left(\sum_{i=1}^{12} x_i\right)\left(\sum_{i=1}^{12} y_i\right)}{n \sum_{i=1}^{12} y_i^2 - \left(\sum_{i=1}^{12} y_i\right)^2}$$

this gives

$$d = \frac{12 \times 365 \cdot 5 - 60 \cdot 5 \times 45}{12 \times 259 \cdot 0 - 45 \times 45} = \frac{4{,}386 - 2{,}722 \cdot 5}{3{,}108 - 2{,}025} = \frac{1{,}663 \cdot 5}{1{,}083}$$

$$= 1 \cdot 54$$

giving

$$c = \bar{x} - d\bar{y}$$
$$c = 5 \cdot 04 - 1 \cdot 54 \times 3 \cdot 75 = 5 \cdot 04 - 5 \cdot 78$$
$$= -0 \cdot 74$$

The regression line of x on y has equation

$$x = -0 \cdot 74 + 1 \cdot 54 y$$

If, for instance, we wanted to know the visibility at A when the visibility at B is 12 km, then putting $y = 12$ in the equation just calculated we get $x = -0 \cdot 74 + 1 \cdot 54 \times 12 = 17 \cdot 74$ km or 17·5 km to the given accuracy.

We have seen that mathematically the two regression lines are both derived by using the method of least squares, but for the regression line of y on x we have measured our errors vertically, and for the regression line of x on y we have measured them horizontally. It is worth noting that the line of best fit could be derived algebraically by minimising the sum of the squares of the perpendicular distances from the points to the line as shown in Fig. 13.7.

Marine Statistics

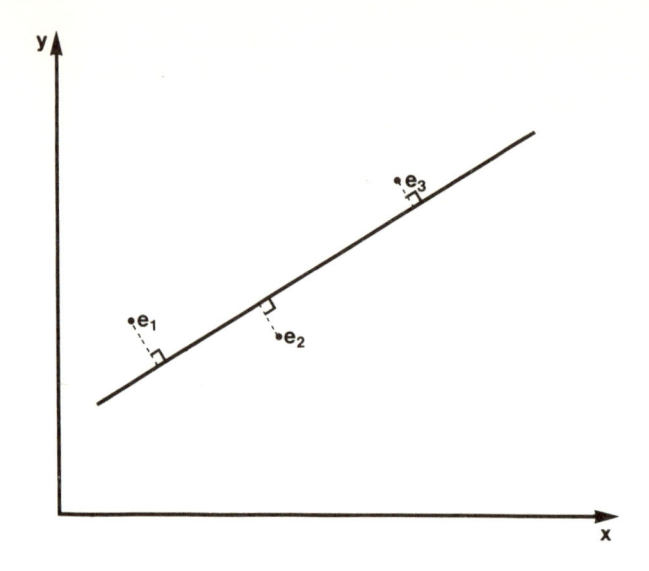

Fig 13·7 The perpendicular minimisation of errors

Hence the existence of three separate lines. It may be shown algebraically that the three lines meet at the point given by the means of the two variables $(\bar{x}, \bar{y})$. We can illustrate this with the visibility data where $\bar{x} = 5{\cdot}04$ and $\bar{y} = 3{\cdot}75$. Putting the coordinates $(5{\cdot}04, 3{\cdot}75)$ in the two lines $y = 1{\cdot}63 + 0{\cdot}42x$ and $x = -0{\cdot}74 + 1{\cdot}54y$, we can see that both equations are satisfied, viz.:

$$1{\cdot}63 + 0{\cdot}42 \times 5{\cdot}04 = 3{\cdot}75 \quad \text{and} \quad -0{\cdot}74 + 1{\cdot}54 \times 3{\cdot}75 = 5{\cdot}04$$

Before leaving this section there is one final point which should be noted and that is that we were trying to estimate x when $y = 12$ km, which is a value of y greater than anything we have observed in our data set. We are thus attempting to extrapolate our line which should always be done with caution as there may be a fundamental change in the relationship. For instance, the underlying relationship may be curvilinear as shown in Fig. 13.8. If we only have data on the portion from A to B then we will probably fit a line as shown. If we now attempt to estimate a value for y when x is at C we will be grossly overestimating the value. The curve drawn is an important one in business as it is the most common form of growth curve. A serious financial disaster could result if one estimated the growth when x is at C from the straight line rather than the curve. Hence care must be taken whenever estimates are made on an extrapolated part of a regression line or curve. Even on a part of the line for which data are available the value estimated must still be treated cautiously because the data will nearly always be sample data.

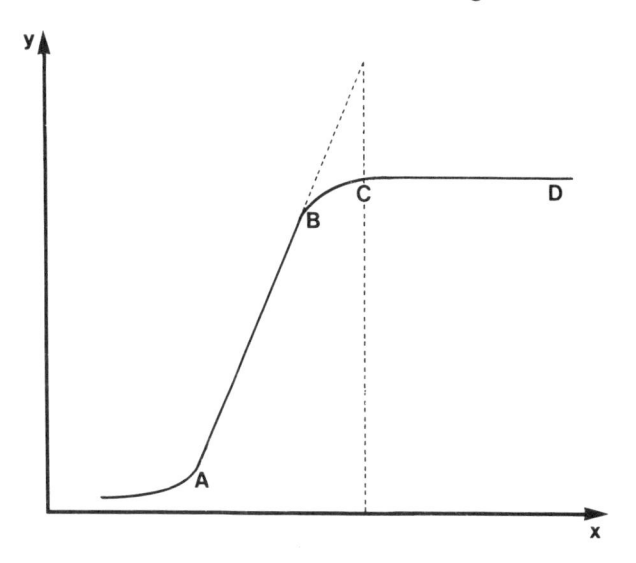

Fig 13·8 The dangers of extrapolation.

13.5 Sampling Aspects of Regression

In an earlier section the form of the population regression line for the regression of y on x was given as $\mu_{y|x} = \alpha + \beta x$, where $\mu_{y|x}$ was the mean of the distribution of y values for a particular x value. We now assume that each of the distributions for $y|x$ is a normal distribution and that the variances of all of them are the same and are denoted by σ^2. From our sample data we calculate estimates of α and β denoted by a and b, but as they are only sample statistics we will often want to construct confidence intervals for α and β and perform hypothesis tests. The derivation of the results in this section is outside the scope of this book and the actual results look rather fearsome. The important thing is for the reader to be aware of the sampling aspects of regression without necessarily following all the details, which are included mainly for reference, if needed.

We must first start by considering how to estimate σ^2. Our measure of error, $E_y = \sum_{i=1}^{n} (y_i - \hat{y}_i)^2$, gives us the variation about the sample regression line. Since $E_y = \sum_{i=1}^{n} (y_i - \hat{y}_i)^2 = \sum_{i=1}^{n} (y_i - a - bx_i)^2$ has two unknown population parameters in the expression our estimate of σ^2, the common variance, must be given by $E_y/(n-2)$, which will have $n-2$ degrees of freedom. Thus

$$\hat{\sigma}^2 = \frac{\sum_{i=1}^{n} (y_i - a - bx_i)^2}{n - 2}$$

This formula is not easy to use computationally but it may be shown algebraically that $\sum_{i=1}^{n} (y_i - a - bx_i)^2 = (n-1)(s_y^2 - b^2 s_x^2)$, where s_x^2 and s_y^2

Marine Statistics

are sample variances of x and y respectively. Thus

$$\hat{\sigma}^2 = \frac{n-1}{n-2}(s_y^2 - b^2 s_x^2)$$

where

$$s_y^2 = \frac{1}{n-1}\left[\sum_{i=1}^{n} y_i^2 - \frac{\left(\sum_{i=1}^{n} y_i\right)^2}{n}\right]$$

and

$$s_x^2 = \frac{1}{n-1}\left[\sum_{i=1}^{n} x_i^2 - \left(\sum_{i=1}^{n} x_i\right)^2\right]$$

Again, the mathematical proofs are outside the scope of this book but the results for the sampling distributions of a and b can now be quoted; a has a normal distribution with mean α and variance estimated by:

$$\hat{\sigma}_a^2 = \left[\frac{\sum_{i=1}^{n} x_i^2}{n(n-1)s_x^2}\right]\hat{\sigma}^2$$

b has a normal distribution with mean β and variance estimated by:

$$\hat{\sigma}_b^2 = \frac{\hat{\sigma}^2}{(n-1)s_x^2}$$

Both these variances depend on σ^2, which will usually be unknown and hence have to be estimated as $\hat{\sigma}^2$. Again, in most situations the sample size will be small so the 't' distribution will have to be used to calculate confidence intervals and perform significance tests.

As an illustration suppose, in the visibility example, we wanted to test whether β could be equal to 0. This is in fact equivalent to testing whether the regression line is significant or not, because a zero slope implies that y does not change as x changes, so any observed changes are purely random effects. Thus

$H_0: \beta = 0$
$H_1: \beta \neq 0$
$\alpha = 0.05$

are our hypotheses and significance level.

A 't' statistic always has the general form

$$t = \frac{\text{sample statistic} - \text{population parameter}}{\text{standard error}}$$

so in this case $t = (b - \beta)/\hat{\sigma}_b = b/\hat{\sigma}_b$ since, under H_0, $\beta = 0$:

$$\hat{\sigma}_b = \frac{\hat{\sigma}}{\sqrt{(n-1)s_x^2}} \quad \text{where} \quad \hat{\sigma}^2 = \frac{n-1}{n-2}[s_y^2 - b^2 s_x^2]$$

The degrees of freedom for t will be $n - 2$, since these are the degrees of freedom for $\hat{\sigma}^2$.

From the earlier calculations:

$$\sum_{i=1}^{12} x_i = 60.5 \qquad \sum_{i=1}^{12} y_i = 45.0 \qquad \sum_{i=1}^{12} x_i^2 = 636.75$$

$$\sum_{i=1}^{12} y_i^2 = 259.0 \qquad b = 0.42 \qquad n = 12$$

$$s_x^2 = \frac{1}{11}\left[\sum_{i=1}^{12} x_i^2 - \frac{\left(\sum_{i=1}^{12} x_i\right)^2}{12}\right] = \frac{1}{11}\left[636.75 - \frac{60.5 \times 60.5}{12}\right]$$

$$= \frac{3{,}980.75}{132} = 30.16$$

$$s_y^2 = \frac{1}{11}\left[\sum_{i=1}^{12} y_i^2 - \frac{\left(\sum_{i=1}^{12} y_i\right)^2}{12}\right] = \frac{1}{11}\left[259 - \frac{45^2}{12}\right]$$

$$= \frac{1{,}083}{132} = 8.21$$

$$\hat{\sigma}^2 = \frac{n-1}{n-2}[s_y^2 - b^2 s_x^2] = \frac{11}{10}[8.21 - (0.42)^2 \times 30.16]$$

$$= \frac{11}{10}[8.21 - 5.32] = 3.18$$

$$\hat{\sigma}_b^2 = \frac{\hat{\sigma}^2}{(n-1)s_x^2} = \frac{3.18}{11 \times 30.16} = 0.0096$$

Thus

$$\hat{\sigma}_b = 0.098$$

This gives

$$t = \frac{0.42}{0.098} = 4.29$$

The degrees of freedom, $n - 2$, is 10, giving as critical region $|t| > t_{10, 2.5\%} = 2.23$; $4.29 > 2.23$, so we have a significant result; hence we may reject H_0. Thus it is likely that the value of β is different from 0 and hence that our regression line is meaningful.

From our calculated regression line we may estimate the mean of the distribution for y given a value of x, $\mu_{y|x}$. Another useful result is to know

Marine Statistics

the form of confidence intervals for this. A 95% confidence interval for $\mu_{y|x=x'}$ is given by the formula:

$$(a + bx') - t_{2.5\%}\hat{\sigma}\sqrt{\frac{1}{n} + \frac{(x' - \bar{x})^2}{(n - 1)s_x^2}} < \mu_{y|x'} < (a + bx')$$

$$+ t_{2.5\%}\hat{\sigma}\sqrt{\frac{1}{n} + \frac{(x' - \bar{x})^2}{(n - 1)s_x^2}}$$

where the t distribution has $n - 2$ degrees of freedom. For any individual value of y when $x = x'$ we have to increase the confidence interval by taking into account the variance σ^2 of the distribution for $y|x'$. Thus a 95% confidence interval is given by the formula:

$$(a + bx') - t_{2.5\%}\hat{\sigma}\sqrt{1 + \frac{1}{n} + \frac{(x' - \bar{x})^2}{(n - 1)s_x^2}} < y < (a + bx')$$

$$+ t_{2.5\%}\hat{\sigma}\sqrt{1 + \frac{1}{n} + \frac{(x' - \bar{x})^2}{(n - 1)s_x^2}}$$

An example may help to illustrate the use of these rather gruesome looking formulae. Using the regression line, we estimated that when the visibility at station A was 12 km, the visibility at B would be 6·67 km. However, a 95% confidence interval for the visibility at B when the visibility at A was 12 km can be calculated using the second formula. The $t_{2.5\%}$ value on 10 degrees of freedom is 2·3. The term

$$t_{2.5\%}\hat{\sigma}\sqrt{1 + \frac{1}{n} + \frac{(x' - \bar{x})^2}{(n - 1)s_x^2}}$$

where $x' = 12$, is equal to:

$$2.23 \times \sqrt{3.17} \times \sqrt{1 + \frac{1}{12} + \frac{(12 - 5.04)^2}{11 \times 30.16}} = 2.23 \times 1.78 \times 1.11$$

$$= 4.40$$

Thus the 95% confidence interval is:

$6.64 - 4.40 < y < 6.64 + 4.40$ i.e. $2.24 < y < 11.04$

Roughly speaking, we can say that although we estimate the visibility at B to be 6·5 km, say, for 95% confidence we have to allow for a discrepancy of 4 km either way. This is the important thing to realise that the estimate from a regression line is not an absolute value but one that is subject to sampling fluctuations.

13.6 Other Forms of Regression

So far we have only dealt with situations with linear relationships between two variables. However these ideas can very readily be extended to cope with more complicated forms of equation.

Suppose firstly we wanted to fit a quadratic curve of the form

Regression and Correlation

$y = a + bx + cx^2$. The method of least squares would be used again to minimise the sum of the errors squared, i.e. $\sum(y - \hat{y})^2$, which in this case is $\sum(y - a - bx - cx^2)$. There are three unknown coefficients, so we will have three normal equations to solve simultaneously, which are:

$$\sum y = na + b\sum x + c\sum x^2$$
$$\sum xy = a\sum x + b\sum x^2 + c\sum x^3$$
$$\sum x^2 y = a\sum x^2 + b\sum x^3 + c\sum x^4$$

The calculations have become much more unwieldy than in the linear case as not only do we have to calculate $\sum y$, $\sum x$, $\sum x^2$ and $\sum xy$ as before, we also have to calculate $\sum x^3$, $\sum x^2 y$ and $\sum x^4$. The burden can be lightened if there is access to a preprogrammed electronic calculator or computer, as these calculations may then be done automatically. However it is perfectly possible to extend the method of least squares to cope with more complicated equations if necessary. There is no point in listing any more sets of normal equations as they can be fairly easily obtained by anyone with a working knowledge of differential calculus.

Another method is to apply a simple transformation on the data so that the required relationship becomes a linear one. Suppose we wanted to fit an equation of the form $y = mx^n$. If we write this equation in logarithmic form it becomes $\log y = \log m + m \log x$. Hence if we look at the linear relationship between $\log y$ and $\log x$ then we will be able to determine the values of m and n. As an illustration of this consider the following data on the ship tonnage on order worldwide at the start of each of the four quarters over a period of two years:

World orderbook date	Million tons d.w.
Year 1 Quarter 1	190
2	155
3	138
4	121
Year 2 Quarter 1	110
2	96
3	85
4	78

We want to fit a curve of the form $y = mx^n$ and hope to use it to forecast the state of the world order book in terms of tonnage on order at the start of the third year. The time variable is therefore labelled x and we must number the quarters consecutively from the start. The tonnage variable becomes y. Our next step is to transform both the variables by writing down their

Marine Statistics

logarithms, and we may then call log x X and log y Y. Using these new variables X and Y we then proceed as before with the calculations shown below.

World orderbook date x	Million tons d.w. y	X (log x)	Y (log y)
1	190	0	2·2788
2	155	0·3010	2·1903
3	138	0·4771	2·1399
4	121	0·6021	2·0828
5	110	0·6990	2·0414
6	96	0·7782	1·9823
7	85	0·8451	1·9294
8	78	0·9031	1·8921

$$\sum X = 4\cdot 6056 \qquad \sum Y = 16\cdot 5370$$
$$\sum X^2 = 3\cdot 3047 \qquad \sum XY = 9\cdot 2431 \qquad n = 8$$

If $Y = a + bX$:

$$b = \frac{n \sum XY - \sum X . \sum Y}{n \sum X^2 - (\sum X)^2} = \frac{8 \times 9\cdot 2431 - 4\cdot 6056 \times 16\cdot 5370}{8 \times 3\cdot 3047 - 4\cdot 6056 \times 4\cdot 6056}$$

$$= \frac{73\cdot 9448 - 76\cdot 1628}{26\cdot 4376 - 21\cdot 2116} = -0\cdot 42$$

$$a = \bar{Y} - b\bar{X} \qquad \bar{Y} = 2\cdot 0671 \qquad \bar{X} = 0\cdot 5757$$
$$a = 2\cdot 0671 + 42 \times 0\cdot 5757 \qquad a = 2\cdot 3089$$

Hence log y = 2·3089 − 0·42 log x or $y = mx^n$, where log m = 2·3089 and $n = -0\cdot 42$. This gives $m = 203\cdot 6$. Thus $y = 204x^{-0\cdot 42}$. For the start of the third year $x = 9$. It is easier to substitute in the logarithmic equation, giving:

$$\log y = 2\cdot 3089 - 0\cdot 42 \times \log 9 = 2\cdot 3089 - 0\cdot 42 \times 0\cdot 9542$$
$$= 2\cdot 3089 - 0\cdot 4008 = 1\cdot 9081$$

Hence $y = 91$ million tons d.w.

It should be noted that this figure is actually higher than the figure for the preceding quarter but this anomaly only arises because the curve takes into account all the points. As a second point it should again be realized that we are extrapolating outside the known data points so the future relationship could be something completely different. However the technique of regressing the variable in question on time is a standard and much used method of forecasting.

In any practical situation it may be considered necessary to fit a variety of curves and a decision then has to be made as to which curve 'fits' the data best. An objective measure for determining this is termed the *Index of*

Determination. This may be defined as the ratio of the variation in the variable y explained by the regression to the total variation in the variable y, assuming that y is the random variable. Symbolically we can write the total variation in y as $\sum_y (y - \bar{y})^2$ and the 'explained' variation in y as $\sum_y (\hat{y} - \bar{y})^2$, where $\hat{y}$ is the estimate of y from the regression curve. It can be shown algebraically that the difference between the total variation and the 'explained' variation is the error sum of squares we have met previously, i.e. $\sum_y (y - \bar{y})^2 = \sum_y (\hat{y} - \bar{y})^2 + \sum_y (y - \hat{y})^2$. Thus the smaller we can make the error sum of squares, the smaller will be the discrepancy between the 'explained' and total variation, and hence the higher our value of the index of determination will be.

13.6.1 The Index of Determination is given by

$$\frac{\text{variation in } y \text{ explained by the regression}}{\text{total variation in } y}$$

Symbolically, it is written as $R^2 = \dfrac{\sum\limits_y (\hat{y} - \bar{y})^2}{\sum\limits_y (y - \bar{y})^2}$

The highest value of R^2 is obviously 1 when there is a perfect algebraic relationship between the variables and the smallest is 0 when no such relationship exists. If one is choosing between a variety of curves then the curve with the highest value of R^2 is taken. This particular situation is most likely to arise if one has access to a package on a computer to perform all the calculations. If one is faced with performing all the computations oneself, one would probably only fit the curve which appeared graphically to be the most suitable.

Non-linear regression is one extension of the basic linear regression which we have been considering. The other common situation which arises is where we have several independent variables, say $x_1, x_2, \ldots, x_K$, and one dependent variable, y. We might then try to fit a line of the form $y = \beta_0 + \beta_1 x_1 + \beta_2 x_2 + \cdots + \beta_K x_K$, or a more complicated form of equation. The whole situation is more difficult to visualise as we are no longer dealing with two dimensions but conceptually it is similar to the regression analysis for two variables which we have considered. This type of regression is known as multiple regression but any treatment of it is outside the scope of this book.

13.7 *Correlation*

If both variables x and y are random variables, then it is theoretically not possible to consider the prediction of values of one of them given a value of the other. However, it is possible to consider how closely related the variables are and, in particular, how closely the variables follow a linear pattern. An

Marine Statistics

analysis of this form is termed a correlation analysis. The basis of it is linked to the index of determination which we defined in the last section. If we have calculated a regression line of *y* on *x*, say, then we could calculate the index of determination:

$$R^2 = \frac{\sum_y (\hat{y} - \bar{y})^2}{\sum_y (y - \bar{y})^2}$$

This may be rewritten as:

$$R^2 = \frac{[n \sum xy - (\sum x)(\sum y)]^2}{[n \sum x^2 - (\sum x)^2][n \sum y^2 - (\sum y)^2]}$$

where *n* is the number of pairs of observations. This formula is symmetrical in *x* and *y* and so is independent of the regression line calculated. If we take the square root of it, we have the correlation coefficient, usually denoted by *r*, which gives a measure of the strength of the linear relationship between two variables *x* and *y*.

13.7.1 The correlation coefficient, *r*, between two variables *x* and *y* is given by the formula

$$r = \frac{n \sum_{i=1}^{n} (x_i y_i) - \left(\sum_{i=1}^{n} x_i\right)\left(\sum_{i=1}^{n} y_i\right)}{\left[n \sum_{i=1}^{n} x_i^2 - \left(\sum_{i=1}^{n} x_i\right)^2\right]^{1/2} \left[n \sum_{i=1}^{n} y_i^2 - \left(\sum_{i=1}^{n} y_i\right)^2\right]^{1/2}}$$

This is sometimes referred to as the Pearson product–moment correlation coefficient. Using the notation of previous sections we can see that *r* is related to the regression coefficients *b* and *d* by the following relationships:

$$r = b \cdot \frac{s_x}{s_y} \quad \text{and} \quad r = d \cdot \frac{s_y}{s_x}$$

If we consider the visibility example again, then

$$n = 12 \quad \sum_{i=1}^{12} x_i = 60.5 \quad \sum_{i=1}^{12} y_i = 45.0 \quad \sum_{i=1}^{12} x_i y_i = 365.5$$

$$\sum_{i=1}^{12} x_i^2 = 636.75 \quad \text{and} \quad \sum_{i=1}^{12} y_i^2 = 259.0$$

Thus

$$r = \frac{12 \times 365 \cdot 5 - 60 \cdot 5 \times 45 \cdot 0}{[12 \times 636 \cdot 75 - (60 \cdot 5)^2]^{1/2}[12 \times 259 \cdot 0 - (45 \cdot 0)^2]^{1/2}} = 0 \cdot 80$$

The formula looks rather more complicated to use than it is, but to overcome this it is often written in an alternative form. If we put $X = x - \bar{x}$ and $Y = y - \bar{y}$ then

$$r = \frac{\sum XY}{\sqrt{\sum X^2 . \sum Y^2}}$$

13.8 Interpretation of the Correlation Coefficient

The next important stage is the interpretation of the correlation coefficient and there are several points to note:

(i) r can take any value between -1 and $+1$.

(ii) If r is negative then we say that the correlation is negative or indirect, indicating that as one variable increases the other variable decreases.

(iii) If r is positive then we say that the correlation is positive or direct, indicating that as one variable increases the other variable also does.

(iv) A value of $r = +1$ or -1 is termed perfect correlation, indicating that there is an exact linear relationship between the two variables, the situation where the two regression lines become coincident.

(v) If $r = 0$, then we may only conclude that there is no *linear* relationship between the variables, not that there is necessarily no relationship at all.

(vi) Since r is the square root of the index of determination, to assess the value of r it is a good idea to square it and consider r^2. Thus if $r = 0.9$, $r^2 = 0.81$, which indicates that 81% of the variation in one variable is explained by the variation in the other. Thus a value of 0.9 indicates a high correlation. However if $r = 0.7$, $r^2 = 0.49$, indicating that only half the variation in one variable is explained by the variation in the other. Thus a value of $r = 0.7$ indicates that there is some correlation but it cannot be termed a high correlation.

If we return to the visibility data then we found a value of r there equal to 0.8. The conclusion would be that there is a direct relationship between the visibility at the two weather stations and that this relationship is reasonably high ($r^2 = 0.64$).

When considering the interpretation of the correlation coefficient there is one important factor to guard against and that is spurious correlation. Consider the following example. During the course of six months a ship was at sea for 20 weeks. Records were kept of the consumption of domestic fresh water, cigarettes and paint, week by week during the period, as follows:

Marine Statistics

Week	Fresh water (tonnes)	Cigarettes (cartons)	Paint (litres)
1	32	20	1
2	25	21	0
3	36	19	4
4	40	15	7
5	38	16	7
6	53	15	10
7	44	12	8
8	48	12	12
9	55	9	14
10	39	10	10
11	60	9	11
12	54	12	12
13	43	15	7
14	51	18	9
15	37	20	6
16	45	15	8
17	38	23	3
18	33	22	2
19	30	20	1
20	27	25	2

If we calculate the correlation coefficient between fresh water and cigarette consumption we obtain a value of -0.78 which suggests a reasonably close association. Is the conclusion therefore that the more water we drink and the more showers we take the lower the cigarette consumption will be? The results certainly seem to suggest a new method for giving up cigarette smoking! Yet again, if we calculate the correlation between fresh water consumption and paint consumption, we obtain a value of 0.90, suggesting a close relationship. Does this mean that we can predict the amount of water likely to be used on a ship simply by knowing the number of litres of paint needed?!

Both these results are examples of the dangers that may be present in attempting to make conclusions from correlation coefficients. It is never possible to establish causal relations by using correlation techniques, only to indicate whether two variables are linked in their movements. Thus we certainly cannot say that increasing water consumption will decrease cigarette smoking, for instance. Furthermore, even if there is shown to be an association between two variables, the circumstances should always be examined carefully to make sure that the association is a meaningful one and not a case of spurious correlation. It is not difficult to find correlations between somewhat surprising pairs of variables, and careful investigation often reveals a third factor to which each of the other two are meaningfully related. The explanation in this case was that week 1 was in April, week 10 in July and week 20 in October. The summer weather in the middle weeks implied higher

Regression and Correlation

consumption of domestic water for showers, etc., but also better conditions for painting so more paint was consumed. Additionally in that period more overtime was worked and hence the crew had less leisure time for smoking.

The main usefulness of correlation analysis is to indicate where interesting relationships between variables may lie. However, if one does require to establish a causal relationship, careful experimentation must then be carried out to show that one variable is the result of the other. As an exploratory technique correlation is very widely used.

13.9 *Significance of the Correlation Coefficient*

It is also necessary to remember that the value of the correlation coefficient which we obtain will most probably be based on a sample of data alone, and hence will only be an estimate of the true population value. Particularly if the sample size is small it is good practice to test whether the correlation is significant, i.e. if the value of r obtained suggests a significant difference from a value of 0. If we call the population correlation coefficient ρ (rho), and we must assume that the distribution of one of the variables is normal, then the test is as follows:

$H_0: \rho = 0$
$H_1: \rho \neq 0$
$\alpha = 0.05$

Test statistic is $t = (r\sqrt{n-2})/\sqrt{1-r^2}$, which is distributed as t on $n-2$ degrees of freedom if $\rho = 0$. Using the visibility data with $n = 12$ and $r = 0.80$, then

$$t = \frac{0.8}{\sqrt{1-0.64}}\sqrt{10} = \frac{0.8}{0.6} \times 3.16 = 4.22$$

Under H_0 we should have a t distribution on 10 degrees of freedom. The critical values are -2.23 and $+2.23$. As $4.22 > 2.23$ we have a significant result and hence we may conclude that it is unlikely that $\rho = 0$.

There are a variety of tests which can be performed to test hypotheses for ρ taking some value other than 0 but these will not be dealt with here.

13.10 *Rank Correlation*

At the start of the chapter it was mentioned that correlations could be calculated for qualitative variables. The only restriction is that it must be possible to rank each variable in some way. Our initial example considered a variable of potential danger in a port from casualties to different types of vessels and it was possible to rank this from low danger to high danger. If we calculate the correlation between two variables given in rank order, then we term it rank correlation. We shall consider here *Spearman's rank correlation coefficient*. There are alternatives but this one is the simplest to calculate. As an illustration we will take the opening example, the data for which are repeated below:

Marine Statistics

Class of ship	Rank for potential danger from casualties	Relative frequency of casualties
B	1	0·025
C	2	0·019
A	3	0·030
E	4	0·028
G	5	0·014
D	6	0·032
F	7	0·016
I	8	0·021
J	9	0·017
M	10	0·024
K	11	0·011
L	12	0·015

The first step is to rank both variables in the same direction, say from 1 — low to n — high, where n is the number of pairs, in this case 12. The second step is to find the difference, d_i, between the ranks for each pair, and the third step is to find d_i^2 in each case.

Spearman's rank correlation coefficient, r_s, is given by the formula:

$$r_s = 1 - \frac{6 \sum_{i=1}^{i=n} d_i^2}{n(n^2 - 1)}$$

The necessary calculations are shown below:

Class of ship	Danger rank	Casualty frequency rank	d_i	d_i^2
B	1	9	−8	64
C	2	6	−4	16
A	3	11	−8	64
E	4	10	−6	36
G	5	2	+3	9
D	6	12	−6	36
F	7	4	+3	9
I	8	7	+1	1
J	9	5	+4	16
H	10	8	+2	4
K	11	1	+10	100
L	12	3	+9	81

$\sum d_i^2 = 436$

$$r_s = 1 - \frac{6 \times 436}{12(144 - 1)} = -0.52$$

Regression and Correlation

0·52 does not indicate a very close relationship but the negative sign does at least confirm the suggestion that, by and large, the risk of casualty decreases as the potential danger of the consequences increases. Strictly speaking, a special test of significance should be carried out on the value −0·52 to see if it is significantly different from 0.

13.11 *Summary*

In this final chapter we have considered methods of examining the relationship between two variables. These conclude the statistical methods and concepts which we have attempted to introduce the reader to in this book. While many more sophisticated techniques exist, the foundations of them have been covered here. We hope that the reader will by this stage at least be aware of the uses and abuses of statistical methods but even more than that he will now be considering applying them to his own practical problems.

Exercises

1. In order to investigate the variation in the breaking load of a particular type of rope, with wear, sixteen samples were subjected to simulated working conditions for various lengths of time up to 1,400 hr and then tested to destruction with results as summarised below:

Sample number	Working time, hours (x)	Breaking load, tonnes (y)
1	0	2·65
2	0	2·78
3	200	2·53
4	200	2·76
5	400	2·70
6	400	2·54
7	600	2·42
8	600	2·40
9	800	2·44
10	800	2·28
11	1,000	2·36
12	1,000	2·20
13	1,200	2·08
14	1,200	2·13
15	1,400	2·09
16	1,400	1·93

(a) Construct a scatter diagram for the relationship between working time and breaking load.

Marine Statistics

    (b) Estimate the equation of the regression line of *y* on *x*.
    (c) Estimate the breaking load (with 95% confidence limits) for a rope with a working life of 900 hr.
    (d) Conduct a significance test, at the 0·05 level, for the slope of the regression line.

2. In order to study the propagation of errors in D.R. navigation, a navigator kept a record of the error in his D.R. position as it accumulated for various distances between fixes. The results are tabulated below:

Distance from last fix (n.m.)	Error in D.R. position (n.m.)
15	0·12
24	0·08
37	0·42
46	0·51
66	0·64
99	1·05
142	0·82
185	1·20
220	1·64
248	0·75
256	1·86
264	2·75

    (a) Estimate the coefficients of the regression line for *y* on *x*, and test the value of the intercept on the *y* axis (*a*) for significance at the 0·05 level.
    (b) Estimate the error in the D.R. position after a run of 200 nautical miles and give the 95% confidence limits.

3. The bunker consumption was noted when a ship made different daily mean speeds as follows:

Mean speed (knots)	Daily consumption (tonnes)
18	43·6
16	29·8
12	14·9
9	8·0
6	3·5
3	2·2

Assuming that we expect a regression line of the form $y = mx^n$, estimate the values to be ascribed to the coefficients *m* and *n*, and hence estimate the daily consumption for a mean speed of 14 knots.

Regression and Correlation

4. The radar detection ranges for end-on contacts in normal atmospheric conditions and for a radar scanner height of 15 m have been quoted as follows:

Size of vessel (gross tons)	Detection range (n.m.)
10	1·9
200	5·5
1,000	6·0
10,000	13·0

If the regression line has the form $y = mx^n$, estimate the coefficients m and n and hence estimate the detection range for an end-on contact of 5,000 tons.

5. Using the data given in Ex. 1, calculate the correlation coefficient for the relationship between the working time and the breaking loads of the ropes. Test your result for significance at the 0·05 level.

6. Using the data given in Ex. 2, calculate the correlation coefficient for the relationship between the D.R. position error and the distance from the previous fix. Test your result for significance at the 0·01 level.

7. In order to investigate the relationship between weather conditions and a ship's speed, log book records were used to find cases where maximum wind forces during the course of a day in the open ocean were recorded as each of the values 1–10. The corresponding daily mean speeds were as follows:

Wind force	Mean speed
1	15·50
2	15·55
3	15·38
4	15·21
5	15·04
6	15·12
7	15·08
8	14·90
9	15·05
10	14·71

Since Beaufort scale wind force does not have a linear relationship with wind speed in knots, a ranking method is considered appropriate for the analysis of these data. Calculate the Spearman rank correlation coefficient for the relationship between wind force and average speed.

8. In a radar simulator experiment, twelve navigators of varying levels of experience were given identical collision avoidance situations to resolve.

Marine Statistics

The closest point of approach achieved by each subject was recorded and these are tabulated below against the ranks of the levels of experience:

Rank of experience	Subjects CPA (n.m.)	Rank of experience	Subjects CPA (n.m.)
1	0·9	7	1·0
2	2·5	8	1·5
3	2·0	9	0·8
4	1·2	10	1·3
5	1·8	11	0·6
6	2·4	12	1·1

Calculate Spearman's rank correlation coefficient for the relationship between these variables.

Answers

1. (b) We tabulate:

Sample number	x	y	xy	x^2	y^2
1	0	2·65	0	0	7·02
2	0	2·78	0	0	7·73
3	200	2·53	506	40,000	6·40
4	200	2·76	552	40,000	7·62
5	400	2·70	1,080	160,000	7·29
6	400	2·54	1,016	160,000	6·45
7	600	2·42	1,452	360,000	5·86
8	600	2·40	1,440	360,000	5·76
9	800	2·44	1,952	640,000	5·95
10	800	2·28	1,824	640,000	5·20
11	1,000	2·36	2,360	1,000,000	5·57
12	1,000	2·20	2,200	1,000,000	4·84
13	1,200	2·08	2,496	1,440,000	4·33
14	1,200	2·13	2,556	1,440,000	4·54
15	1,400	2·09	2,926	1,960,000	4·37
16	1,400	1·93	2,702	1,960,000	3·72
	11,200	38·29	25,062	11,200,000	92·65

$$\bar{x} = \frac{11{,}200}{16} = \underline{700} \qquad \bar{y} = \frac{38·29}{16} = \underline{2·393}$$

$$s_x^2 = \underline{224{,}000} \qquad s_y^2 = \underline{0·0678}$$

Regression and Correlation

The regression coefficient is calculated as:

$$b = \frac{n \sum xy - (\sum x)(\sum y)}{n \sum x^2 - (\sum x)^2} = \frac{16 \times 25{,}062 - 11{,}200 \times 38 \cdot 29}{16 \times 11{,}200{,}000 - 11{,}200^2}$$

$$= \frac{-27{,}856}{53{,}760{,}000} = \underline{-0 \cdot 000518}$$

Then

$$a = \bar{y} - b\bar{x} = 2 \cdot 393 + 0 \cdot 000518 \times 700$$
$$= \underline{2 \cdot 756}$$

(c) To estimate the breaking load, y, for a working life $x = 900$ hr, we use:

$$y = a + bx = 2 \cdot 756 - 0 \cdot 000518 \times 900$$
$$= \underline{2 \cdot 29 \text{ tonnes}}$$

The 95% confidence interval is given by:

$$(a + bx') - t_{n-2, 0 \cdot 025}\hat{\sigma}\sqrt{1 + \frac{1}{n} + \frac{(x' - \bar{x})^2}{(n-1)s_x^2}} < y < (a + bx')$$

$$+ t_{0 \cdot 025, n-2}\hat{\sigma}\sqrt{1 + \frac{1}{n} + \frac{(x' - \bar{x})^2}{(n-1)s_x^2}}$$

We calculate:

$$\hat{\sigma} = \sqrt{\frac{n-1}{n-2}[s_y^2 - b^2 s_x^2]} = \sqrt{\frac{15}{14}[0 \cdot 0678 - 0 \cdot 000518^2 \times 224{,}000]}$$

$$= \underline{0 \cdot 0908}$$

We find $t_{14, 0 \cdot 025} = 2 \cdot 14$.

We next calculate the term:

$$t_{14, 0 \cdot 025}\hat{\sigma}\sqrt{1 + \frac{1}{n} + \frac{(x' - \bar{x})^2}{(n-1)s_x^2}}$$

$$= 2 \cdot 14 \times 0 \cdot 0908 \sqrt{1 + \frac{1}{16} + \frac{(900 - 700)^2}{15 \times 224{,}000}} = 0 \cdot 201$$

Thus the limits within which we are 95% certain that the true value of y will lie are:

$$2 \cdot 29 - 0 \cdot 20 < y < 2 \cdot 29 + 0 \cdot 20 \quad \text{i.e.} \quad \underline{2 \cdot 09 < y < 2 \cdot 49}$$

(d) The slope of the regression line for the sample is b, which was found to be $-0 \cdot 000518$. To decide whether this is significant, we set up:

$$\left.\begin{array}{l} H_0: \beta = 0 \\ H_1: \beta \neq 0 \end{array}\right\} \text{ where } \beta \text{ is the slope of the population regression line}$$

$$\alpha = 0 \cdot 05$$

Marine Statistics

We calculate:

$$\hat{\sigma}_b = \frac{\hat{\sigma}}{\sqrt{(n-1)s_x^2}} = \frac{0.0908}{\sqrt{15 \times 224,000}} = \underline{0.0000495}$$

We may then calculate:

$$t = \frac{b}{\hat{\sigma}_b} = \frac{0.000518}{0.0000495} = \underline{10.46}$$

The degrees of freedom, $n - 2 = 14$, and using this we find that the rejection region is given by:

$$|t| > t_{0.025, 14} = 2.14$$

The calculated value is thus in the rejection region. Hence we reject H_0 and conclude that the value of β is significantly different from zero.

2. (a) We tabulate:

Distance run (x)	D.R. error (y)	xy	x^2	y^2
15	0·12	1·80	225	0·0144
24	0·08	1·92	576	0·0064
37	0·42	15·54	1,369	0·1764
46	0·51	23·46	2,116	0·2601
66	0·64	42·24	4,356	0·4096
99	1·05	103·95	9,801	1·1025
142	0·82	116·44	20,164	0·6724
185	1·20	222·00	34,225	1·4400
220	1·64	360·80	48,400	2·6896
248	0·75	186·00	61,504	0·5625
256	1·86	476·16	65,536	3·4596
264	2·75	726·00	69,696	7·5625
1,602	11·84	2,276·31	317,968	18·3560

$$\bar{x} = \frac{1,602}{12} = \underline{133.5} \qquad \bar{y} = \underline{0.987}$$

$$s_x^2 = \underline{9,463.7} \qquad s_y^2 = \underline{0.607}$$

We calculate the regression coefficient as:

$$b = \frac{n \sum xy - (\sum x)(\sum y)}{n \sum x^2 - (\sum x)^2} = \frac{12 \times 2,276.31 - 1,602 \times 11.84}{12 \times 317,968 - 1,602 \times 1,602}$$

$$= \frac{8,348.04}{1,249,212} = \underline{0.00668}$$

$$a = \bar{y} - b\bar{x} = 0.987 - 0.00668 \times 133.5$$
$$= \underline{0.0952}$$

Regression and Correlation

The intercept of the regression line on the y axis, $a = 0.095$, is tested for significance by setting up:

$H_0: \alpha = 0$ (where α is the intercept for the population regression
$H_0: \alpha \neq 0$ line)
$\alpha = 0.05$

We calculate:

$$\sigma^2 = \frac{n-1}{n-2}(s_y^2 - b^2 s_x^2) = \frac{11}{10}(0.607 - 0.00668^2 \times 9{,}463.7)$$

$$= \underline{0.203}$$

We calculate:

$$\hat{\sigma}_a^2 = \frac{\sum x^2}{n(n-1)s_x^2}\hat{\sigma}^2 = \frac{317{,}968 \times 0.203}{12 \times 11 \times 9{,}463.7}$$

$$= \underline{0.0517}$$

We calculate:

$$t = \frac{a}{\hat{\sigma}_a} = \frac{0.0952}{0.0517} = \underline{1.84}$$

We find that the rejection region is given by:

$$|t| > t_{10, 0.025} = 2.23$$

Hence we accept H_0 and conclude that the value of α is not significantly different from zero. This result corresponds with a commonsense interpretation since there must always be a zero D.R. error corresponding to a zero elapsed distance which implies that the regression line should pass through the origin.

(b) We estimate the error in the D.R. position after a run of 200 nautical miles by the equation:

$$y = a + bx' = 0.0952 + 0.00668 \times 200$$
$$= 1.43 \text{ nautical miles}$$

The 95% confidence limits are given by:

$$(a + bx') - t_{n-2, 0.025}\hat{\sigma}\sqrt{1 + \frac{1}{n} + \frac{(x' - \bar{x})^2}{(n-1)s_x^2}} < y < (a + bx')$$

$$+ t_{n-2, 0.025}\hat{\sigma}\sqrt{1 + \frac{1}{n} + \frac{(x' - \bar{x})^2}{(n-1)s_x^2}}$$

We find:

$t_{10, 0.025} = \underline{2.23}$

Marine Statistics

We calculate:

$$t_{10, 0.025} \hat{\sigma} \sqrt{1 + \frac{1}{n} + \frac{(x' - \bar{x})^2}{(n-1)s_x^2}}$$

$$= 2 \cdot 23 \times 0 \cdot 451 \sqrt{1 + \frac{1}{12} + \frac{(200 - 133 \cdot 5)^2}{11 \times 9{,}463 \cdot 7}} = \underline{1 \cdot 07}$$

Thus the limits within which the value of y will lie on 95% of occasions are:

$$(1 \cdot 43 - 1 \cdot 07) < y < (1 \cdot 43 + 1 \cdot 07) \quad \text{i.e. } \underline{0 \cdot 36 < y < 2 \cdot 50}$$

3. We re-tabulate the data to include logarithmic values thus:

x Speed	y Consumption	X Log speed	Y Log consumption
18	43·6	1·255	1·639
16	29·8	1·204	1·474
12	14·9	1·079	1·173
9	8·0	0·954	0·903
6	3·5	0·778	0·544
3	2·2	0·477	0·342

$$\sum X = 5 \cdot 747 \qquad \sum Y = 6 \cdot 075 \qquad n = 6$$
$$\sum X^2 = 5 \cdot 932 \qquad \sum XY = 6 \cdot 545$$

For the equation $Y = a + bX$, we find:

$$b = \frac{n \sum XY - (\sum X)(\sum Y)}{n \sum X^2 - (\sum X)^2} = \frac{6 \times 6 \cdot 545 - 5 \cdot 747 \times 6 \cdot 075}{6 \times 5 \cdot 932 - (5 \cdot 747)^2}$$

$$= \frac{4 \cdot 357}{2 \cdot 564} = 1 \cdot 699$$

$$a = \bar{Y} - b\bar{X} = \frac{6 \cdot 075}{6} - \frac{1 \cdot 699 \times 5 \cdot 747}{6} = \underline{-0 \cdot 615}$$

The equation thus becomes:

$$Y = -0 \cdot 615 + 1 \cdot 699 X$$

or

$$\log y = -0 \cdot 615 + 1 \cdot 699 \log x \quad \text{i.e. } y = mx^n$$

where $\log m = -0 \cdot 615 \Rightarrow m = 0 \cdot 243$ and $n = 1 \cdot 699$.

To estimate the consumption for a speed of 14 knots, we put $x = 14$ in:

$$y = 0 \cdot 243 x^{1 \cdot 699} = 0 \cdot 243 \times 14^{1 \cdot 7}$$
$$= \underline{21 \cdot 6 \text{ tonnes}}$$

Regression and Correlation

4. We re-tabulate the data to include logarithmic values thus:

x Size	y Detection range	X Log size	Y Log detection range
10	1·9	1·000	0·279
200	5·5	2·301	0·740
1,000	6·0	3·000	0·778
10,000	13·0	4·000	1·114

$$\sum X = \underline{10 \cdot 301} \qquad \sum Y = \underline{2 \cdot 911}$$
$$\sum X^2 = \underline{31 \cdot 295} \qquad \sum XY = \underline{8 \cdot 772}$$

For the equation $Y = a + bX$, we find:

$$b = \frac{n \sum XY - (\sum X)(\sum Y)}{n \sum X^2 - (\sum X)^2} = \frac{4 \times 8 \cdot 772 - 10 \cdot 301 \times 2 \cdot 911}{4 \times 31 \cdot 295 - 10 \cdot 301^2}$$

$$= \frac{5 \cdot 102}{19 \cdot 069} = \underline{0 \cdot 268}$$

$$a = \bar{Y} - b\bar{X} = \frac{2 \cdot 911}{4} - \frac{0 \cdot 268 \times 10 \cdot 301}{4} = \underline{0 \cdot 0376}$$

The equation thus becomes:

$$Y = 0 \cdot 0376 + 0 \cdot 268 X$$

or

$\log y = 0 \cdot 0376 + 0 \cdot 268 \log x$ i.e. $y = mx^n$

where $\log m = 0 \cdot 0376 \Rightarrow m = 1 \cdot 09$ and $n = 0 \cdot 268$.

To estimate the detection range for a 5,000-ton vessel, we put $x = 5{,}000$ in the equation:

$$y = 1 \cdot 09 \times 5{,}000^{0 \cdot 268}$$
$$= \underline{10 \cdot 7 \text{ nautical miles}}$$

5. We have:

$$n = 16 \qquad \sum x = 11{,}200 \qquad \sum y = 38 \cdot 29 \qquad \sum xy = 25{,}062$$
$$\sum x^2 = 11{,}200{,}000 \qquad \sum y^2 = 92 \cdot 65$$

$$r = \frac{n \sum (xy) - (\sum x)(\sum y)}{[n \sum x^2 - (\sum x)^2]^{1/2}[n \sum y^2 - (\sum y)^2]^{1/2}}$$

$$= \frac{16 \times 25{,}062 - 11{,}200 \times 38 \cdot 29}{(16 \times 11{,}200{,}000 - 11{,}200^2)^{1/2}(16 \times 92 \cdot 65 - 38 \cdot 29^2)^{1/2}}$$

$$= \underline{-0 \cdot 94}$$

Marine Statistics

To test for significance, we set up:

$H_0: \rho = 0$
$H_1: \rho \neq 0$
$\alpha = 0.05$

The test statistic is:

$$t = \frac{r\sqrt{n-2}}{\sqrt{1-r^2}} = \frac{-0.94\sqrt{14}}{\sqrt{1-0.94^2}}$$
$$= -10.31$$

Under H_0, the critical values of $t_{0.05}$ for 14 d.f. are ± 2.12. Hence, since our calculated value of $-10.31 < -2.12$, it is in the rejection region. We thus reject H_0 and conclude that there is a significant negative correlation between the two variables.

6. We have:

$n = 12 \quad \sum x = 1,602 \quad \sum y = 11.84 \quad \sum xy = 2,276.31$
$\sum x^2 = 317,968 \quad \sum y^2 = 18.356$

$$r = \frac{n \sum xy - \sum x \sum y}{[n \sum x^2 - (\sum x)^2]^{1/2}[n \sum y^2 - (\sum y)^2]^{1/2}}$$
$$= \frac{12 \times 2,276.31 - 1,602 \times 11.84}{(12 \times 317,968 - 1,602^2)^{1/2}(12 \times 18.356 - 11.84^2)^{1/2}}$$

To test for significance, we set up:

$H_0: \rho = 0$
$H_1: \rho \neq 0$
$\alpha = 0.01$

We calculate:

$$t = \frac{r\sqrt{n-2}}{\sqrt{1-r^2}} = \frac{0.83\sqrt{10}}{\sqrt{1-0.83^2}}$$
$$= 4.71$$

The critical values of $t_{0.01}$ for 10 degrees of freedom are ± 3.17. Hence, since our calculated value of $4.71 > 3.17$, it is in the rejection region. We thus reject H_0 and conclude that there is a significant positive correlation between the two variables.

7. We re-tabulate the data with ranks assigned to the mean speeds as well as to the wind force:

Wind force	Mean speed	Speed rank	d	d^2
1	15·50	9	−8	64
2	15·55	10	−8	64
3	15·38	8	−5	25
4	15·21	7	−3	9
5	15·04	3	2	4
6	15·12	6	0	0
7	15·08	5	2	4
8	14·90	2	6	36
9	15·05	4	5	25
10	14·71	1	9	81

$$\sum d^2 = 312$$

$$r_s = 1 - \frac{6 \sum d^2}{n(n^2 - 1)} = 1 - \frac{6 \times 312}{10 \times 99}$$
$$= \underline{-0.89}$$

8. We re-tabulate the data to include ranks for the achieved CPAs thus:

Experience rank	CPA (n.m.)	CPA rank	d	d^2
1	0·9	3	−2	4
2	2·5	12	−10	100
3	2·0	10	−7	49
4	1·2	6	−2	4
5	1·8	9	−4	16
6	2·4	11	−5	25
7	1·0	4	3	9
8	1·5	8	0	0
9	0·8	2	7	49
10	1·3	7	3	9
11	0·6	1	10	100
12	1·1	5	7	49

$$\sum d^2 = 414$$

$$r_s = 1 - \frac{6 \sum d^2}{n(n^2 - 1)} = 1 - \frac{6 \times 414}{12 \times 143}$$
$$= \underline{-0.45}$$

We thus conclude that there appears to be an inverse correlation between length of experience and achieved CPAs, but that such correlation is rather weak.

Summary of notation and formulae

Arithmetic mean:

Sample: Population

$$\mu = \frac{\sum_i f_i x_i}{n}$$ where f_i is the frequency with which the value x_i occurs and $n = \sum_i f_i$

Sample

$$\bar{x} = \frac{\sum_i f_i x_i}{n}$$

Median: $\tilde{x}$

Standard deviation:

Population

$$\sigma = \sqrt{\frac{1}{n}\left[\sum_i f_i x_i^2 - \frac{(\sum_i f_i x_i)^2}{n}\right]}$$

Sample

$$s = \sqrt{\frac{1}{n-1}\left[\sum_i f_i x_i^2 - \frac{(\sum_i f_i x_i)^2}{n}\right]}$$

Addition law of probability:

$$p(A \cup B) = p(A) + p(B) - p(A \cap B)$$

Multiplication law of probability:

$$p(A \cap B) = p(A)p(B|A) \quad \text{or} \quad p(B)p(A|B)$$

Regression line of y on x ($y = a + bx$):

Slope

$$b = \frac{n\sum_{i=1}^{n}(x_i y_i) - \left(\sum_{i=1}^{n} x_i\right)\left(\sum_{i=1}^{n} y_i\right)}{n\sum_{i=1}^{n} x_i^2 - \left(\sum_{i=1}^{n} x_i\right)^2}$$

Intercept

$$a = \bar{y} - b\bar{x}$$

Marine Statistics

Regression line of x on y ($x = c + dy$):

Slope

$$d = \frac{n \sum_{i=1}^{n} (x_i y_i) - \left(\sum_{i=1}^{n} x_i\right)\left(\sum_{i=1}^{n} y_i\right)}{n \sum_{i=1}^{n} y_i^2 - \left(\sum_{i=1}^{n} y_i\right)^2}$$

Intercept

$$c = \bar{x} - d\bar{y}$$

Correlation coefficient:

$$r = \frac{n \sum_{i=1}^{n} (x_i y_i) - \left(\sum_{i=1}^{n} x_i\right)\left(\sum_{i=1}^{n} y_i\right)}{\left[n \sum_{i=1}^{n} x_i^2 - \left(\sum_{i=1}^{n} x_i\right)^2\right]^{1/2} \left[n \sum_{i=1}^{n} y_i^2 - \left(\sum_{i=1}^{n} y_i\right)^2\right]^{1/2}}$$

Tables

Table 1. Normal Curve Areas

z	·00	·01	·02	·03	·04	·05	·06	·07	·08	·09
0·0	·0000	·0040	·0080	·0120	·0160	·0199	·0239	·0279	·0319	·0359
0·1	·0398	·0438	·0478	·0517	·0557	·0596	·0636	·0675	·0714	·0753
0·2	·0793	·0832	·0871	·0910	·0948	·0987	·1026	·1064	·1103	·1141
0·3	·1179	·1217	·1255	·1293	·1331	·1368	·1406	·1443	·1480	·1517
0·4	·1554	·1591	·1628	·1664	·1700	·1736	·1772	·1808	·1844	·1879
0·5	·1915	·1950	·1985	·2019	·2054	·2088	·2123	·2157	·2190	·2224
0·6	·2257	·2291	·2324	·2357	·2389	·2422	·2454	·2486	·2517	·2549
0·7	·2580	·2611	·2642	·2673	·2704	·2734	·2764	·2794	·2823	·2852
0·8	·2881	·2910	·2939	·2967	·2995	·3023	·3051	·3078	·3106	·3133
0·9	·3159	·3186	·3212	·3238	·3264	·3289	·3315	·3340	·3365	·3389
1·0	·3413	·3438	·3461	·3485	·3508	·3531	·3554	·3577	·3599	·3621
1·1	·3643	·3665	·3686	·3708	·3729	·3749	·3770	·3790	·3810	·3830
1·2	·3849	·3869	·3888	·3907	·3925	·3944	·3962	·3980	·3997	·4015
1·3	·4032	·4049	·4066	·4082	·4099	·4115	·4131	·4147	·4162	·4177
1·4	·4192	·4207	·4222	·4236	·4251	·4265	·4279	·4292	·4306	·4319
1·5	·4332	·4345	·4357	·4370	·4382	·4394	·4406	·4418	·4429	·4441
1·6	·4452	·4463	·4474	·4484	·4495	·4505	·4515	·4525	·4535	·4545
1·7	·4554	·4564	·4573	·4582	·4591	·4599	·4608	·4616	·4625	·4633
1·8	·4641	·4649	·4656	·4664	·4671	·4678	·4686	·4693	·4699	·4706
1·9	·4713	·4719	·4726	·4732	·4738	·4744	·4750	·4756	·4761	·4767
2·0	·4772	·4778	·4783	·4788	·4793	·4798	·4803	·4808	·4812	·4817
2·1	·4821	·4826	·4830	·4834	·4838	·4842	·4846	·4850	·4854	·4857
2·2	·4861	·4864	·4868	·4871	·4875	·4878	·4881	·4884	·4887	·4890
2·3	·4893	·4896	·4898	·4901	·4904	·4906	·4909	·4911	·4913	·4916
2·4	·4918	·4920	·4922	·4925	·4927	·4929	·4931	·4932	·4934	·4936
2·5	·4938	·4940	·4941	·4943	·4945	·4946	·4948	·4949	·4951	·4952
2·6	·4953	·4955	·4956	·4957	·4959	·4960	·4961	·4962	·4963	·4964
2·7	·4965	·4966	·4967	·4968	·4969	·4970	·4971	·4972	·4973	·4974
2·8	·4974	·4975	·4976	·4977	·4977	·4978	·4979	·4979	·4980	·4981
2·9	·4981	·4982	·4982	·4983	·4984	·4984	·4985	·4985	·4986	·4986
3·0	·4987	·4987	·4987	·4988	·4988	·4989	·4989	·4989	·4990	·4990

Table 2. Values of t

d.f.	$t_{.100}$	$t_{.050}$	$t_{.025}$	$t_{.010}$	$t_{.005}$	d.f.
1	3·078	6·314	12·706	31·821	63·657	1
2	1·886	2·920	4·303	6·965	9·925	2
3	1·638	2·353	3·182	4·541	5·841	3
4	1·533	2·132	2·776	3·747	4·604	4
5	1·476	2·015	2·571	3·365	4·032	5
6	1·440	1·943	2·447	3·143	3·707	6
7	1·415	1·895	2·365	2·998	3·499	7
8	1·397	1·860	2·306	2·896	3·355	8
9	1·383	1·833	2·262	2·321	3·250	9
10	1·372	1·812	2·228	2·764	3·169	10
11	1·363	1·796	2·201	2·718	3·106	11
12	1·356	1·782	2·179	2·681	3·055	12
13	1·350	1·771	2·160	2·650	3·012	13
14	1·345	1·761	2·145	2·624	2·977	14
15	1·341	1·753	2·131	2·602	2·947	15
16	1·337	1·746	2·120	2·583	2·921	16
17	1·333	1·740	2·110	2·567	2·898	17
18	1·330	1·734	2·101	2·552	2·878	18
19	1·328	1·729	2·093	2·539	2·861	19
20	1·325	1·725	2·086	2·528	2·845	20
21	1·323	1·721	2·080	2·518	2·881	21
22	1·321	1·717	2·074	2·508	2·819	22
23	1·319	1·714	2·069	2·500	2·807	23
24	1·318	1·711	2·064	2·492	2·797	24
25	1·316	1·708	2·060	2·485	2·787	25
26	1·315	1·706	2·056	2·479	2·779	26
27	1·314	1·703	2·052	2·473	2·771	27
28	1·313	1·701	2·048	2·467	2·763	28
29	1·311	1·699	2·045	2·462	2·756	29
inf.	1·282	1·645	1·960	2·326	2·576	inf.

Taken from Table IV of R. A. Fisher, *Statistical Methods for Research Workers*. Copyright © 1970 University of Adelaide, published by Hafner Press, New York, by permission of the author and publisher.

Table 3. Percentage Points of the Chi-Square Distribution

ν	0·995	0·99	0·975	0·95	0·05	0·025	0·01	0·005
1	0·04393	0·0³157	0·0³982	0·02393	3·841	5·024	6·635	7·879
2	0·0100	0·0201	0·0506	0·103	5·991	7·378	9·210	0·597
3	0·0717	0·115	0·216	0·352	7·815	9·348	11·345	12·838
4	0·207	0·297	0·484	0·711	9·488	11·143	13·277	14·860
5	0·412	0·554	0·831	1·145	11·070	12·832	15·086	16·750
6	0·676	0·872	1·237	1·635	12·592	14·449	16·812	18·548
7	0·989	1·239	1·690	2·167	14·067	16·013	18·475	20·278
8	1·344	1·646	2·180	2·733	15·507	17·535	20·090	21·955
9	1·735	2·088	2·700	3·325	16·919	19·023	21·666	23·589
10	2·156	2·558	3·247	3·940	18·307	20·483	23·209	25·188
11	2·603	3·053	3·816	4·575	19·675	21·920	24·725	26·757
12	3·074	3·571	4·404	5·226	21·026	23·337	26·217	28·300
13	3·565	4·107	5·009	5·892	22·362	24·736	27·688	29·819
14	4·075	4·660	5·629	6·571	23·685	26·119	29·141	31·319
15	4·601	5·229	6·262	7·261	24·996	27·488	30·578	32·801
16	5·142	5·812	6·908	7·962	26·296	28·845	32·000	34·267
17	5·697	6·408	7·564	8·672	27·587	30·191	33·409	35·718
18	6·265	7·015	8·231	9·390	28·869	31·526	34·805	37·156
19	6·844	7·633	8·907	10·117	30·144	32·852	36·191	38·582
20	7·434	8·260	9·591	10·851	31·410	34·170	37·566	39·997
21	8·034	8·897	10·283	11·591	32·671	35·479	38·932	41·401
22	8·643	9·542	10·982	12·338	33·924	36·781	40·289	42·796
23	9·260	10·196	11·689	13·091	35·172	38·076	41·638	44·181
24	9·886	10·856	12·401	13·848	36·415	39·364	42·980	45·558
25	10·520	11·524	13·120	14·611	37·652	40·646	44·314	46·928
26	11·160	12·198	13·844	15·379	38·885	41·923	45·642	48·290
27	11·808	12·879	14·573	16·151	40·113	43·194	46·963	49·645
28	12·461	13·565	15·308	16·928	41·337	44·461	48·278	50·993
29	13·121	14·256	16·047	17·708	42·557	45·722	49·588	52·336
30	13·787	14·953	16·791	18·493	43·773	46·979	50·892	53·672

Taken from Table 8 of *Biometrika Tables for Statisticians*, Vol. I, by permission of the Biometrika Trustees.

Table 4(a). 5 Per Cent Points of the F-Distribution

v_2 \ v_1	1	2	3	4	5	6	7	8	10	12	24	∞
1	161·4	199·5	215·7	224·6	230·2	234·8	236·8	238·9	241·9	243·9	249·0	254·3
2	18·5	19·0	19·2	19·2	19·3	19·3	19·4	19·4	19·4	19·4	19·5	19·5
3	10·13	9·55	9·28	9·12	9·01	8·94	8·89	8·85	8·79	8·74	8·64	8·53
4	7·71	6·94	6·59	6·39	6·26	6·16	6·09	6·04	5·96	5·91	5·77	5·63
5	6·61	5·79	5·41	5·19	5·05	4·95	4·88	4·82	4·74	4·68	4·53	4·36
6	5·99	5·14	4·76	4·53	4·39	4·28	4·21	4·15	4·06	4·00	3·84	3·67
7	5·59	4·74	4·35	4·12	3·97	3·87	3·79	3·73	3·64	3·57	3·41	3·23
8	5·32	4·46	4·07	3·84	3·69	3·58	3·50	3·44	3·35	3·28	3·12	2·93
9	5·12	4·26	3·86	3·63	3·48	3·37	3·29	3·23	3·14	3·07	2·90	2·71
10	4·96	4·10	3·71	3·48	3·33	3·22	3·14	3·07	2·98	2·91	2·74	2·54
11	4·84	3·98	3·59	3·36	3·20	3·09	3·01	2·95	2·85	2·79	2·61	2·40
12	4·75	3·89	3·49	3·26	3·11	3·00	2·91	2·85	2·75	2·69	2·51	2·30
13	4·67	3·81	3·41	3·18	3·03	2·92	2·83	2·77	2·67	2·60	2·42	2·21
14	4·60	3·74	3·34	3·11	2·96	2·85	2·76	2·70	2·60	2·53	2·35	2·13
15	4·54	3·68	3·29	3·06	2·90	2·79	2·71	2·64	2·54	2·48	2·29	2·07
16	4·49	3·63	3·24	3·01	2·85	2·74	2·66	2·59	2·49	2·42	2·24	2·01
17	4·45	3·59	3·20	2·96	2·81	2·70	2·61	2·55	2·45	2·38	2·19	1·96
18	4·41	3·55	3·16	2·93	2·77	2·66	2·58	2·51	2·41	2·34	2·15	1·92
19	4·38	3·52	3·13	2·90	2·74	2·63	2·54	2·48	2·38	2·31	2·11	1·88
20	4·35	3·49	3·10	2·87	2·71	2·60	2·51	2·45	2·35	2·28	2·08	1·84
21	4·32	3·47	3·07	2·84	2·68	2·57	2·49	2·42	2·32	2·25	2·05	1·81
22	4·30	3·44	3·05	2·82	2·66	2·55	2·46	2·40	2·30	2·23	2·03	1·78
23	4·28	3·42	3·03	2·80	2·64	2·53	2·44	2·37	2·27	2·20	2·00	1·76
24	4·26	3·40	3·01	2·78	2·62	2·51	2·42	2·36	2·25	2·18	1·98	1·73
25	4·24	3·39	2·99	2·76	2·60	2·49	2·40	2·34	2·24	2·16	1·96	1·71
26	4·23	3·37	2·98	2·74	2·59	2·47	2·39	2·32	2·22	2·15	1·95	1·69
27	4·21	3·35	2·96	2·73	2·57	2·46	2·37	2·31	2·20	2·13	1·93	1·67
28	4·20	3·34	2·95	2·71	2·56	2·45	2·36	2·29	2·19	2·12	1·91	1·65
29	4·18	3·33	2·93	2·70	2·55	2·43	2·35	2·28	2·18	2·10	1·90	1·64
30	4·17	3·32	2·92	2·69	2·53	2·42	2·33	2·27	2·16	2·09	1·89	1·62
32	4·15	3·29	2·90	2·67	2·51	2·40	2·31	2·24	2·14	2·07	1·86	1·59
34	4·13	3·28	2·88	2·65	2·49	2·38	2·29	2·23	2·12	2·05	1·84	1·57
36	4·11	3·26	2·87	2·63	2·48	2·36	2·28	2·21	2·11	2·03	1·82	1·55
38	4·10	3·24	2·85	2·62	2·46	2·35	2·26	2·19	2·09	2·02	1·81	1·53
40	4·08	3·23	2·84	2·61	2·45	2·34	2·25	2·18	2·08	2·00	1·79	1·51
60	4·00	3·15	2·76	2·53	2·37	2·25	2·17	2·10	1·99	1·92	1·70	1·39
120	3·92	3·07	2·68	2·45	2·29	2·18	2·09	2·02	1·91	1·83	1·61	1·25
∞	3·84	3·00	2·60	2·37	2·21	2·10	2·01	1·94	1·83	1·75	1·52	1·00

Taken from Table 18 of *Biometrika Tables for Statisticians*, Vol. I, by permission of the Biometrika Trustees.

Table 4(b). 2½ Per Cent Points of the F-Distribution

v_2 \ v_1 =	1	2	3	4	5	6	7	8	10	12	24	∞
1	648	800	864	900	922	937	948	957	969	977	997	1018
2	38·5	39·0	39·2	39·2	39·3	39·3	39·4	39·4	39·4	39·4	39·5	39·5
3	17·4	16·0	15·4	15·1	14·9	14·7	14·6	14·5	14·4	14·3	14·1	13·9
4	12·22	10·65	9·98	9·60	9·36	9·20	9·07	8·98	8·84	8·75	8·51	8·26
5	10·01	8·43	7·76	7·39	7·15	6·98	6·85	6·76	6·62	6·52	6·28	6·02
6	8·81	7·26	6·60	6·23	5·99	5·82	5·70	5·60	5·46	5·37	5·12	4·85
7	8·07	6·54	5·89	5·52	5·29	5·12	4·99	4·90	4·76	4·67	4·42	4·14
8	7·57	6·06	5·42	5·05	4·82	4·65	4·53	4·43	4·30	4·20	3·95	3·67
9	7·21	5·71	5·08	4·72	4·48	4·32	4·20	4·10	3·96	3·87	3·61	3·33
10	6·94	5·46	4·83	4·47	4·24	4·07	3·95	3·85	3·72	3·62	3·37	3·08
11	6·72	5·26	4·63	4·28	4·04	3·88	3·76	3·66	3·53	3·43	3·17	2·88
12	6·55	5·10	4·47	4·12	3·89	3·73	3·61	3·51	3·37	3·28	3·02	2·72
13	6·41	4·97	4·35	4·00	3·77	3·60	3·48	3·39	3·25	3·15	2·89	2·60
14	6·30	4·86	4·24	3·89	3·66	3·50	3·38	3·29	3·15	3·05	2·79	2·49
15	6·20	4·76	4·15	3·80	3·58	3·41	3·29	3·20	3·06	2·96	2·70	2·40
16	6·12	4·69	4·08	3·73	3·50	3·34	3·22	3·12	2·99	2·89	2·63	2·32
17	6·04	4·62	4·01	3·66	3·44	3·28	3·16	3·06	2·92	2·82	2·56	2·25
18	5·98	4·56	3·95	3·61	3·38	3·22	3·10	3·01	2·87	2·77	2·50	2·19
19	5·92	4·51	3·90	3·56	3·33	3·17	3·05	2·96	2·82	2·72	2·45	2·13
20	5·87	4·46	3·86	3·51	3·29	3·13	3·01	2·91	2·77	2·68	2·41	2·09
21	5·83	4·42	3·82	3·48	3·25	3·09	2·97	2·87	2·73	2·64	2·37	2·04
22	5·79	4·38	3·78	3·44	3·22	3·05	2·93	2·84	2·70	2·60	2·33	2·00
23	5·75	4·35	3·75	3·41	3·18	3·02	2·90	2·81	2·67	2·57	2·30	1·97
24	5·72	4·32	3·72	3·38	3·15	2·99	2·87	2·78	2·64	2·54	2·27	1·94
25	5·69	4·29	3·69	3·35	3·13	2·97	2·85	2·75	2·61	2·51	2·24	1·91
26	5·66	4·27	3·67	3·33	3·10	2·94	2·82	2·73	2·59	2·49	2·22	1·88
27	5·63	4·24	3·65	3·31	3·08	2·92	2·80	2·71	2·57	2·47	2·19	1·85
28	5·61	4·22	3·63	3·29	3·06	2·90	2·78	2·69	2·55	2·45	2·17	1·83
29	5·59	4·20	3·61	3·27	3·04	2·88	2·76	2·67	2·53	2·43	2·15	1·81
30	5·57	4·18	3·59	3·25	3·03	2·87	2·75	2·65	2·51	2·41	2·14	1·79
32	5·53	4·15	3·56	3·22	3·00	2·84	2·72	2·62	2·48	2·38	2·10	1·75
34	5·50	4·12	3·53	3·19	2·97	2·81	2·69	2·59	2·45	2·35	2·08	1·72
36	5·47	4·09	3·51	3·17	2·94	2·79	2·66	2·57	2·43	2·33	2·05	1·69
38	5·45	4·07	3·48	3·15	2·92	2·76	2·64	2·55	2·41	2·31	2·03	1·66
40	5·42	4·05	3·46	3·13	2·90	2·74	2·62	2·53	2·39	2·29	2·01	1·64
60	5·29	3·93	3·34	3·01	2·79	2·63	2·51	2·41	2·27	2·17	1·88	1·48
120	5·15	3·80	3·23	2·89	2·67	2·52	2·39	2·30	2·16	2·05	1·76	1·31
∞	5·02	3·69	3·12	2·79	2·57	2·41	2·29	2·19	2·05	1·94	1·64	1·00

Table 4(c). 1 Per Cent Points of the F-Distribution

$v_1 =$	1	2	3	4	5	6	7	8	10	12	24	∞
$v_2 = 1$	4052	5000	5403	5625	5764	5859	5928	5981	6056	6106	6235	6366
2	98·5	99·0	99·2	99·2	99·3	99·3	99·4	99·4	99·4	99·4	99·5	99·5
3	34·1	30·8	29·5	28·7	28·2	27·9	27·7	27·5	27·2	27·1	26·6	26·1
4	21·2	18·0	16·7	16·0	15·5	15·2	15·0	14·8	14·5	14·4	13·9	13·5
5	16·26	13·27	12·06	11·39	10·97	10·67	10·46	10·29	10·05	9·89	9·47	9·02
6	13·74	10·92	9·78	9·15	8·75	8·47	8·26	8·10	7·87	7·72	7·31	6·88
7	12·25	9·55	8·45	7·85	7·46	7·19	6·99	6·84	6·62	6·47	6·07	5·65
8	11·26	8·65	7·59	7·01	6·63	6·37	6·18	6·03	5·81	5·67	5·28	4·86
9	10·56	8·02	6·99	6·42	6·06	5·80	5·61	5·47	5·26	5·11	4·73	4·31
10	10·04	7·56	6·55	5·99	5·64	5·39	5·20	5·06	4·85	4·71	4·33	3·91
11	9·65	7·21	6·22	5·67	5·32	5·07	4·89	4·74	4·54	4·40	4·02	3·60
12	9·33	6·93	5·95	5·41	5·06	4·82	4·64	4·50	4·30	4·16	3·78	3·36
13	9·07	6·70	5·74	5·21	4·86	4·62	4·44	4·30	4·10	3·96	3·59	3·17
14	8·86	6·51	5·56	5·04	4·70	4·46	4·28	4·14	3·94	3·80	3·43	3·00
15	8·68	6·36	5·42	4·89	4·56	4·32	4·14	4·00	3·80	3·67	3·29	2·87
16	8·53	6·23	5·29	4·77	4·44	4·20	4·03	3·89	3·69	3·55	3·18	2·75
17	8·40	6·11	5·18	4·67	4·34	4·10	3·93	3·79	3·59	3·46	3·08	2·65
18	8·29	6·01	5·09	4·58	4·25	4·01	3·84	3·71	3·51	3·37	3·00	2·57
19	8·18	5·93	5·01	4·50	4·17	3·94	3·77	3·63	3·43	3·30	2·92	2·49
20	8·10	5·85	4·94	4·43	4·10	3·87	3·70	3·56	3·37	3·23	2·86	2·42
21	8·02	5·78	4·87	4·37	4·04	3·81	3·64	3·51	3·31	3·17	2·80	2·36
22	7·95	5·72	4·82	4·31	3·99	3·76	3·59	3·45	3·26	3·12	2·75	2·31
23	7·88	5·66	4·76	4·26	3·94	3·71	3·54	3·41	3·21	3·07	2·70	2·26
24	7·82	5·61	4·72	4·22	3·90	3·67	3·50	3·36	3·17	3·03	2·66	2·21
25	7·77	5·57	4·68	4·18	3·86	3·63	3·46	3·32	3·13	2·99	2·62	2·17
26	7·72	5·53	4·64	4·14	3·82	3·59	3·42	3·29	3·09	2·96	2·58	2·13
27	7·68	5·49	4·60	4·11	3·78	3·56	3·39	3·26	3·06	2·93	2·55	2·10
28	7·64	5·45	4·57	4·07	3·75	3·53	3·36	3·23	3·03	2·90	2·52	2·06
29	7·60	5·42	4·54	4·04	3·73	3·50	3·33	3·20	3·00	2·87	2·49	2·03
30	7·56	5·39	4·51	4·02	3·70	3·47	3·30	3·17	2·98	2·84	2·47	2·01
32	7·50	5·34	4·46	3·97	3·65	3·43	3·26	3·13	2·93	2·80	2·42	1·96
34	7·45	5·29	4·42	3·93	3·61	3·39	3·22	3·09	2·90	2·76	2·38	1·91
36	7·40	5·25	4·38	3·89	3·58	3·35	3·18	3·05	2·86	2·72	2·35	1·87
38	7·35	5·21	4·34	3·86	3·54	3·32	3·15	3·02	2·83	2·69	2·32	1·84
40	7·31	5·18	4·31	3·83	3·51	3·29	3·12	2·99	2·80	2·66	2·29	1·80
60	7·08	4·98	4·13	3·65	3·34	3·12	2·95	2·82	2·63	2·50	2·12	1·60
120	6·85	4·79	3·95	3·48	3·17	2·96	2·79	2·66	2·47	2·34	1·95	1·38
∞	6·63	4·61	3·78	3·32	3·02	2·80	2·64	2·51	2·32	2·18	1·79	1·00

Table 5. Critical Values of W in the Wilcoxon Test for Paired Observations

n	One-sided $\alpha = 0.01$ Two-sided $\alpha = 0.02$	One-sided $\alpha = 0.025$ Two-sided $\alpha = 0.05$	One-sided $\alpha = 0.05$ Two-sided $\alpha = 0.10$
5			1
6		1	2
7	0	2	4
8	2	4	6
9	3	6	8
10	5	8	11
11	7	11	14
12	10	14	17
13	13	17	21
14	16	21	26
15	20	25	30
16	24	30	36
17	28	35	41
18	33	40	47
19	38	46	54
20	43	52	60
21	49	59	68
22	56	66	75
23	62	73	83
24	69	81	92
25	77	90	101
26	85	98	110
27	93	107	120
28	102	117	130
29	111	127	141
30	120	137	152

Reproduced from F. Wilcoxon and R. A. Wilcox, *Some Rapid Approximate Statistical Procedures*, American Cyanamid Company, Pearl River, N.Y., 1964, by permission of the American Cyanamid Company.

Table 6. $\Pr(U \leq u | H_0 \text{ is true})$ in the Wilcoxon Two-Sample Test

$n_2 = 3$

	\multicolumn{3}{c}{n_1}		
u	1	2	3
0	0·250	0·100	0·050
1	0·500	0·200	0·100
2	0·750	0·400	0·200
3		0·600	0·350
4			0·500
5			0·650

$n_2 = 4$

	n_1			
u	1	2	3	4
0	0·200	0·067	0·028	0·014
1	0·400	0·133	0·057	0·029
2	0·600	0·267	0·114	0·057
3		0·400	0·200	0·100
4		0·600	0·314	0·171
5			0·429	0·243
6			0·571	0·343
7				0·443
8				0·557

$n_2 = 5$

	n_1				
u	1	2	3	4	5
0	0·167	0·047	0·018	0·008	0·004
1	0·333	0·095	0·036	0·016	0·008
2	0·500	0·190	0·071	0·032	0·016
3	0·667	0·286	0·125	0·056	0·028
4		0·429	0·196	0·095	0·048
5		0·571	0·286	0·143	0·075
6			0·393	0·206	0·111
7			0·500	0·278	0·155
8			0·607	0·365	0·210
9				0·452	0·274
10				0·548	0·345
11					0·421
12					0·500
13					0·579

Table 6. $\Pr(U \leqslant u | H_0 \text{ is true})$ in the Wilcoxon Two-Sample Test (continued)

$n_2 = 6$

				n_1			
u	1	2	3	4	5	6	
0	0·143	0·036	0·012	0·005	0·002	0·001	
1	0·286	0·071	0·024	0·010	0·004	0·002	
2	0·428	0·143	0·048	0·019	0·009	0·004	
3	0·571	0·214	0·083	0·033	0·015	0·008	
4		0·321	0·131	0·057	0·026	0·013	
5		0·429	0·190	0·086	0·041	0·021	
6		0·571	0·274	0·129	0·063	0·032	
7			0·357	0·176	0·089	0·047	
8			0·452	0·238	0·123	0·066	
9			0·548	0·305	0·165	0·090	
10				0·381	0·214	0·120	
11				0·457	0·268	0·155	
12				0·545	0·331	0·197	
13					0·396	0·242	
14					0·465	0·294	
15					0·535	0·350	
16						0·409	
17						0·469	
18						0·531	

$n_2 = 7$

				n_1			
u	1	2	3	4	5	6	7
0	0·125	0·028	0·008	0·003	0·001	0·001	0·000
1	0·250	0·056	0·017	0·006	0·003	0·001	0·001
2	0·375	0·111	0·033	0·012	0·005	0·002	0·001
3	0·500	0·167	0·058	0·021	0·009	0·004	0·002
4	0·625	0·250	0·092	0·036	0·015	0·007	0·003
5		0·333	0·133	0·055	0·024	0·011	0·006
6		0·444	0·192	0·082	0·037	0·017	0·009
7		0·556	0·258	0·115	0·053	0·026	0·013
8			0·333	0·158	0·074	0·037	0·019
9			0·417	0·206	0·101	0·051	0·027
10			0·500	0·264	0·134	0·069	0·036
11			0·583	0·324	0·172	0·090	0·049
12				0·394	0·216	0·117	0·064
13				0·464	0·265	0·147	0·082
14				0·538	0·319	0·183	0·104
15					0·378	0·223	0·130
16					0·438	0·267	0·159
17					0·500	0·314	0·191
18					0·562	0·365	0·228
19						0·418	0·267
20						0·473	0·310
21						0·527	0·355
22							0·402
23							0·451
24							0·500
25							0·549

Table 6. $\Pr(U \leq u | H_0$ is true) in the Wilcoxon Two-Sample Test (continued)

$n_2 = 8$

u	n_1=1	2	3	4	5	6	7	8
0	0.111	0.022	0.006	0.002	0.001	0.000	0.000	0.000
1	0.222	0.044	0.012	0.004	0.002	0.001	0.000	0.000
2	0.333	0.089	0.024	0.008	0.003	0.001	0.001	0.000
3	0.444	0.133	0.042	0.014	0.005	0.002	0.001	0.001
4	0.556	0.200	0.067	0.024	0.009	0.004	0.002	0.001
5		0.267	0.097	0.036	0.015	0.006	0.003	0.001
6		0.356	0.139	0.055	0.023	0.010	0.005	0.002
7		0.444	0.188	0.077	0.033	0.015	0.007	0.003
8		0.556	0.248	0.107	0.047	0.021	0.010	0.005
9			0.315	0.141	0.064	0.030	0.014	0.007
10			0.387	0.184	0.085	0.041	0.020	0.010
11			0.461	0.230	0.111	0.054	0.027	0.014
12			0.539	0.285	0.142	0.071	0.036	0.019
13				0.341	0.177	0.091	0.047	0.025
14				0.404	0.217	0.114	0.060	0.032
15				0.467	0.262	0.141	0.076	0.041
16				0.533	0.311	0.172	0.095	0.052
17					0.362	0.207	0.116	0.065
18					0.416	0.245	0.140	0.080
19					0.472	0.286	0.168	0.097
20					0.528	0.331	0.198	0.117
21						0.377	0.232	0.139
22						0.426	0.268	0.164
23						0.475	0.306	0.191
24						0.525	0.347	0.221
25							0.389	0.253
26							0.433	0.287
27							0.478	0.323
28							0.522	0.360
29								0.399
30								0.439
31								0.480
32								0.520

Reproduced from H. B. Mann and D. R. Whitney, 'On a test of whether one of two random variables is stochastically larger than the other', *Ann. Math. Statist.*, vol. 18, pp. 52–54 (1947), by permission of the authors and the publisher.

Table 7. Critical Values of U in the Wilcoxon Two-Sample Test

One-Tailed Test at $\alpha = 0.001$ or Two-Tailed Test at $\alpha = 0.002$

n_1	n_2=9	10	11	12	13	14	15	16	17	18	19	20
1												
2												
3									0	0	0	0
4		0	0	0	1	1	1	2	2	3	3	3
5	1	1	2	2	3	3	4	5	5	6	7	7
6	2	3	4	4	5	6	7	8	9	10	11	12
7	3	5	6	7	8	9	10	11	13	14	15	16
8	5	6	8	9	11	12	14	15	17	18	20	21
9	7	8	10	12	14	15	17	19	21	23	25	26
10	8	10	12	14	17	19	21	23	25	27	29	32
11	10	12	15	17	20	22	24	27	29	32	34	37
12	12	14	17	20	23	25	28	31	34	37	40	42
13	14	17	20	23	26	29	32	35	38	42	45	48
14	15	19	22	25	29	32	36	39	43	46	50	54
15	17	21	24	28	32	36	40	43	47	51	55	59
16	19	23	27	31	35	39	43	48	52	56	60	65
17	21	25	29	34	38	43	47	52	57	61	66	70
18	23	27	32	37	42	46	51	56	61	66	71	76
19	25	29	34	40	45	50	55	60	66	71	77	82
20	26	32	37	42	48	54	59	65	70	76	82	88

One-Tailed Test at $\alpha = 0.01$ or Two-Tailed Test at $\alpha = 0.02$

n_1	n_2=9	10	11	12	13	14	15	16	17	18	19	20
1												
2					0	0	0	0	0	0	1	1
3	1	1	1	2	2	2	3	3	4	4	4	5
4	3	3	4	5	5	6	7	7	8	9	9	10
5	5	6	7	8	9	10	11	12	13	14	15	16
6	7	8	9	11	12	13	15	16	18	19	20	22
7	9	11	12	14	16	17	19	21	23	24	26	28
8	11	13	15	17	20	22	24	26	28	30	32	34
9	14	16	18	21	23	26	28	31	33	36	38	40
10	16	19	22	24	27	30	33	36	38	41	44	47
11	18	22	25	28	31	34	37	41	44	47	50	53
12	21	24	28	31	35	38	42	46	49	53	56	60
13	23	27	31	35	39	43	47	51	55	59	63	67
14	26	30	34	38	43	47	51	56	60	65	69	73
15	28	33	37	42	47	51	56	61	66	70	75	80
16	31	36	41	46	51	56	61	66	71	76	82	87
17	33	38	44	49	55	60	66	71	77	82	88	93
18	36	41	47	53	59	65	70	76	82	88	94	100
19	38	44	50	56	63	69	75	82	88	94	101	107
20	40	47	53	60	67	73	80	87	93	100	107	114

Marine Statistics

One-Tailed Test at $\alpha = 0.025$ or Two-Tailed Test at $\alpha = 0.05$

n_1	9	10	11	12	13	14	15	16	17	18	19	20
1												
2	0	0	0	1	1	1	1	1	2	2	2	2
3	2	3	3	4	4	5	5	6	6	7	7	8
4	4	5	6	7	8	9	10	11	11	12	13	13
5	7	8	9	11	12	13	14	15	17	18	19	20
6	10	11	13	14	16	17	19	21	22	24	25	27
7	12	14	16	18	20	22	24	26	28	30	32	34
8	15	17	19	22	24	26	29	31	34	36	38	41
9	17	20	23	26	28	31	34	37	39	42	45	48
10	20	23	26	29	33	36	39	42	45	48	52	55
11	23	26	30	33	37	40	44	47	51	55	58	62
12	26	29	33	37	41	45	49	53	57	61	65	69
13	28	33	37	41	45	50	54	59	63	67	72	76
14	31	36	40	45	50	55	59	64	67	74	78	83
15	34	39	44	49	54	59	64	70	75	80	85	90
16	37	42	47	53	59	64	70	75	81	86	92	98
17	39	45	51	57	63	67	75	81	87	93	99	105
18	42	48	55	61	67	74	80	86	93	99	106	112
19	45	52	58	65	72	78	85	92	99	106	113	119
20	48	55	62	69	76	83	90	98	105	112	119	127

One-Tailed Test at $\alpha = 0.05$ or Two-Tailed Test at $\alpha = 0.10$

n_1	9	10	11	12	13	14	15	16	17	18	19	20
1											0	0
2	1	1	1	2	2	2	3	3	3	4	4	4
3	3	4	5	5	6	7	7	8	9	9	10	11
4	6	7	8	9	10	11	12	14	15	16	17	18
5	9	11	12	13	15	16	18	19	20	22	23	25
6	12	14	16	17	19	21	23	25	26	28	30	32
7	15	17	19	21	24	26	28	30	33	35	37	39
8	18	20	23	26	28	31	33	36	39	41	44	47
9	21	24	27	30	33	36	39	42	45	48	51	54
10	24	27	31	34	37	41	44	48	51	55	58	62
11	27	31	34	38	42	46	50	54	57	61	65	69
12	30	34	38	42	47	51	55	60	64	68	72	77
13	33	37	42	47	51	56	61	65	70	75	80	84
14	36	41	46	51	56	61	66	71	77	82	87	92
15	39	44	50	55	61	66	72	77	83	88	94	100
16	42	48	54	60	65	71	77	83	89	95	101	107
17	45	51	57	64	70	77	83	89	96	102	109	115
18	48	55	61	68	75	82	88	95	102	109	116	123
19	51	58	65	72	80	87	94	101	109	116	123	130
20	54	62	69	77	84	92	100	107	115	123	130	138

Adapted and abridged from Tables 1, 3, 5 and 7 of D. Auble, 'Extended tables for the Mann–Whitney statistic', *Bulletin of the Institute of Educational Research at Indiana University*, vol. 1, no. 2 (1953), by permission of the director.

Table 8. Random Numbers

99 70	44 70	23 38	38 61	24 72	91 17	42 28	05 69	08 72	87 59
91 23	66 34	42 60	53 70	51 82	20 22	04 49	45 60	95 97	68 97
73 82	92 69	37 01	29 65	78 44	01 28	49 31	37 11	37 99	51 09
56 79	56 07	34 53	32 54	77 80	47 65	78 15	23 74	05 79	91 14
90 77	93 75	34 11	91 87	33 62	22 18	12 18	94 15	55 85	59 05
19 13	22 21	40 51	52 95	38 57	05 20	77 97	26 04	67 28	01 51
94 85	19 32	56 30	48 49	60 58	46 22	99 49	30 21	85 86	21 82
95 93	69 20	59 64	89 65	45 05	24 32	91 02	48 97	40 10	04 50
66 45	00 48	59 44	70 51	90 91	23 44	65 04	57 66	94 55	82 40
78 32	36 65	90 50	24 71	19 40	41 39	17 72	90 42	11 63	84 87
91 21	64 86	92 52	06 41	73 05	49 24	59 07	98 24	64 00	23 17
08 15	89 57	94 49	31 83	79 47	73 84	00 90	40 37	50 94	03 04
52 21	18 47	48 64	44 88	69 57	58 25	95 86	23 21	66 98	38 67
26 52	36 20	38 37	09 96	96 84	87 44	90 49	28 73	66 91	69 84
47 16	56 37	12 53	06 71	73 88	29 03	33 85	25 96	33 58	27 30
01 20	24 25	85 56	60 53	81 40	68 40	49 79	28 28	52 49	36 14
36 89	45 98	08 51	24 28	56 56	88 46	74 48	11 89	74 98	89 39
48 51	27 19	26 24	42 23	96 61	75 85	37 25	66 88	50 26	19 96
96 03	72 28	99 34	08 12	85 65	79 62	22 43	45 93	49 46	65 65
95 67	58 84	08 92	08 79	16 09	39 45	77 96	63 76	19 65	03 61
20 88	05 30	39 08	27 07	95 46	02 04	87 46	00 14	64 17	42 95
74 68	89 76	27 47	64 71	34 13	45 58	98 62	65 30	18 43	67 86
05 32	05 69	33 76	46 12	76 25	69 88	57 31	39 31	65 58	98 36
97 07	39 59	01 79	88 71	30 62	19 35	58 37	65 17	79 90	28 74
21 68	86 76	85 79	29 01	67 39	36 00	63 42	33 41	07 23	52 40
16 15	55 86	93 14	03 85	59 66	14 00	96 25	86 10	90 08	34 34
11 92	48 07	26 35	86 43	10 33	01 88	94 78	48 63	53 82	10 55
28 88	44 75	03 66	63 13	23 42	66 81	60 09	78 71	98 17	97 26
88 02	64 78	97 85	26 89	32 52	43 15	89 48	54 02	08 91	88 08
94 10	65 67	97 26	62 19	30 55	20 40	77 77	58 51	37 21	43 39
92 69	19 28	04 67	94 62	99 45	33 19	26 04	03 83	20 17	75 73
15 48	00 97	20 99	10 60	94 60	12 85	13 23	64 72	74 49	17 27
96 27	23 33	72 10	34 59	47 59	54 84	37 96	30 92	94 70	70 40
99 95	89 25	96 23	60 90	82 42	64 54	42 83	61 47	22 15	85 33
29 62	11 74	26 11	21 29	66 26	97 37	12 18	88 33	93 29	00 79
44 25	51 55	08 19	50 90	81 39	02 10	40 16	92 34	45 04	68 36
57 38	94 60	06 87	96 92	93 48	11 54	93 99	67 68	44 91	32 59
06 57	37 18	75 62	54 45	14 03	40 88	54 30	33 40	16 23	05 88
12 93	01 37	20 19	28 34	21 04	15 12	61 89	22 03	04 50	90 80
96 40	49 36	14 09	62 03	60 07	50 57	13 44	14 29	32 70	23 20
22 37	51 87	03 61	72 61	32 22	69 85	47 67	33 26	03 64	54 55
25 57	59 26	70 96	57 95	35 25	06 77	97 37	35 66	62 49	78 41
23 66	63 37	17 85	15 53	33 94	27 84	60 87	59 58	61 00	44 13
82 37	88 83	68 70	31 75	19 83	55 04	32 00	65 63	89 03	60 77
73 49	17 78	85 43	94 42	76 93	22 77	00 15	54 55	01 72	02 07
21 42	51 90	27 25	30 32	00 11	20 81	14 24	30 85	27 56	72 40
52 03	49 03	84 41	71 03	31 42	19 07	62 02	32 75	49 95	78 30
68 57	06 86	93 44	20 33	46 55	51 07	26 15	36 74	49 74	01 88
33 81	52 80	56 34	45 50	75 18	60 45	12 44	54 28	20 26	33 79
31 96	55 24	67 90	85 54	91 22	89 27	46 77	51 40	48 87	78 39

Table 9. Values of e^{-x}

x	e^{-x}	x	e^{-x}	x	e^{-x}	x	e^{-x}
0·0	1·000	2·5	0·082	5·0	0·0067	7·5	0·00055
0·1	0·905	2·6	0·074	5·1	0·0061	7·6	0·00050
0·2	0·819	2·7	0·067	5·2	0·0055	7·7	0·00045
0·3	0·741	2·8	0·061	5·3	0·0050	7·8	0·00041
0·4	0·670	2·9	0·055	5·4	0·0045	7·9	0·00037
0·5	0·607	3·0	0·050	5·5	0·0041	8·0	0·00034
0·6	0·549	3·1	0·045	5·6	0·0037	8·1	0·00030
0·7	0·497	3·2	0·041	5·7	0·0033	8·2	0·00028
0·8	0·449	3·3	0·037	5·8	0·0030	8·3	0·00025
0·9	0·407	3·4	0·033	5·9	0·0027	8·4	0·00023
1·0	0·368	3·5	0·030	6·0	0·0025	8·5	0·00020
1·1	0·333	3·6	0·027	6·1	0·0022	8·6	0·00018
1·2	0·301	3·7	0·025	6·2	0·0020	8·7	0·00017
1·3	0·273	3·8	0·022	6·3	0·0018	8·8	0·00015
1·4	0·247	3·9	0·020	6·4	0·0017	8·9	0·00014
1·5	0·223	4·0	0·018	6·5	0·0015	9·0	0·00012
1·6	0·202	4·1	0·017	6·6	0·0014	9·1	0·00011
1·7	0·183	4·2	0·015	6·7	0·0012	9·2	0·00010
1·8	0·165	4·3	0·014	6·8	0·0011	9·3	0·00009
1·9	0·150	4·4	0·012	6·9	0·0010	9·4	0·00008
2·0	0·135	4·5	0·011	7·0	0·0009	9·5	0·00008
2·1	0·122	4·6	0·010	7·1	0·0008	9·6	0·00007
2·2	0·111	4·7	0·009	7·2	0·0007	9·7	0·00006
2·3	0·100	4·8	0·008	7·3	0·0007	9·8	0·00006
2·4	0·091	4·9	0·007	7·4	0·0006	9·9	0·00005

Table 10.

n	n^2	$\sqrt{n}$	$\sqrt{10n}$	n	n^2	$\sqrt{n}$	$\sqrt{10n}$
1·0	1·00	1·000	3·162	5·5	30·25	2·345	7·416
1·1	1·21	1·049	3·317	5·6	31·36	2·366	7·483
1·2	1·44	1·095	3·464	5·7	32·49	2·387	7·550
1·3	1·69	1·140	3·606	5·8	33·64	2·408	7·616
1·4	1·96	1·183	3·742	5·9	34·81	2·429	7·681
1·5	2·25	1·225	3·873	6·0	36·00	2·449	7·746
1·6	2·56	1·265	4·000	6·1	37·21	2·470	7·810
1·7	2·89	1·304	4·123	6·2	38·44	2·490	7·874
1·8	3·24	1·342	4·243	6·3	39·69	2·510	7·937
1·9	3·61	1·378	4·359	6·4	40·96	2·530	8·000
2·0	4·00	1·414	4·472	6·5	42·25	2·550	8·062
2·1	4·41	1·449	4·583	6·6	43·56	2·569	8·124
2·2	4·84	1·483	4·690	6·7	44·89	2·588	8·185
2·3	5·29	1·517	4·796	6·8	46·24	2·608	8·246
2·4	5·76	1·549	4·899	6·9	47·61	2·627	8·307
2·5	6·25	1·581	5·000	7·0	49·00	2·646	8·367
2·6	6·76	1·612	5·099	7·1	50·41	2·665	8·426
2·7	7·29	1·643	5·196	7·2	51·84	2·683	8·485
2·8	7·84	1·673	5·292	7·3	53·29	2·702	8·544
2·9	8·41	1·703	5·385	7·4	54·76	2·720	8·602
3·0	9·00	1·732	5·477	7·5	56·25	2·739	8·660
3·1	9·61	1·761	5·568	7·6	57·76	2·757	8·718
3·2	10·24	1·789	5·657	7·7	59·29	2·775	8·775
3·3	10·89	1·817	5·745	7·8	60·84	2·793	8·832
3·4	11·56	1·844	5·831	7·9	62·41	2·811	8·888
3·5	12·25	1·871	5·916	8·0	64·00	2·828	8·944
3·6	12·96	1·897	6·000	8·1	65·61	2·846	9·000
3·7	13·69	1·924	6·083	8·2	67·24	2·864	9·055
3·8	14·44	1·949	6·164	8·3	68·89	2·881	9·110
3·9	15·21	1·975	6·245	8·4	70·56	2·898	9·165
4·0	16·00	2·000	6·325	8·5	72·25	2·915	9·220
4·1	16·81	2·025	6·403	8·6	73·96	2·933	9·274
4·2	17·64	2·049	6·481	8·7	75·69	2·950	9·327
4·3	18·49	2·074	6·557	8·8	77·44	2·966	9·381
4·4	19·36	2·098	6·633	8·9	79·21	2·983	9·434
4·5	20·25	2·121	6·708	9·0	81·00	3·000	9·487
4·6	21·16	2·145	6·782	9·1	82·81	3·017	9·539
4·7	22·09	2·168	6·856	9·2	84·64	3·033	9·592
4·8	23·04	2·191	6·928	9·3	86·49	3·050	9·644
4·9	24·01	2·214	7·000	9·4	88·36	3·066	9·695
5·0	25·00	2·236	7·071	9·5	90·25	3·082	9·747
5·1	26·01	2·258	7·141	9·6	92·16	3·098	9·798
5·2	27·04	2·280	7·211	9·7	94·09	3·114	9·849
5·3	28·09	2·302	7·280	9·8	96·04	3·130	9·899
5·4	29·16	2·324	7·348	9·9	98·01	3·146	9·950

Index

Acceptance region 174
Addition law of probability 80
Alternative hypothesis 172
  one-sided 173
  two-sided 173
Analysis of variance 243
  one way classification 247, 255
  two way classification
    several observations per cell 263
    single observations per cell 259
Average 40

Bar chart
  component 31
  multiple 31
  simple 31
Bayes' rule 87
Binomial distribution 99
  mean 102
  normal approximation to 132
  Poisson approximation to 106
  variance 102

Chi-squared distribution 199
  goodness of fit test 202
Class 17
  interval 17, 46
  midpoint 18, 43
Coefficient of variation 61
Combinations 88
Complement of a set 79
Complementary probability 79
Component bar chart 31
Conditional probability 83
Confidence interval 150
  difference between means 157
  difference between proportions 160
  mean 152
  proportion 159
  ratio of two variances 246
  regression estimates 294

  regression parameters 292
  variance 201
Confidence limits 151
Contingency table 208
Continuous probability distribution 119
Correlation 279, 297
Correlation coefficient
  product-moment 298
  rank 301
Critical region 174
Critical value 175

Deciles 51
Degrees of freedom
  analysis of variance 250, 256, 260, 265
  chi-squared distribution 199
  contingency table 210
  F distribution 244
  goodness of fit 203
  t distribution 155
Difference between means
  confidence interval for 157
  sampling distribution 157
  testing hypotheses 181, 231
Difference between proportions
  confidence interval for 160
  sampling distribution of 160
  testing hypotheses 187
Discrete probability distribution 95
Distribution
  binomial 99
  chi-squared 199
  double exponential 134
  exponential 124
  F 244
  geometric 109
  hypergeometric 110
  negative binomial 108
  normal 126
  Poisson 104, 213

Marine Statistics

Distribution—*continued*
   t 155
   uniform
      continuous 123
      discrete 98
Distribution-free test 226
Double exponential distribution 134

Empirical probability 77
Error
   type I and type II 173
Error sum of squares
   one way classification 250, 255
   regression 286
   two way classification 259
      several observations per cell 264
      single observations per cell 260
Estimate
   interval 150
   point 150
Events
   complementary 79
   independent 85
   mutually exclusive 81
Expectation 97
Expected value 97
Exponential distribution 124

F distribution 244
Factorial notation 88
Frequency distribution 14
   cumulative 25
   relative 24
Frequency polygon 22

Gaussian distribution 126
Geometric distribution 109
Goodness of fit test 202
   degrees of freedom 203
Grouped data
   deciles 51
   mean 42
   median 45
   mode 47
   percentiles 52
   quantiles 51
   range 57
   standard deviation 61
   variance 61

Histogram
   frequency 21
   probability 98
   relative frequency 24

Historigram 26
Hypergeometric distribution 110
   mean 110
   variance 110

Independent events 85
Interaction 213
Intersection of sets 80
Interval estimate 150

Least significant difference 253
Least squares, method of 287
Level of significance 178
Linear regression 282
Lorenz curve 28

Mann-Whitney test 231
Mean (arithmetic) 40, 50
   confidence interval for 152
   population 41
   sample 41
   sampling distribution of 144
   testing hypotheses 179, 223
   geometric 40, 49
   harmonic 49
   mean deviation 64
Measure of central tendency 46
   of dispersion 57
Median 45, 50
Median class 46
Modal class 47
Mode 47, 50
Multiple bar chart 31
Multiple regression 297
Multiplication law of probability 84
Mutually exclusive events 81

Negative binomial distribution 108
Nonparametric tests 225
Normal curve 127
   areas under 129
Normal distribution 126
   approximation to the binomial 132
   standard 128
Null hypothesis 172

Ogive 25
One-sided alternative 172
One-tailed test 172
One way classification
   analysis of variance table 251, 256
   sum of squares identity 250

Index

Paired observations
   sampling distribution of 184
   tests for 184
   Wilcoxon test for 228
Parameter 142
Pearson product-moment correlation coefficient 298
Percentile 52
Permutation 88
Pictogram 29
Pie chart 32
Point estimate 150
Poisson distribution 104, 213
   approximation to the binomial 106
   mean 106
   variance 108
Polygon frequency 22
Population 13
   mean 41
   standard deviation 63
Power 178
Prediction 291
Probability
   addition law of 80
   of complementary events 79
   conditional 83
   of independent events 85
   multiplication law of 84
   of mutually exclusive events 81
Probability density function 122
Probability distribution
   continuous 119
   discrete 95
Probability histogram 98
Probability sample 143
Proportion
   confidence interval for 159
   sampling distribution of 159
   testing hypotheses 185

Quartiles 51

Random sample 143
Random variable 97
Range 57, 65
Regression 279
   coefficient 286, 289
   line 283
   linear 283
   multiple 297
   non-linear 295
   parameters
      confidence intervals for 292

      estimation of 292
      tests of hypotheses 292
Relative frequency 24

Sample 13
   mean 142
   size 160
   variance 142
Sampling distribution 144
   of differences of means 157
   of differences of proportions 160
   of the mean 152, 153, 154
   of proportions 159
Scatter diagram 279
Semi-interquartile range 64, 66
Semi-logarithmic graph 27
Sets
   intersection 80
   union of 80
Sign test 226
Significance
   level 178
   region 174
Simple random sample 143
Skewness 66
   Pearsonian coefficient of 67
Standard deviation
   of a population 59, 66
   of a sample 63
Standard error
   of the difference of means 157
   of the difference of proportions 160
   of the mean 146
   of proportions 159
Standard normal distribution 128
Standardising transformation 131
Statistic 142
Statistical hypothesis
   alternative 172
   null 172
Student-t distribution 155
Sum of squares
   column means 250
      computational formula 251, 255, 261, 263
   error 250
   interaction 263
      computational formula 265
   row means 261
      computational formula 261, 263
   total 250
      computational formula 251, 222, 261, 263

Sum-of-squares identity
  one way classification 250
  two way classification
    several observations per cell 264
    single observation per cell 260
Summation notation 41

t distribution 155
Tests of hypotheses
  analysis of variance 243
  difference between means 181
  difference between proportions 187
  distribution-free 226
  goodness of fit 202
  independence 208
  Mann-Whitney test 231
  means 179
  non-parametric 225
  one-tailed 172
  paired observations 183
  proportions 185
  regression parameters 291
  sign test 226
  two variances 247
  two-tailed 172
  variance 201
  Wilcoxon test 228

Total sum of squares
  one way classification 251, 257
  two way classification 261
    several observations per cell 265
    single observation per cell 261
Two-sided test
Two way classification
  several observations per cell 263
  analysis of variance table 265
  single observation per cell
  analysis of variance table 261
Type I and type II errors 173

Uniform distribution
  continuous 123
  discrete 98
Union of sets 80

Variance 60
  analysis of 243
  confidence interval for 202
  testing hypothesis 201
  Venn diagram 80

Wilcoxon test 228

Yates' correction 211